Semantic Web Technologies

Semantic web technologies (SWTs) offer the richest machine-interpretable (rather than just machine-processable) and explicit semantics that are being extensively used in various domains and industries. This book provides a roadmap for semantic web technologies (SWTs) and highlights their role in a wide range of domains including cloud computing, Internet of Things, big data, sensor network, and so forth. It also explores the prospects of these technologies including different data interchange formats, query languages, ontologies, Linked Data, and notations. The role of SWTs in 'epidemic Covid-19', 'e-learning platforms and systems', 'block chain', 'open online courses', and 'visual analytics in healthcare' is described as well. This book:

- Explores all the critical aspects of semantic web technologies (SWTs)
- Discusses the impact of SWTs on cloud computing, Internet of Things, big data, and sensor network
- Offers a comprehensive examination of the emerging research in the areas of SWTs and their related domains
- Provides a template to develop a wide range of smart and intelligent applications
- Includes latest applications and examples with real data

This book is aimed at researchers and graduate students in computer science, informatics, web technology, cloud computing, and Internet of Things.

T0321115

Computational Intelligence in Engineering Problem Solving

Series Editor
Nilanjan Dey

Computational Intelligence (CI) can be framed as a heterogeneous domain that harmonized and coordinated several technologies, such as probabilistic reasoning, artificial life, multi-agent systems, neuro-computing, fuzzy systems, and evolutionary algorithms. Integrating several disciplines, such as Machine Learning (ML), Artificial Intelligence (AI), Decision Support Systems (DSS), and Database Management Systems (DBMS) increases the CI power and impact in several engineering applications. This book series provides a well-standing forum to discuss the characteristics of CI systems in engineering. It emphasizes the development of CI techniques and their role as well as the state-of-the-art solutions in different real world engineering applications. The book series is proposed for researchers, academics, scientists, engineers, and professionals who are involved in the new techniques of CI. CI techniques including artificial fuzzy logic and neural networks are presented for biomedical image processing, power systems, and reactor applications.

Applied Intelligent Decision Making in Machine Learning
Himansu Das, Jitendra Kumar Rout, Suresh Chandra Moharana,
and Nilanjan Dey

Machine Learning and IoT for Intelligent Systems and Smart Applications
Madhumathy P, M Vinoth Kumar, and R. Umamaheswari

Industrial Power Systems
Evolutionary Aspects
Amitava Sil and Saikat Maity

Fuzzy Optimization Techniques in the Areas of Science and Management
Edited by Santosh Kumar Das and Massimiliano Giacalone

Semantic Web Technologies: Research and Applications
Edited by Archana Patel, Narayan C. Debnath, and Bharat Bhushan

For more information about this series, please visit: www.routledge.com/ Computational-Intelligence-in-Engineering-Problem-Solving/book-series/ CIEPS

Semantic Web Technologies
Research and Applications

Edited by
Archana Patel, Narayan C. Debnath, and
Bharat Bhushan

CRC Press
Taylor & Francis Group
Boca Raton London New York

CRC Press is an imprint of the
Taylor & Francis Group, an **informa** business

First edition published 2023
by CRC Press
6000 Broken Sound Parkway NW, Suite 300, Boca Raton, FL 33487-2742

and by CRC Press
4 Park Square, Milton Park, Abingdon, Oxon, OX14 4RN

CRC Press is an imprint of Taylor & Francis Group, LLC

© 2023 selection and editorial matter, Archana Patel, Narayan C. Debnath, and Bharat Bhushan; individual chapters, the contributors

Library of Congress Cataloging-in-Publication Data
Names: Patel, Archana (Lecturer in software engineering), editor. |
Debnath, N. C. (Narayan C.), editor. | Bhushan, Bharat, 1989- editor.
Title: Semantic web technologies : research and applications /
edited by Archana Patel, Narayan C. Debnath, Bharat Bhushan.
Description: First edition. | Boca Raton : CRC Press, 2023. |
Series: Computational intelligence in engineering problem solving |
Includes bibliographical references and index.
Subjects: LCSH: Semantic Web.
Classification: LCC TK5105.88815 .S4565 2023 (print) |
LCC TK5105.88815 (ebook) |
DDC 025.042/7—dc23/eng/20220803
LC record available at https://lccn.loc.gov/2022017870
LC ebook record available at https://lccn.loc.gov/2022017871

ISBN: 978-1-032-31369-6 (hbk)
ISBN: 978-1-032-31370-2 (pbk)
ISBN: 978-1-003-30942-0 (ebk)

DOI: 10.1201/9781003309420

Typeset in Times New Roman
by codeMantra

Contents

Editors

Archana Patel is currently a faculty of the Department of Software Engineering, School of Computing and Information Technology, Binh Duong Province, Vietnam. She has completed Postdoc from the Freie Universität Berlin, Berlin, Germany. She has filed a patent titled "Method and System for Creating Ontology of Knowledge Units in a Computing Environment" in November 2019. She has received Doctor of Philosophy (Ph.D.) in Computer Applications and postgraduate degree both from the National Institute of Technology (NIT) Kurukshetra, India in 2020 and 2016 respectively. She has qualified GATE and UGC-NET/JRF exams in 2017. Dr. Patel has also contributed to a research project funded by Defence Research and Development Organization (DRDO) for the period of two years. Her contribution in teaching is also remarkable. Dr. Patel is an author or co-author of more than 30 publications in numerous refereed journals and conference proceedings. She has been awarded best paper award (three times) in international conferences. Dr. Patel has received various awards for presentation of research work at various international conferences, teaching and research institutions. Her primary research interests are in Ontological Engineering, Semantic Web, Big Data, Expert System and Knowledge Warehouse.

Narayan C. Debnath is the Founding Dean of the School of Computing and Information Technology at Eastern International University, Vietnam. He is also serving as the Head of the Department of Software Engineering at Eastern International University, Vietnam. Dr. Debnath has been the Director of the International Society for Computers and their Applications (ISCA), USA, since 2014. Formerly, Dr. Debnath served as a Full Professor of Computer Science at Winona State University, Minnesota, USA for 28 years (1989–2017). He was elected as the Chairperson of the Computer Science Department at Winona State University for three consecutive terms and assumed the role of the Chairperson of the Computer Science Department at Winona State University for seven years (2010–2017).

Dr. Debnath earned a Doctor of Science (D.Sc.) degree in Computer Science and also a Doctor of Philosophy (Ph.D.) degree in Physics. In the past, he served as the elected President for two separate terms, Vice President, and Conference Coordinator of the International Society for Computers and their Applications, and has been a member of the ISCA Board of Directors since 2001. Before being elected as the Chairperson of the Department of Computer Science in 2010 at Winona State University, he served as the Acting Chairman of the Department. Dr. Debnath received numerous Honors and Awards while serving as a Professor of Computer Science during the period 1989–2017. During 1986–1989, Dr. Debnath served as an Assistant Professor in the Department of Mathematics and Computer Systems at the University of Wisconsin-River Falls, USA, where he was nominated for the prestigious US National Science Foundation (NSF) Presidential Young Investigator Award in 1989.

Professor Debnath has made significant contributions to teaching, research, and services across the academic and professional communities. He has made original research contributions in Software Engineering Models, Metrics and Tools, Software Testing, Software Management, and Information Science, Technology and Engineering. Dr. Debnath is an author or co-author of over 450 publications in numerous refereed journals and conference proceedings in Computer Science, Information Science, Information Technology, System Sciences, Mathematics, and Electrical Engineering.

Professor Debnath has made numerous teaching, research and invited keynote presentations at various international conferences, industries, and teaching and research institutions in Africa, Asia, Australia, Europe, North America, and South America. He has offered courses and workshops on Software Engineering and Software Testing at universities in Asia, Africa, the Middle East, and South America. Dr. Debnath has been a visiting professor at universities in Argentina, China, India, Sudan, and Taiwan. He has been maintaining active research and professional collaborations with many universities, faculty, scholars, professionals, and practitioners across the globe.

Dr. Debnath is an editor of several books published by Springer, Elsevier, CRC Press, and Bentham Science Press on emerging fields of computing. His recently published books on IoT Devices and Sensor Networks, IOT Security paradigm, Blockchain technology for data privacy, and Security and Privacy issues in IOT devices made significant impacts in new developments on these topics. Dr. Debnath has been actively involved with the IEEE International Conference on Industrial Technology (ICIT) and IEEE International Conference on Industrial Informatics (INDIN), flagship annual conferences of the IEEE Industrial Electronics Society (IES). For the last many years, Dr. Debnath has been organizing highly successful Special Sessions attracting a large number of quality submissions at both IEEE INDIN and IEEE ICIT. He also served as a guest editor of the Journal of Computational Methods in Science and Engineering (JCMSE) published by the IOS Press, the Netherlands. Dr. Debnath has been an active member of the ACM, IEEE Computer Society, Arab Computer Society, and a senior member of the ISCA.

Bharat Bhushan is an Assistant Professor in the Department of Computer Science and Engineering (CSE) at School of Engineering and Technology, Sharda University, Greater Noida, India. He received his Undergraduate Degree (B-Tech in Computer Science and Engineering) with Distinction in 2012, received his Postgraduate Degree (M-Tech in Information Security) with Distinction in 2015 and Doctorate Degree (PhD in Computer Science and Engineering) in 2021 from Birla Institute of Technology, Mesra, India. He earned numerous international certifications such as CCNA, MCTS, MCITP, RHCE, and CCNP. He has Published more than 80 research papers in various renowned International conferences and SCI indexed journals including Journal of Network and Computer

Applications (Elsevier), Wireless Networks (Springer), Wireless Personal Communications (Springer), Sustainable Cities and Society (Elsevier), and Emerging Transactions on Telecommunications (Wiley). He has contributed more than 25 book chapters in various books and has edited 11 books from the most famed publishers like Elsevier, IGI Global, and CRC Press. He has served as Keynote Speaker (resource person) in numerous reputed international conferences held in different countries including India, Morocco, China, Belgium, and Bangladesh. He has served as a Reviewer/Editorial Board Member for several reputed international journals. In the past, he worked as an assistant professor at HMR Institute of Technology and Management, New Delhi and Network Engineer in HCL Infosystems Ltd., Noida. He has qualified GATE exams for successive years and gained the highest percentile of 98.48 in GATE 2013. He is also a member of numerous renowned bodies including IEEE, IAENG, CSTA, SCIEI, IAE and UACEE.

Contributors

Zaenal Akbar
Research Center for Informatics
National Research and Innovation
  Agency
Jakarta, Indonesia

Ayesha Ameen
Department of I.T
Deccan College of Engineering and
  Technology
Hyderabad, India

Ayush Kumar A
Department of Chemical Engineering
National institute of Technology
Tiruchirappalli, India

P. Shanthi Bala
Department of Computer Science
Pondicherry University
Kalapet, India

Ayesha Banu
Department of CSE
Vaagdevi College of Engineering
Bollikunta, India

Houda El Bouhissi
LIMED Laboratory, Faculty of Exact
  Sciences
University of Bejaia
Bejaia, Algeria

Michael DeBellis
155 Gardenside Dr. #24
San Francisco, California

Narayan C. Debnath
Department of Software Engineering
Eastern International University
Bình Dương, Vietnam

Gerard Deepak
Department of Computer Science and
  Engineering
National institute of
  Technology
Tiruchirappalli, India

Tutie Djarwaningsih
Research Center for Informatics
National Research and Innovation
  Agency
Jakarta, Indonesia

Esingbemi P. Ebietomere
Department of Computer
  Science
University of Benin
Benin City, Nigeria

Godspower O. Ekuobase
Department of Computer Science
University of Benin
Benin City, Nigeria

Gina George
School of Computer Science and
  Engineering
VIT
Vellore, India

Ariani Indrawati
Research Center for
  Informatics
National Research and Innovation
  Agency
Jakarta, Indonesia

Noman Islam
Karachi Institute of Economics and
  Technology
Karachi, Pakistan

G. Jeyakodi
Department of Computer Science
Pondicherry University
Kalapet, India

Yulia A. Kartika
Research Center for Informatics
National Research and Innovation
    Agency
Jakarta, Indonesia

Khaleel Ur Rahman Khan
ACE Engineering College
Hyderabad, India

Anisha M. Lal
School of Computer Science and
    Engineering
VIT
Vellore, India

Shiva Shankar Mahato
Srinivasa Ramanujan Library
Indian Institute of Science Education
    and Research
Pune, India

Taufik Mahendra
Research Center for Informatics
National Research and Innovation
    Agency
Jakarta, Indonesia

Lindung P. Manik
Research Center for Informatics
National Research and Innovation
    Agency
Jakarta, Indonesia

Hani F. Mustika
Research Center for Informatics
Indonesian Institute of Sciences
Jarkarta, Indonesia

Olaide N. Oyelade
Department of Computer Science
Ahmadu Bello University
Zaria, Nigeria

Archana Patel
Department of Software Engineering
Eastern International University
Bình Dương, Vietnam

Halyna Pidnebesna
International Research and
    Training Centre for Information
    Technologies and Systems of the
    NAS and MES of Ukraine
Ukraine

B. Padmaja Rani
Department of CS
JNTUH
Hyderabad, India

Asif Raza
NED University
Karachi, Pakistan

Dadan R. Saleh
Research Center for Informatics
National Research and Innovation
    Agency
Jakarta, Indonesia

Ika A. Satya
Research Center for Informatics
National Research and Innovation
    Agency
Jakarta, Indonesia

Foni A. Setiawan
Research Center for Informatics
National Research and Innovation
    Agency
Jakarta, Indonesia

Prashant Kumar Sinha
Documentation Research and Training
  Centre
Indian Statistical Institute
Bangalore, India
Department of Library and
  Information Science
University of Calcutta
Kolkata, India

Volodymyr Stepashko
International Research and
  Training Centre for Information
  Technologies and Systems of the
  NAS and MES of Ukraine
Ukraine

Darakhshan Syed
Iqra University
Karachi, Pakistan

Wita Wardani
Research Center for Informatics
National Research and Innovation
  Agency
Jakarta, Indonesia

Aris Yaman
Research Center for Informatics
National Research and Innovation
  Agency
Jakarta, Indonesia

Mariz Zafar
NED University
Karachi, Pakistan

Preface

Gone are the days when data was interlinked with related data by humans and then by means of human interpretation put it in a coherent manner to find insights. We are in such an era where data is no more just data; it has more to do with. Data of a domain of discourse has been considered now, a Thing or Entity or Concept- to bring the meaning to it, so that the machine could not only understand the concept but also extrapolate the way humans do. This becomes possible in today's world with the help of semantic web technologies (SWTs) that adds semantics to the data and transform it into machine-understandable and processable form. For the same reasons, SWTs are highly valuable tools to simplify the existing problems in various domains leading to new opportunities. However, there are some challenges in various domains that need to be addressed to make industrial applications and machines smarter. These challenges can be mitigated by the utilization of SWTs and therefore, it is essential to find the solution to the existing challenges to fully realize the vision of SWTs in other domains. The proposed book aims to provide a roadmap for SWTs and highlights the role of these technologies in a wide range of domains. The book also explores the present and future prospects of these technologies, including different data interchange formats, query languages, ontologies, linked data, knowledge graph, and web ontology language. Moreover, the book potentially serves as an important guide towards the latest applications of semantic web technologies for the upcoming generation. Along with this, the book provides the answers to various questions that guide the researchers about the opportunities and challenges of these technologies.

1 Semantic Web Technologies

Esingbemi P. Ebietomere and
Godspower O. Ekuobase

CONTENTS

DOI: 10.1201/9781003309420-1

1.1 INTRODUCTION

The advent of the internet came with several mechanisms which have and is still revolutionizing it. One such outstanding machinery is the web—a complex system of interconnected components which has become a global repository of vast distributed heterogeneous data and services. The web has metamorphosed through generations from web 1.0—which allows for published contents to be searched and read (read only), through web 2.0—which allows for dynamism and collaborations (read and write) to web 3.0—which extends collaboration by allowing machine participation (read, write, and execute). Web 3.0 also referred to as the semantic web became incumbent as a result of the need to efficiently and effectively exploit the massive information and services on the web [1]. The semantic web began as a vision in about the mid-90s but was popularized by Tim Berners-Lee, Jim Hendler, and Ora Lassila's seminal paper presented at the www conference held in May 2001; published in Scientific American [2–7].

Also, the semantic web is undoubtedly indistinct and misconstruing plausibly due to the use of the word "semantics" and its connectedness with other terminologies like Web of Data, Linked Data, and Web of Linked Data [6,8,9]. This has made it even more challenging for a consensus definition of the semantic web. One of the goals of this chapter is to explicate the schools of thought on the semantic web to facilitate comprehension—particularly by new entrants and enthusiasts. It is pertinent to mention—irrespective of the school of thought—that there seems to be a consensus that the semantic web is dependent on many technologies with some predating it, some birthed with its advent, and others transforming it [2,3,10,11]. These technologies that are critical to the implementation, development, and realization of the semantic web are what we refer to in the context of this work as semantic web technologies (SWTs).

These SWTs consist of standards, methods, and tools. By standards, we mean formalized technologies that enforce consistency in the realization of the semantic web. These manifest in the form of format, language, or vocabularies—for example, Uniform Resource Identifier (URI)/Internationalized Resource Identifier (IRI), Extensible Markup Language (XML), Resource Description Framework (RDF)/Resource Description Framework Schema (RDFS), Simple Knowledge Organization System (SKOS), Web Ontology Language (OWL), SPARQL Protocol and RDF Query Language (SPARQL), and Rule Interchange Format (RIF) [12–14]. Methods refer to a systematic approach, practice, or technique followed in the realization of the semantic web—examples of methods include contextual analysis, reasoning engines, natural language understanding, ontology, knowledge graph, linked data, and their associated methodologies [11,15–18]. Tools are a set of software development environments, components, and plug-ins that aid developers in the efficient production and maintenance of programs—which in this work are semantic applications. These tools include annotation tools—knowtator, semantator, and GATE; ontology rendering tools—Protégé, OntoEdit, and OntoGen; and reasoners—Pellet, RacerPro, HermiT, Fact++. The primary goal of this chapter is to espouse these technologies along the convened dimensions. It is pertinent to make clear that the list of the technologies that will be discussed is not exhaustive but sure guarantees a good place to build from.

This chapter begins with the concept of the semantic web. This will then be followed by an exposé of the several technologies associated with it and their categorizations. Subsequently, some pragmatic issues associated with the implementation and deployment of SWTs were discussed.

1.2 THE CONCEPT OF SEMANTIC WEB

The semantic web has several appellations and definitions. The instability in its descriptions is due to the several perspectives of the concept, resulting in fuzziness on the topic. Some of the descriptions that have been given to the semantic web include a vision, a program, a technology, a project, an extension, a movement, a framework, a roadmap, an artifact, and a field of research [1–6,8–10,19,20]. What is clear about the semantic web at provenance is it being a vision as evident in [2], where the author attempted to give a roadmap, and supposed architectural blueprint of the concept. This was re-echoed by the same author, in the company of others including Jim Hendler, and Ora Lassila in 2001. From [3], two notable perspectives are evident in their vision, stated thus;

i. "The Semantic Web will bring structure to the meaningful content of Web pages, creating an environment where software agents roaming from page to page can readily carry out sophisticated tasks for users." [3] (p. 1)

ii. "The Semantic Web is not a separate Web but an extension of the current Web in which information is given well-defined meaning, better enabling computers and people to work in cooperation." [3] (p. 1)

Following the two statements, it is palpable that the first perspective envisioned a scenario where software agents crawl through the sea of information on the web—negotiating with each other intelligently, and performing complex and repetitive cognitive and decision support tasks such as searching, and scheduling on behalf of users. The second perspective envisioned annotating the already existing web with meaning to allow for seamless comprehension of information both by humans and machines. For the vision to come to fruition, it is apposite to permeate the web with meaning—that is, have a contraption for defining semantics about resources and links—and as well create structure on the web.

The authors [6] exposed how the description of the semantic web has evolved over a period of about two decades, as depicted in Figure 1.1.

Figure 1.1 reiterates a consensus on the concept starting out as a vision, and the fact that it is not a separate web but an extension of an existing web.

The work [5] exposed three major views of the semantic web—as a vision, program, and technology. As a vision, reference was to its use at provenance. The concept became a program with the emergence of the World Wide Web Consortium (W3C) as a body to enforce standards on the web—as a result

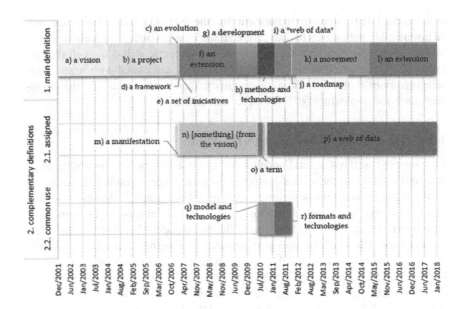

FIGURE 1.1   Several descriptions of the semantic web [6].

of the fear of the web dissociating due to the burden imposed by competitive commercial interests. Besides, the term technology is used to refer to methods, tools, and standards that could help realize a meaningfully annotated web. Also, Hitzler [9] exposed two broad perspectives from which the semantic web can be perceived—as an artifact, and a field of research. The author's inclination was toward the semantic web as a field of research rather than a concrete artifact which according to the author may not exist despite the obvious argument in support of the existence of the artifact already. Interestingly, the author's [9] view of the semantic web as a field of research does not negate every other view but rather subsumes it. Besides, it is not uncommon to see a vision become a field of research as evident in the field of software engineering, artificial intelligence, and computational linguistics which all started as a vision and are now fields of research. Thus, it is from this perspective we viewed the semantic web—a vision-turned research field.

Another very important aspect to mention is that of the associated concept with the semantic web—web of data, linked data, and knowledge graph. These concepts appear misleading due to different definitions and descriptions adduced to them, resulting in the concepts being misconstrued by many as the semantic web. It is pertinent to mention that efforts have also been made in the literature to show the dichotomy between these concepts and the semantic web as evident in [6,15,17,18]. While the work agrees subtly with the semantic web being referred to as web of data, it is obvious that the concepts "linked data" and "knowledge graph" are just ways of achieving the semantic web, and as such can be rightly categorized as SWTs—and will therefore be discussed suitably as SWTs.

The quest for the realization of the semantic web no doubt has attracted several fields including linguistics, artificial intelligence (AI), and data mining. These fields though have become even more interwoven with the semantic web in recent times so much so that many are beginning to confuse them altogether. These domains are different but undeniably complementary as evident in [21–24]. Natural Language Processing (NLP), for instance, is originally a sub-field of linguistics from which computational linguistics (CL) emerged. In as much as it cannot be claimed that the advent of the semantic web birthed the field of CL—as the field predates the semantic web, it is imperative to mention that the rapid advancement of the domain is greatly influenced by the advent of the semantic web, as evident with the emergence of several platforms and tools for text engineering and annotation (e.g. GATE and Knowtator) and has found practical use both in academia and industry.

It is also important to appreciate that both AI and CL graced the world at about the same period in the 1950s with the vision of AI being to make machines perform cognitive tasks as if they were humans, and that of CL was for machines to automatically recognize and manipulate texts in human language. CL has over the years become vital to both the field of AI and the semantic web—and has become a great approach to realizing the visions of both fields [25]. Again, data mining, which is often referred to as knowledge discovery on databases (KDD) [23,24], became popular at about the 1990s, owing to the problem of information overload—the inability of users to deal with a large volume of data called "big data" and discern the value in the data that meet their needs for prompt and accurate decision making. The kernel and the end product of data mining is "knowledge". Combining data mining with SWTs can help enrich data mining processes with domain knowledge and as well ease knowledge discovery, representation, and reuse [23,24].

Overall, while, on the one hand, the semantic web and its associated technologies are helping these fields realize their visions, on the other hand, these fields are helping to realize the semantic web's vision too; a clear indication that these fields are strongly intertwined and complementary. Having described the transformation of the semantic web and some of the key fields associated with it, we next describe the SWTs.

1.3 SEMANTIC WEB TECHNOLOGIES

When referring to SWTs what traditionally comes to mind are the entities described and represented in the semantic web layer cake [2,3] as christened by the pioneers of the field of the semantic web. The semantic web layer cake has received several modifications over the years to align with current realities, particularly with the proliferation of several disruptive technologies. The standard layer cake is depicted in Figure 1.2.

From Figure 1.2, while some experts have described the middle slice of the architecture—that is, the portion above the XML and below the Unifying Logic—as the core SWTs, this work, in addition, sees the technologies below the middle slice—technologies predating the semantic web—as part of SWTs also.

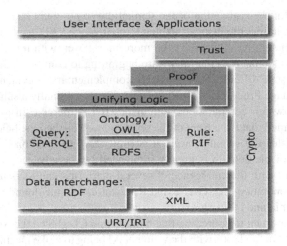

FIGURE 1.2 A variant of the semantic web layered cake [26].

TABLE 1.1
Categorization of the Semantic Web Technologies

	Category	Description	Instances
Semantic Web Technologies	Standards	Formats, languages, or vocabularies	URI/IRI, XML, RDF/RDFS, SKOS, OWL, SPARQL, RIF
	Methods	Approaches, practices, or techniques	Contextual analysis, reasoning engine, natural language understanding, knowledge graph, linked data, and ontology
	Tools	Editors, development environments, or plug-ins	Annotation tools (e.g. Knowtator), acquisition tools (e.g. OntoEdit), ontology rendering tools (e.g. Protege), and reasoners (e.g. Pellet)

Though SWTs have received several categorizations and descriptions in literature [2,5,9,11,27], this work has convened these technologies along a tripod dimension of standards, methods, and tools; for clarity purposes. Thus, this work views SWTs as collections of standards, methods and tools that aid the actualization of the vision of machines' seamless comprehension of data or information and performance of cognitive tasks as surrogates. This categorization of the SWTs is described in Table 1.1.

1.3.1 SEMANTIC WEB STANDARDS

The focus of this subsection is to explicate the standards listed in Table 1.1. To achieve this, their definition, description, benefits, and flaws where applicable were exposed.

1.3.1.1 URI/IRI

Uniform Resource Identifier (URI) may be viewed as a sequence of characters chosen from a subgroup of ASCII characters that identifies an abstract or physical resource [28,29]. These characters are representative of words in natural languages. URI is decked by three entities—uniformity, resource, and identifier—with each contributing a unique adornment to the entire concept of URI. It is evident from [28] that while uniformity enforces consistency, resource depicts anything abstract or physical that is identifiable by URI, and identifier holds vital information to differentiate an identified resource from every other within its scope of identification. The adoption of this simple and extensible mechanism presents huge benefits which includes allowing (i) usage of different types of resource identifiers in the same contexts even with differences in access mechanism deployed, (ii) reusability of identifiers in diverse contexts, (iii) existence of a consistent semantic interpretation of common syntactic protocols across disparate types of resource identifiers, and (iv) introduction of new types of resource identifier without interfering with the way that existing identifiers are used [28,29]. Figure 1.3 is a representation of a typical URI and its components.

From Figure 1.3, the upper URI shows a scenario with all components; scheme, authority, path, query, and fragment present, whereas the lower URI represents a scenario with some missing components. For a more detailed description of these components, see [28].

The Internationalized Resource Identifier (IRI) became incumbent due to the problem of ambiguity introduced by the constraint of having to transcribe non-Latin scripts to natural scripts using Latin scripts as evident in URI [29]. The quest to get out of the dilemma led to creating a new protocol known as IRI. The possibility of mapping IRIs to URIs makes them complementary and thus allows IRIs and URIs to be used interchangeably where and when necessary [29]. IRI serves as a global identifier that can be used to identify any resource and an elemental constituent of the resource description framework data model [30,31]— which is key to the success of both the present web and the semantic web.

1.3.1.2 XML

The Extensible Markup Language (XML)—a progeny of the Standard Generalized Markup Language (SGML—ISO8879:1986)—is a formal meta-markup language and an open standard for information interchange between conceivably two dissimilar systems. Its emergence is a result of the recognition of the inability of the

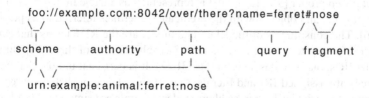

FIGURE 1.3 URI and its component parts [28].

```
1.  <?xml version="1.0" encoding="ISO-8859-1"?>
2.  <book>
3.       <title lang="en">Semantic Web Technologies</title>
4.       <author>Dr. Esingbemi Princewill Ebietomere</author>
5.       <author>Prof. Godspower O. Ekuobase</author>
6.  </book>
```

FIGURE 1.4 XML snippet.

prior infrastructures (e.g. HTML and SGML) of the original web to scale up to meet the need of the semantic web—occasioned by the restrictiveness and the complexity inherent with the HyperText Markup Language (HTML) and SGML, respectively. Consequently, XML incorporates the simplicity of the HTML and the extensibility prowess of the SGML. Besides, while the XML was designed to store and transport data, it was fabricated also to be software and hardware independent. The paper [32] identified the following usages of the XML: (i) it allows users to merge several files to form a whole, (ii) it allows users to identify where illustrations are to be incorporated into text files, and the format used for encoding such illustrations, (iii) it allows users provide processing control information to support programs, including, document validators and browsers, and (iv) it allows users add comments to files. More so, a typical XML consists of two major parts: prolog and body. The prolog which is used for the XML declaration carries information such as version, encoding, schema, and style sheets. Moreover, the body consists of elements with each of the elements consisting of data and markups (tags) which are meta-information about the data it describes. For example, consider the simple XML snippet depicted in Figure 1.4.

From Figure 1.4, the first line represents a prolog with information on the XML version and its encoding. The other parts (lines 2–6) of the snippet collectively referred to as the body, contain the element "book" which is the root element and two other markups—"title" and "author" as branches. Thus, for line 2, the data is "Semantic Web Technologies". Also, line 2 contains additional information like attribute name (lang) and value (en). It is worthy to note also, that every open tag on the left of each of the elements has an associated closing tag as evident in the snippet.

1.3.1.3   RDF/RDFS

The Resource Description Framework (RDF) is a data model that came with the advent of the semantic web and is consequently referred to as the first layer of the semantic web proper [5]. This model is represented as a triple in the form of subject, predicate, and object—with the predicate sometimes referred to as property [33–36]. The constituents of the RDF data model are the RDF terms that can be used in reference to resources. This term in itself consists of three disjoint subsets; the IRIs, blank nodes, and literals [31,37]. It is pertinent to note that not all resources are assigned IRI and literals, thus, the existence of such resources are denoted with variables known as blank nodes. Literals are simply a set of lexical values. For clarity purposes, let us assume I is the set of all IRIs, B is a set of blank nodes, L is a set of literals, and t denotes a triple. Then, the RDF triple may be defined as $t = <s, p, o>$, where $s \in I \cup B$, $p \in I$, and $o \in I \cup B \cup L$. Furthermore, a collection of RDF triples may be viewed as a labeled multi-graph where the subjects and the objects are the nodes in the graph and the properties posing as connectors between the nodes to form an edge shown as $s \xrightarrow{p} o$ [36].

The Resource Description Framework Schema (RDFS) rather than being a data model is a language consisting of vocabularies for describing the classes of resources as well as the properties describing the resources [5,26,38,39]. More succinctly, RDFS may be seen as a vocabulary for RDF data modeling [35,40]. The advent of the RDFS is one of the foremost means of realizing ontologies, allowing RDF resources to be described in terms of classes, properties, and instances hierarchically to bring proper structure and meaning to the web. The RDFS also offers a reasoning framework for inference. An application of RDF and RDFS is the popular Friend of a Friend project (FOAF)—which describes people and the relations between them. Besides, there are several variations and extensions of RDFS (e.g. Resource Description Framework in Attribute (RDFa), and RDFS++). RDFa is a microformat that allows structured data to be added to HTML pages to make it machine-readable [41], while RDFS++ is RDFS with some constructs in OWL to further enrich it. Readers interested in learning the constructs of RDF and RDFS can begin with the following materials: [26,38–40].

1.3.1.4 SKOS

Knowledge has become the soul of every organization particularly in recent times so much so that every economy is becoming knowledge-driven [42]. The quest for organizing this knowledge for ease of use has resulted in several artifacts for controlling vocabularies such as thesauri, taxonomies, folksonomies, classification schemes, and subject heading lists. Though all these controlled vocabulary systems are similar in types but sharing and linking data therefrom on the web comes with a high cost. A way to reduce this cost and allow for easy porting of existing knowledge organization systems on the web at least requires a common data model for these knowledge organization systems which SKOS provides. SKOS is an RDF-based lightweight conceptual modeling language for building and sharing knowledge among organizational systems.

According to Isaac and Summers [43], "SKOS allows concepts to be composed and published on the World Wide Web, linked with data on the web and integrated into other concept schemes". Overall, the crux of SKOS is to allow for wider reuse and interoperability among knowledge organization systems. Besides, it is important to note that, SKOS can be used independently or in combination with more expressive languages like OWL depending on the application need. For a comprehensive knowledge of SKOS [43], is a good text, to begin with.

1.3.1.5 OWL

The Web Ontology Language (OWL)—an extension of the RDFS—is the most formal and expressive knowledge representation language, strongly rooted in classical logic—e.g. descriptive logic. OWL came to existence due to the expressive restrictiveness of the RDFS [34] to completely represent knowledge in a domain of discourse. Its potency in rendering ontologies has seen it become a de-facto language, and as such made it found application across several domains. OWL has three sub-languages—OWL lite, OWL DL, and OWL full—and each with its strength and weakness. Their level of expressivity and formality increases from OWL lite, through OWL DL to OWL full [44]. If expressivity without restriction is the quest, OWL full is the choice, but if automated reasoning and consistency check is considered, then OWL DL is the choice. There are several versions of OWL with the most recent being OWL 2 which has RDF/XML as its primary exchange syntax. The work [44] gave an excellent description of the OWL syntax.

1.3.1.6 SPARQL

Querying is an important process and an integral part of any system that is dependent on a repository (database or knowledgebase or web). It allows relevant data or knowledge to be extracted from its store as may be required by the users—using a defined standard language/format. SPARQL is a recursive acronym for SPARQL Protocol and RDF Query Language. It is a standard query language designed to work with RDF-based repositories [36,45] and has become a de-facto language recommended by W3C. Consequently, SPARQL can be used to manipulate ontologies developed or stored in RDFS and OWL formats. Just like the Structured Query Language is for manipulating databases, so is SPARQL for manipulating knowledge bases.

1.3.1.7 RIF

The use of rules in knowledge representation and business modeling is a popular phenomenon. These rule systems differ along several dimensions particularly semantically, thus making it difficult for these systems to seamlessly interact. The aim of making rule exchange from disparate rule-based systems interoperable, occasioned by the need to have a support mechanism to conceal the limitations of ontology rendering languages like OWL led to the birthing of the Rule Interchange Format (RIF) [46,47]. RIF consists of a family of languages often referred to as dialects—with well-defined syntax and semantics—and designed with the focus of ensuring consistency across several dialects and ease of extensibility. Broadly, three types of rule systems can be identified—first order, logic programming, or action rule [47], but the focus of RIF Working Group of W3C has been on the logic-based, and rules with actions dialects. The logic-based systems have first-order logic (e.g. Horns), and families of non-first order logic as the foundation while the rule with action emphasized production rules.

Overall, it is important to mention there are several tools in existence leveraging these standards and allowing knowledge to be represented in a format like OWL and facilitating its conversion to another format such as XML or RDF as may be required. It also allows for reasoning, and querying in standard format (e.g. Pellet and SPARQL respectively). It is pertinent to mention that designers and developers of web applications, particularly semantic web applications must of necessity pay attention to the discussed standards considering their strengths, limitations, and the type of semantic application they want to realize so as to be certain of which to adopt. If they are employing several standards, they must also be sure of the compatibility of these standards to ensure seamless interaction.

1.3.2 SEMANTIC WEB METHODS

In this section, the focus is to explicate the approaches, practices, or techniques for realizing the semantic web as listed in Table 1.1. To achieve this, a clear definition or description, benefits, and flaws where necessary were explored.

1.3.2.1 Contextual Analysis

Contextual analysis is a popular concept that has been extensively explored in AI [48,49] and has become one of the ways of realizing the semantic web [16]. The semantic web, being a web of heterogeneous linked data requires mechanisms that can support disambiguation and effective delivery of data and knowledge on the semantic web. Contextual analysis is one of such mechanisms and its primary goal is to aid in making data and knowledge on the web unambiguous. Context analysis has been particularly useful in semantic information retrieval [16,50] because of the lacuna that exists between queries and the repository (the web). For example, if a user issues a query with the word "right" which could mean; to be correct, a position, or an entitlement, in a syntactic scenario, all documents containing the word "right" irrespective of its context of use would be retrieved and the number can be overwhelming. This challenge is often referred to as "polysemy" and "synonymy". To overcome the challenge, it is incumbent to make the system understand the context in which the word is used by providing formal and explicit representations for contexts [48,51].

Oftentimes, context modeling is being handled at the level of application programs which is still a problem due to the vastness of data and knowledge on the web and as such, there are proposals on creating frameworks that can go beyond the level of application programs and mere community-specific agreements [48]. This obviously will facilitate linking, reuse, and integration of data and knowledge on the semantic web. Very importantly, we must mention that while contextual analysis, on the one hand, is facilitating the realization of the semantic web, the semantic web on the other hand is revolutionizing contextual modeling.

1.3.2.2 Reasoning Engine

For machines (software agents) to be able to perform repetitive cognitive tasks on behalf of the users, it is imperative to make these agents intelligent as the web is being permeated with data and knowledge since intelligence is conditioned on knowledge [52,53]. At the center of the aforementioned is reasoning—a mechanism that facilitates the deduction of new knowledge from existing ones—which can make machines think as if they were human. Reasoning has been so valuable in different fields including AI, linguistics, philosophy, and computer science and has become an integral part of knowledge-based systems [53,54]. Interestingly, the basic foundation for reasoning is logic and the knowledge reasoned through is usually formally represented using logic [52]. Consequently, while reasoning engines have become one of the drivers of the semantic web, the SWTs are helping in turn to strengthen reasoning engines [55]. This is evident as standards like OWL have found great use in knowledge representation and reasoning. Furthermore, the reasoning engine has found usage in semantic search and expert systems [16,56].

1.3.2.3 Natural Language Understanding

As earlier implied, natural language processing (NLP) and semantic web have different but complementary goals. Thus, while NLP has found application in the semantic web, the semantic web has also helped revolutionize NLP. Natural language understanding (NLU) also referred to as semantic analysis [34] aims to represent the semantics (meaning) of a given text toward enabling machine readability and comprehension [25,34]. NLU has become one of the approaches to the semantic web, particularly in the area of knowledge management. It has found its role in automating knowledge acquisition tasks [57] thereby solving the problem of knowledge acquisition bottleneck – effort and time [56] that bedeviled explicit knowledge description of ontology at the early stage [58,59]. Also, the impact of NLU is visible in semantic web search, information visualization, and connecting text to linked open data [34,57]. Some of the SWTs that support NLU are highlighted under the section dedicated to semantic web tools.

1.3.2.4 Knowledge Graph

The term knowledge graph (KG) is not new [60] but a buzzword supposedly reinvented by Google on the semantic web and has been adopted by several companies and the academia—and due to the diverse perspectives of the concept, it has been misconstrued for other concepts like ontology, knowledgebase, and knowledgebase systems [17]. Besides, its different use has seen it devoid of a single fit in its definition and description [17,60,61]. One notable view of KG is that of [61], which took a path that seemingly aggregates the major views of the concept in the literature under three distinct categories—describing KGs as a knowledge representation tool, a knowledge management system, and a knowledge application service. However, we defined KG as an autonomous knowledge representation technique that access, extract, and aggregate data and knowledge from heterogeneous and distributed data sources in real-time. Irrespective of the views, what is obvious from most of the descriptions of KG is the fact that it is an approach to the semantic web. Consequently, the semantic web interestingly has been described as the broadest KG, or on the other hand—a KG that is capable of crawling the entire web could be construed as a self-contained semantic web [17]. KGs have found usage in semantic search, question answering, and recommender systems [17,18,61–64]. It is important to mention that standards exist for creating KGs. These standards are RDF-based.

1.3.2.5 Linked Data

The concept—linked data—has been used in literature by some authors to refer to data published on the web in such a way to allow for easy consumption and comprehension by machines [41]. Interestingly, this description could as well fit perfectly for open data, thereby creating an impression that linked data is a synonym for open data. In this work, linked data is seen as it should be from the perspective of [15,65,66], where, linked data is succinctly described as a set of best practices or guiding principles for publishing and associating possibly unstructured data or a mix on the web for ease of readability and comprehension by machine. If linked data can be openly and freely shared or distributed, it becomes a linked open data.

The goal of linked data is to achieve proper linking, integration, and reuse of vast and heterogeneous data or knowledge in a repository (e.g. web) [65–67]. Thus just like contextual analysis, reasoning engine, natural language understanding, knowledge graphs, linked data is also a method of realizing the semantic web vision. Linked data has found usage in digital libraries, security, meteorology, etc. Furthermore, techniques and standards exist for publishing linked data—e.g. RDF and SPARQL [14]—which have been discussed.

1.3.2.6 Ontology

The concept of ontology emanated from the field of philosophy and has diffused to the field of AI and computer science—culminating in its several definitions and descriptions [59,68]. There is confusion in the field as to what is considered an ontology and the confusion ranges from the issue of structure and formalism (complexity) to granularity—and this has contributed greatly to the several classifications of the concept in literature [69,70]. Amidst this confusion is the consensus on the potential the concept holds in guaranteeing sharing of common understanding, reusability of knowledge, disambiguation of domain knowledge, and separation of domain knowledge from operational knowledge [71]. Ontology at provenance according to [72] is described as "the science of what is, of the kinds and structures of objects, properties, events, processes and relations in every area of reality"—from which the term was borrowed to be used in AI and computer science. It is worth mentioning that the advent of the semantic web saw the concept redefined with great emphasis on its potential toward the realization of the semantic web vision—so much so—the concept has been labeled as the cornerstone of the semantic web [68,73].

One notable definition that is practicable, easily relatable, and as well resonates with the common perception of the concept is that of [74], which describes the concept thus;

> an ontology is a formal explicit description of concepts in a domain of discourse (classes (sometimes called concepts)), properties of each concept describing various features and attributes of the concept (slots (sometimes called roles or properties)), and restrictions on slots (facets (sometimes called role restrictions)).

Ontology has been used in a myriad of applications across several domains including; knowledge representation, semantic search, interoperability, and decision support [68,71,75].

To conclude this section, the methods for semantic web obviously are dependent on several standards. For example in KG implementation—RDF Mapping Language (RDFML) and its family (R2RML, xR2RML) [61] are popular standards but are dependent on RDF. It is obvious therefore that this work has only discussed the core standards for the realization of the semantic web as specified in the semantic web layer cake. Furthermore, for proper implementation of these methods, there are laid down guidelines and methodologies that must be imbibed by developers. These methodologies are particularly evident in ontology crafting because of the quest for quality (due to its proliferation), and its criticality toward realizing the semantic web. Some of the methodologies include; Methontology, Gruninger and Fox, KATUS, and Noy and McGuiness [59,68]. It is important to note that none of these methodologies is adjudged superior to the other, for they can all lead to building quality ontologies. Consequently, the methodology to employ is a matter of preference.

1.3.3 SEMANTIC WEB TOOLS

Semantic Web tools in this work refer to software components that facilitate the efficient implementation of semantic web methods. These tools include those born as a result of the fundamental challenges that came with the advent of the semantic web, i.e. how to semantically annotate knowledge, acquire knowledge, and represent knowledge. Some of these tools are multi-purpose in nature and thus have overlapping functionalities, for example, many of the semantic knowledge representation tools double up as semantic knowledge acquisition tools; the same can also be said about some knowledge annotation tools acting as knowledge acquisition tools. Furthermore, at the heart of these tools is NLP, acting as a common base and creating a strong interlink among these semantic web tools and as a result further blurring the fine lines among these tools. Very importantly, since the number of tools in existence for annotation, acquisition, and representation of knowledge is huge and can hardly be exhausted by a single text talk less of being part of a book chapter, this work restricts itself to the following semantic web tools: (i) open source—freely available for use, (ii) supported by the standards discussed in this chapter, and (iii) with visualization capability. Again, this section focuses on discussing these semantic web tools along the following dimensional space—semantic knowledge annotation tools, semantic knowledge acquisition tools, semantic knowledge representation tools, and reasoners.

1.3.3.1 Semantic Knowledge Annotation Tools

There are several definitions in literature as to what annotation is [76–80]. These definitions and descriptions observably are perspective-dependent. In any case, the essence of annotation is to enrich content (e.g. data or knowledge) with additional information to advance machine comprehension of the content—that is, associating metadata with content. This work particularly refers to those annotation tools that imbibe principles that suggest the use of semantic web standards (e.g. RDF, and OWL) and methods (e.g. ontology, and NLU) as a basic mechanism or allows for the binding of such to existing web resources, as semantic annotation tools. The content (resources) to be annotated can be texts, images, speech, and videos [76,80–84].

Several tools abound for semantic annotation with these tools supporting manual, semi-automatic to automatic processes. Manual annotation though yields a great result, is impracticable with vast and heterogeneous content to annotate. Automatic annotation though the desired as it can cope with vast and heterogeneous content, is bedeviled with the inaccurate result. Thus, to achieve a balance, the semiautomatic process which combines the strength of both the manual and the automatic processes is usually preferred. Manual annotation tools include Annotea [85] and Knowtator [76]; while Semantator [86] and GATE [87] allow for both manual and automatic annotation. Also, some of these tools can be general purpose—in which case they can be used across several domains e.g. Knowtator and GATE—or domain-specific, that is, built for a specific domain e.g. Semantator (biomedical). It is imperative to mention which of the annotation tools to employ at any point in time is contingent on several scenarios including; nature of problem, domain of problem, and functionality desired.

1.3.3.2   Semantic Knowledge Acquisition Tools

Semantic knowledge acquisition is the process of mining meaningful knowledge from a repository. Knowledge acquisition is often a very costly activity with vast and heterogeneous data and knowledge, particularly when employing the ontological approach to knowledge representation; both from the point of knowledge engineers and domain experts. Acquisition of knowledge can be done manually, semi-automatically, or automatically. The manual style yields great results which are usually the gold standard but very costly (effort and time) because it compels knowledge engineers and domain experts to create ontology from scratch. The automatic style though the desired—as it requires less effort and time— often results in less accurate result (the results are often spurious) as it allows ontology to be bootstrapped from a repository without human intervention. The semi-automatic style allows ontology to be bootstrapped from a repository and refined by the ontology engineer/domain expert. This often yields great results as it leverages the strength of both manual and automatic styles. Some popular knowledge acquisition tools include Protégé [88], OntoEdit [89], OntoLT [90], OntoGen [91] and GATE [87]. All these tools have found use in different applications across several domains with Protégé being the most prominent for manual knowledge acquisition. It is pertinent to note that the act of bootstrapping ontology from a repository is sometimes referred to as ontology learning.

1.3.3.3   Semantic Knowledge Representation Tools

Knowledge representation is a popular term both in the field of AI and the semantic web largely because of its role in allowing knowledge to be encoded and organized in such a way that it can be machine and human-comprehensible [92]. Questions arising in semantic knowledge representation include those of the trade-offs between representational adequacy, fidelity, acquisition cost, and computational cost. Consequently, the following semantic knowledge representation ensues—bag of words, taxonomy, thesaurus, and ontology [58]—with formality, complexity, and explicit description increasing from a bag of words, through taxonomy, and thesaurus to ontology. For obvious reasons, most of the knowledge representation tools are geared toward ontology construction with facilities for graphical visualization (e.g., Owl-viz, Jambalaya, and Ontograf) and storage in triple (RDF) forms—using such stores as Sesame. Several knowledge representation tools often referred to as ontology rendering tools exist in the literature supporting, either manual, semi-automatic, or automatic knowledge representation. These tools include; Protégé, WebODE [93], OntoEdit, and OntoGen.

1.3.3.4 Reasoners

Reasoning is an integral part of knowledge-based systems and a key to the realization of the semantic web vision. Its criticality has led to the proliferation of software components referred to as reasoners. A reasoner typically consists of knowledge representation formalism and inference mechanism. Two basic paradigms are underlining the workings of reasoners viz: the use of classical logic (e.g., description logic), and the datalog (rule) [94]. It is not uncommon to find a hybrid of the two approaches. Some popular reasoners include Pellet [95], RacerPro [96], HermiT [97], FACT++ [98], and JFact. These reasoners are already incorporated into many knowledge representation editors.

To conclude this section, it is important to reiterate that there are many semantic web tools in existence and for pedagogical reasons; these tools have been discussed along the dimensional spaces of annotation, acquisition, representation, and reasoning giving instances of the tools falling under each of the dimensions. Furthermore, some of these tools have become a complete platform incorporating the processes of annotation, acquisition, representation, and reasoning, and thus are being used to build complete semantic applications.

1.4   A PEEP INTO THE PRAGMATICS OF SEMANTIC WEB TECHNOLOGIES

The semantic web and its associated technologies have recorded successes across several domains—enabling machine comprehension of vast heterogeneous data from multiple independent sources toward performing cognitive and decision support tasks; and in the vision of industry 4.0, extends decision support to decision dictate. The sustenance of these successes is dependent on the pragmatics of deploying these technologies. Already, many of the semantic web tools glitch with increasing data or prolonged high-velocity use—scalability problem. The interoperability challenge of semantic web technologies is being worsened by the demands of smart environment and industry 4.0 [99]. It is easy to appreciate that data sources can easily be compromised to generate, contain or send false data thereby forcing the machines to make wrong decisions—data quality/integrity problem. Thus, security compromises on real-time semantic web applications including denial of service attacks may emanate from sensors and other data sources. Linked data and knowledge graph also pose the challenge of data privacy. Multilingual and cultural issues are challenges that must be addressed for the anticipated seamless global comprehension of data by machines. Also critical is the issue of data availability—guaranteed freshness of data. As an emerging field of study, experts and personnel with the right technical know-how remain scarce. Despite these challenges, the application of the semantic web and its associated technologies is fast diffusing into more areas because of the benefits already enjoyed by those areas that have embraced it, a trend that will continue, particularly as gilt-edge solutions are being churned out in the field. Although, efforts at resolving some of these challenges have begun [100], attenuating the impacts of these challenges is the semantic web desired "breadth of life"—until this happens, the destination is still far ahead of us.

1.5 CONCLUSION

This chapter began with the exploration of the semantic web concept and concluded that the semantic web is a vision turned research field—dedicated to the seamless comprehension of vast heterogeneous and distributed data and the performance of complex and repetitive cognitive and decision support tasks as surrogates. The work exposed the relationship and disparity between the semantic web and some conflicting concepts such as AI, CL, and data mining; and established that they are different fields that reinforce one another. Also, for ease of appreciation of the various SWTs the work examined them under three basic categories: standards, methods and tools. Terminological conflicts associated with some instances in the various categories (e.g. RDF/RDFS, linked data/linked open data, ontology/knowledge graph) were resolved. It was noted that some SWT tools can perform more than one task classification of knowledge annotation, acquisition, representation and reasoning; under which they were discussed in the work. Such tools are often referred to as platforms (e.g. GATE, Protégé).

Finally, the chapter made evident that with the increasing burden of data overload and heterogeneity coupled with distributed data sources; the promising future of semantic web may be stifled by the problem of data security, privacy, integrity, scalability, and availability. More stifling is the problem of integration and interoperability which the SWTs and their associated technologies and resultant applications are presently struggling with. Many readers, particularly new entrants into the field of semantic web, may expect the chapter to practically expose them to the SWTs but considering the myriads of the technologies, this is not feasible, particularly with a chapter in book. It is however hoped that this introduction to SWTs will equip new entrants into semantic web with an excellent cognitive stand on SWTs.

REFERENCES

1. Horrocks, I., "Semantic Web: The Story So Far", *In the Proceedings of International Cross-Disciplinary Conference on Web Accessibility* (W4A2007), ACM, pp. 120–125, 2007.
2. Berners-Lee, T., "Semantic Web Road Map", W3C, 1998. https://www.w3.org/DesignIssues/Semantic.html.

3. Berners-Lee, T., Hendler, J. and Lassila, O., "The Semantic Web: A New Form of Web Content That is Meaningful to Computers Will Unleash a Revolution of New Possibilities", Scientific American, 3pp., 2001.

4. Hayes, P., "The SW vision", *In the International Conference on Information and Knowledge Management* (CIKM'04), ACM, pp. 416–416, 2004.

5. Matthews, B., "Semantic Web Technologies", *JISC Technology and Standards Watch*, 20pp., 2005.

6. Machado, L. M. O., Borges, M. M. and Souza, R. R., "The Evolution of the Concept of Semantic Web in the Context of Wikipedia: An Exploratory Approach to Study the Collective Conceptualization in a Digital Collaborative Environment", *Publications*, vol. 6, no. 44, 2018. doi:10.3390/publications6040044.

7. Barati, M., "The vision of Semantic Web", *The Vision of Semantic Web*. In *IEEE/ ACM 12th International Conference on Utility and Cloud Computing Companion (UCC '19 Companion)*, ACM, pp. 175–176, 2019. doi:10.1145/3368235.3370268.

8. Machado, L. M. O., Souza, R. R. and Simoes, M. G., "Semantic Web or Web of Data? A Diachronic Study (1999 to 2017) of the Publications of Tim Berners-Lee and the World Wide Web Consortium", *Journal of the Association for Information Science and Technology*, vol. 70, no. 7, pp. 701–714, 2019. doi:10.1002/asi.24111.

9. Hitzler, P., "A Review of the Semantic Web Field", *Communications of the ACM*, vol. 64, no. 2, pp. 76–83, 2021. doi:10.1145/3397512.

10. Cardiff, J., "The Evolution of the Semantic Web", 2009. https://www.researchgate. net/publication/228533759_The_Evolution_of_the_Semantic_Web.

11. Hogan, A., "Linked Data and the Semantic Web Standards", In A. Harth, K. Hose, and R. Schenkel (Eds.), *Linked Data Management*, Chapman and Hall/CRC Press, New York, 53pp., 2013.

12. Gutierrez, C., Hurtado, C. and Mendelzon, A. O., "Foundations of Semantic Web Databases", In *Proceedings of Database systems (PODS'04)*, ACM, pp. 95–106, 2004.

13. Skritek, S., "Foundational Aspects of Semantic Web Optimization", *In the International Conference on Management of Data (SIGMOD/PODS'12)*, ACM, pp. 45–49, 2012.

14. Yumusak, S., Kamilaris, A. and Dogdu, E., Kodaz, H., Uysal, E. and Aras, R. E., "A Discovery and Analysis Engine for SemanticWeb", *International Workshop on Profiling and Searching Data on the Web, WWW'18*, ACM, pp. 1497–1505, 2018. doi:10.1145/3184558.3191599.

15. Bizer, C., Heath, T. and Berners-Lee, T., "Linked Data—The Story So Far", In T. Heath, M. Hepp, and C. Bizer (Eds.), *Special Issue on Linked Data*, International Journal on Semantic Web and Information Systems (IJSWIS), vol. 5, no. 3, 22pp., 2009. doi:10.4018/jswis.2009081901.

16. Sudeepthi, G., Anuradha, G. and Babu, M. S. P., "A Survey on Semantic Web Search Engine", *International Journal of Computer Science Issues (IJCSI)*, vol. 9, no 1, pp. 241–245, 2012.

17. Ehrlinger, L. and WoB, W., "Towards a Definition of Knowledge Graphs", *SEMANTICS: Posters and Demos Track*, 4pp., 2016. http://ceur-ws.org/Vol-1695/ paper4.pdf.

18. Hao, X., Ji, Z., Li, X., Yin, L., Liu, L., Sun, M., Liu, Q. and Yang, R., "Construction and Application of Knowledge Graph", *Remote Sensing*, vol. 13, no. 2511, 2021. doi:10.3390/rs13132511.

19. Parsia, B. and Patel-Schneider, P. F., "Meaning and Semantic Web", *In World Wide Web conference (WWW'04)*, ACM, pp.306–307, 2004.

20. Choudhury, N., "World Wide Web and Its Journey from Web 1.0 to Web 4.0", *International Journal of Computer Science and Information Technologies (IJCSIT)*, vol. 5, no. 6, pp. 8096–8100, 2014.

21. Aqel, D. and Vadera, S., "A Framework for Employee Appraisals Based on Sentiment Analysis", *In the Proceedings of the 1st International Conference on Intelligent Semantic Web-Services and Applications (ISWSA 2010)*, ACM, 6pp., 2010.

22. Kumar, A. and Joshi, A., "Ontology Driven Sentiment Analysis on Social Web for Government Intelligence", *In the Proceedings of the 10th International Conference on Theory and Practice of Electronic Governance*, ACM, pp. 134–139, 2017. doi:10.1145/3055219.3055229.

23. Ristoski, P. and Paulheim, H., "Semantic Web in Data Mining and Discovery: A Comprehensive Survey", *Web Semantics: Science, Services and Agents on the World Wide Web*, vol. 36, no. 2016, pp. 1–22, 2016. doi:10.1016/j.websem.2016.01.001.

24. Petrova, E., Pauwels, P., Svidt, K. and Jensen, R. L., "Semantic Data Mining and Linked Data for a Recommender System in the AEC Industry", *European Conference on Computing in Construction (EC³)*, pp. 173–181, 2019.

25. Navigli, R., "Natural language Understanding: Instructions for (Present and Future) Use", *Proceedings of the Twenty-Seventh International Joint Conference on Artificial Intelligence (IJCAI-18)*, pp. 5697–5702, 2018.

26. Sheth, A. P., "Semantic Web: Technologies and Applications for the Real-World", *CORE Scholar*, Wright State University, 128pp., 2007. https://corescholar.libraries.wright.edu/knoesis/640.

27. Patel, A., and Jain, S., "Present and Future of Semantic Web Technologies: A Research Statement", *International Journal of Computers and Applications*, vol. 43, no. 5, pp. 413–422, 2019. doi:10.1080/1206212X.2019.1570666.

28. Berners-Lee, T., Fielding, R. and Masinter, L., "Internet Standard", *RFC 3986, The Internet Society*, 2005. https://datatracker.ietf.org/doc/html/rfc3986#section-3.

29. Duerst, M. and Suignard, M., "Internationalized Resource Identifiers (IRIs)", *The Internet Society*, 46pp., 2005.

30. Masnter, L. and Duerst, M., "Equivalence and Canonicalization of Internationalized Resource Identifiers (IRIs)", *TOC, IETF*, 8pp., 2011.

31. Tomaszuk, D., "Inference Rules for RDF(S) and OWL in N3Logic", 5pp., 2016. arXiv:1601.02650v1.

32. Bryan, M., "An Introduction to the Extensible Markup Language (XML)", *Bulletin of the American Society for Information Science*, vol.25, no.1, pp. 11–14, 1998.

33. Horrocks, I., "Ontologies and the Semantic Web", *Communications of the ACM*, vol.51, no.12, pp. 58–67, 2008. doi:10.1145/1409360.1409377.

34. Habernal, I., "Semantic Web Search Using Natural Language Processing", Doctoral Thesis, Department of Computer Science and Engineering, Faculty of Applied Sciences, University of Bohemia, Plzen, 120pp., 2012.

35. Ciobanu, G., Hornea, R. and Sassoned, V., "A Descriptive Type Foundation for RDF Schema", *Journal of Logical and Algebraic Methods in Programming, ScienceDirect*, 32pp., 2016. doi:10.1016/j.jlamp.2016.02.006.

36. Wylot, M., Hauswirth, M., Cudré-Mauroux, P. and Sakr, S., "RDF Data Storage and Query Processing Schemes: A Survey", *ACM Computing Survey*, vol. 51, no. 4, 36pp., 2018 doi:10.1145/3177850.

37. Fensel, D. and Facca, F., "Semantic Web: RDF and RDF Schema", 2008. https://www.sti-innsbruck.at/sites/default/files/courses/fileadmin/documents/semweb09-10/03_SW-RDF_and_RDFS.pdf.

38. Manola, F. and Miller, E., "RDF Primer", W3C, 2004. http://www.w3.org/TR/2004/REC-rdf-primer-20040210/.

39. Manola, F., Miller, E., Beckett, D. and Herman, I., "RDF Primer—Turtle Version", W3C, 2007. http://www.w3.org/2007/02/turtle/primer/.

40. Brickley, D. and Guha, R. V., "RDF Schema 1.1", W3C, 2014. http://www.w3.org/TR/2014/REC-rdf-schema-20140225/.
41. Rocha, A. and Prazeres, C., "LDoW-PaN: Linked Data on the Web—Presentation and Navigation", *ACM Transactions on the Web*, vol. 11, no. 4, 42pp., 2017. doi:10.1145/2983643.
42. Zakharova, E. N., Chistova, M. V., Abesalashvili, M. Z. and Gonenko, D. V., "The Role of Intellectual Services Sector in the Development of Innovative Processes of the Modern Knowledge Economy", *Revista ESPA CIOS*, vol. 40, no. 40, 11pp., 2019.
43. Isaac, A. and Summers, E., "SKOS Simple Knowledge Organization System Primer", W3C, 2009. http://www.w3.org/TR/2009/NOTE-skos-primer-20090818/.
44. Horridge, M., Jupp, S., Moulton, G., Rector, A., Stevens, R. and Wroe, C., "A Practical Guide To Building OWL Ontologies Using Protégé 4 and CO-ODE Tools Edition 1.1", University of Manchester, UK, 103pp., 2007.
45. Abdelaziz, I., Mansour, E., Ouzzani, M., Aboulnaga, A. and Kalnis, P., "Lusail: A System for Querying Linked Data at Scale", *Proceedings of the VLDB Endowment*, vol. 11, no. 4, pp. 485–498, 2017. doi:10.1145/3164135.3164144.
46. Kifer, M., "Rule Interchange Format: The Framework", In D. Calvanese, and G. Lausen (Eds), *Web Reasoning and Rule Systems*, LNCS, Springer, vol. 5341, 2008. doi:10.1007/978-3-540-88737-9_1.
47. Kifer, M. and Boley, H., "RIF Overview (Second Edition)", W3C Working Group Note 5, 2013. http://www.w3.org/TR/2013/NOTE-rif-overview-20130205/.
48. Bao, J., Tao, J., McGuiness, D. L. and Smart, P. R., "Context Representation for the Semantic Web", *In Proceedings of WebSci10 (2010)*, 8pp., 2010.
49. Patton, D. U., Frey, W. R., McGregor, K. A., Lee, F., McKeown, K. and Moss, E., "Contextual Analysis of Social Media: The Promise and Challenge of Eliciting Context in Social Media Posts with Natural Language Processing", *In the Proceedings of the AAAI/ACM Conference on AI, Ethics and Society (AIES'20)*, 6pp., 2020. doi:10.1145/3375627.3375841.
50. Ruas, T. and Grosky, W., "Keyword Extraction through Contextual Semantic Analysis of Documents", *In the Proceedings of the 2017 MEDES Conference (MEDES'17)*, Bangkok, Thailand, pp.150–156, 2017. doi:10.1145/3167020.3167043.
51. Luo, Z., "Contextual Analysis of Word Meaning in Type-Theoretical Semantics", *In the Proceeding of the 2011 LACL Conference, LNAI 6736*, Springer-Verlag, pp. 159–174, 2011. doi:10.1007/978-3-642-22221-4_11.
52. Brachman, R. J. and Levesque, H. J., *Knowledge Representation and Reasoning*, Morgan Kaufman Publishers (Elsevier), San Francisco, CA, United States, 413pp., 2004.
53. Fan, T. and Liau, C., "Reason-maintenance Belief Logic with Uncertain Information", *ACM Transactions on Computational Logic (TOCL)*, vol. 21, no. 1, 32pp., 2019. doi:10.1145/3355608.
54. Mitsch, S., Platzer, A., Retschitzegger, W. and Schwinger, W., "Logic-Based Modeling Approaches for Qualitative and Hybrid Reasoning in Dynamic Spatial Systems", *ACM Computing Survey*, vol. 48, no. 1, 40 pp., 2015.
55. Maarala, A. L., Su, X. and Riekki, J., "Semantic Reasoning for Context-aware Internet of Things Applications", *Internet of Things Journal, IEEE*, pp.1–12, 2016.
56. Gutierrez, C. and Sequeda, J. F., "Knowledge Graph", *Communications of the ACM*, vol. 64, no. 3, pp. 96–104, 2021. doi:10.1145/3418294.
57. Maynard, D., Bontcheva, K. and Augenstein, I., *Natural Language Processing for the Semantic Web*, Morgan & Claypool Publishers, 194pp., 2016. doi:10.2200/S00741ED1V01Y201611WBE015.

58. Sanchez, M. F., "Semantically Enhanced Information Retrieval: An Ontology-based Approach", Ph.D. Thesis, Escuela Politécnica Superior, Universidad Autonoma De Madrid, 254pp., 2009.

59. Ebietomere, E. P., "A semantic Retrieval System for Nigerian Case Law", Doctoral Thesis. University of Benin, Benin City, 260pp., 2018.

60. Hogan, A., Cochez, M., D'amato, C., Melo, G. D., Gutierrez, C., Kirrane, S., Gayo, J. E. L., Navigli, R., Neumaier, S., Ngomo, A. N., Polleres, A., Rashid, S. M., Rula, A., Schmelzeisen, L., Sequeda, J., Staab, S. and Zimmermann, A., "Knowledge Graphs", *ACM Computing Surveys*, vol. 54, no. 4, 37pp., 2021. doi:10.1145/3447772.

61. Bellomarini, L., Sallinger, E. and Vahdati, S., "Knowledge Graph: The Layered Perspective", In V. Janev, D. Graux, H. Jabeen, and E. Sallinger (Eds.), *Knowledge Graph and Big Data Processing*, LNCS 12072, Springer, Cham, Switzerland, 212pp., 2020.

62. Gonzalez, L. and Hogan, A., "Modeling Dynamics in Semantic Web Knowledge Graphs with Formal Concept Analysis", *In the World Wide Web Conference (WWW'18)*, ACM, 10pp., 2018. doi:10.1145/3178876.3186016.

63. Li, Z., Xu, Q., Jiang, Y., Cao, X. and Huang, Q., "Quaternion-Based Knowledge Graph Network for Recommendation", *In Proceedings of the 28th ACM International Conference on Multimedia (MM'20)*, ACM, pp. 880–888, 2020. doi:10.1145/3394171.3413992.

64. Li, J., Qu, K., Li, K., Chen, Z., Fang, S. and Yan, J., "Knowledge Graph Question Answering Based on TE-BiLTM and Knowledge Graph Embedding", *In the 5th International Conference on Innovation in Artificial Intelligence (ICIAI'21)*, ACM, 6pp., 2021. doi:10.1145/3461353.3461366.

65. Mountantonakis, M. and Tzitzikas, Y., "Large-Scale Semantic Integration of Linked Data: A Survey", *ACM Computing Survey*, vol. 52, no. 5, 40pp., 2019. doi:10.1145/3345551.

66. Haller, A. Fernandez, J. D., Kamdar, M. R. and Polleres, A., "What Are Links in Linked Open Data? A Characterization and Evaluation of Links between Knowledge Graphs on the Web" *ACM Journal of Data and Information Quality*, vol. 12, no. 2, 34pp., 2020. doi:10.1145/3369875.

67. Mountantonakis, M. and Tzitzikas, Y., "Scalable Methods for Measuring the Connectivity and Quality of Large Numbers of Linked Datasets", *ACM Journal of Data and Information Quality*, vol. 9, no. 3, 49 pp., 2018. doi:10.1145/3165713.

68. Ekuobase, G. O. and Ebietomere, E. P., "Ontology for alleviating poverty among farmers in Nigeria", *The 10th International Conference on Informatics and Systems, Association for Computing Machinery*, pp.28–34, 2016. doi. org/10.1145/2908446.2908465.

69. Uschold, M., "Building Ontologies: Towards a Unified Methodology", *Proceedings of Expert Systems, the 16th Annual Conference of British Computer Society Specialist Group on Expert Systems*, 20pp., 1996.

70. Roussey, C., Pinet, F., Ah Kang, M. and Corcho, O., "Introduction to Ontologies and Ontology Engineering", In G. Falquet, C. Metral, J. Teller and C. Tweed, *Ontologies in Urban Development Projects*, Springer, Heidelberg, New York, pp. 9–38, 2011.

71. Ebietomere, E. P., Nse, U., Ekuobase, Betty U. and Ekuobase, G. O., "Crafting Electronic Medical Record Ontology for Interoperability", *The African Journal of Information Systems*, vol. 13, no. 3, pp. 296–315, 2021.

72. Smith, B., "Ontology", In L. Floridi (Eds.), *Blackwell Guide to the Philosophy of Computing and Information*, Oxford: Blackwell, pp. 155–166, 2003.

73. Xue, X., Wu, X. and Chen, J., "Optimizing Ontology Alignment Through an Interactive Compact Genetic Algorithm", *ACM Transactions on Management Information Systems*, vol. 12, no. 2, 17pp., 2021. doi:10.1145/3439772.

74. Noy, N. F. and McGuinness, D. L., *Ontology Development 101: A guide to Creating Your First Ontology*, Stanford, CA: Stanford University, 25pp., 2001.

75. Ebietomere, E. P., and Ekuobase, G. O., "A Semantic Retrieval System for Case Law", *Applied Computer Systems*, vol. 24, no.1, pp. 38–48, 2019. doi:10.2478/acss-2019-0006.

76. Ogren, P. V., "Knowtator: A Protégé Plug-in for Annotated Corpus Construction", *In the Proceedings of the Human Language Technology Conference of the NAACL, Companion Volume*, pp. 273–275, 2006.

77. Oren, E., Möller, K. H., Scerri, S., Handschuh, S. and Sintek M., *What Are Semantic Annotations*, Galway, Ireland: Technical Report, 2006.

78. Belloze, K. T., Monteiro, D. I. S. B., Lima, T. F., Silva, Jr., F. P. S., and Cavalcanti, M. C., "An Evaluation of Annotation Tools for Biomedical Texts", *In The Joint V Seminar on Ontology Research in Brazil and VII International Workshop on Metamodels, Ontologies and Semantic Technologies, CEUR-WS*, vol. 938, pp. 108–119, 2012.

79. Gayoso-Cabada, J., Sarasa-Cabezuelo, A. and Sierra-Rodríguez, J., "A Review of Annotation Classification Tools in the Educational Domain", *Open Computing Science, De Gruyter*, vol. 9, pp. 299–307, 2019. doi:10.1515/comp-2019-0021.

80. Liao, X. and Zhao, Z., "Unsupervised Approaches for Textual Semantic Annotation, A Survey", *ACM Computing Survey*, vol. 52, no. 4, 45pp., 2019. doi:10.1145/3324473.

81. Tetreault, J., Swift, M., Prithviraj, P., Dzikovska, M. and Allen, J., "Discourse Annotation in the Monroe Corpus", *In the Proceedings of the ACL Workshop on Discourse Annotation (DiscAnnotation'04)*, ACL, pp. 103–109, 2004.

82. Leong, C. W. and Mihalcea, R., "Explorations in Automatic Image Annotation Using Textual Features", *In the Proceedings of the Third Linguistic Annotation Workshop (ACL-IJCNLP'09)*, ACL, pp. 56–59, 2009.

83. Marrero, M., Urbano, J., Morato, J. and Sánchez-Cuadrado, S., "On the Definition of Patterns for Semantic Annotation", *In the Proceedings of the third workshop on Exploiting semantic annotations in information retrieval (ESAIR'10)*, ACM, pp. 15–16, 2010. doi:10.1145/1871962.1871972.

84. Schoning, J., Faion, P., and Heidemann, G., "Semi-Automatic Ground Truth Annotation in Videos: An Interactive Tool for Polygon-Based Object Annotation and Segmentation", *In the Proceeding of the 8th International Conference on Knowledge Capture (K-CAP)*, ACM, no. 17, pp. 1–4, 2015. doi:10.1145/2815833.2816947.

85. Kahan, J., Koivunen, M., Prud'Hommeaux, E. and Swick, R. R. "Annotea: An Open RDF Infrastructure for Shared Web Annotations", *In the Proceedings of the WWW10 International Conference*, Hong Kong, 2001.

86. Tao, C., Song, D., Sharma, D. and Chute, C. G., "Semantator: Semantic Annotator for Converting Biomedical Text to Linked Data", *Journal of Biomedical Informatics*, vol. 46, no. 2013, pp.882–893. doi:10.1016/j.jbi.2013.07.003.

87. Cunningham, H., Maynard, D., Bontcheva, K., Tablan, V., Aswani, N., Roberts, I., Gorrell, G., Funk, A., Roberts, A., Dalmjanovic, D., Heitz, T., Greenwood, M. A., Saggion, H., Petrak, J., Li, Y., Peters, W., Derczynski, L. et al., "Developing Language Processing Components with GATE Version 8 (A User Guide)", The University of Sheffield, Department of Computer Science, 632pp., 2020.

88. Noy, N. F. Sintek, M. Decker, S. Crubézy, M. Fergerson, R. W. and Musen, M. A., "Creating Semantic Web Contents with Protégé-2000", *IEEE Intelligent Systems*, vol. 16, pp. 60–71, 2001.

89. Sure, Y., Staab, S., Erdmann, M., Angele, J., Studer, R. and Wenke, D., "OntoEdit: Collaborative Ontology Development for the Semantic Web", *In the Proceeding of International Semantic Web Conference (ISWC'02)*, Springer, pp. 221–235, 2002.

90. Buitelaar, P., Olejnik, D. and Sintek, M., "A Protégé Plug-In for Ontology Extraction from Text Based on Linguistic Analysis", *In Proceedings of the 1st European Semantic Web Symposium (ESWS)*, 14pp., 2004.

91. Fortuna, B., Grobelnik, M. and Mladenić, D., "Semi-Automatic Data-Driven Ontology Construction System", *In Proceedings of the 9th international multi-conference Information Society IS, Ljubljana*, Slovenia, pp. 223–226, 2006.

92. Hayes, P. J., "Knowledge Representation", *Encyclopedia of Computer Science*, vol. 3, pp. 947–95, 2003.

93. Corcho, O., Fernandez-Lopez, M., Gomez-Perez, A. and Vicente, O., "WebODE: An Integrated Workbench for Ontology Representation, Reasoning and Exchange", *In the Proceedings of 13th International conference on Knowledge Engineering and Knowledge Management (EKAW'02)*, Springer, LNAI, vol. 2473, pp. 138–153, 2002.

94. Meditskos, G. and Bassiliades, N., "Rule-based OWL Reasoning Systems: Implementations, Strengths and Weaknesses", In A. Giurca, D. Gasevic, and K. Taveter (Eds.), *Handbook of Research on Emerging Rule-Based Languages and Technologies: Open Solutions and Approaches*, Aristotle University of Thessaloniki, Greece, pp. 124–148, 2008.

95. Sirin, E., Parsia, B., Grau, B. C., Kalyanpur, A., and Katz, Y., "Pellet: A Practical OWL-DL Reasoner", *Web Semantics: Science, Services and Agents on the World Wide Web*, vol. 5, no. 2, pp. 51–53, 2007.

96. Haarslev, V. and Moller, R., "Racer: A Core Inference Engine for the Semantic Web", *In the Proceedings of the International Workshop on Evaluation of Ontology-based Tools*, pp. 27–36, 2003.

97. Glimm, B., Horrocks, I., Motik, B., Stoilos, G. and Wang, Z., "HermiT: An OWL_2 Reasoner", *Journal of Automated Reasoning*, vol. 53, no. 3, pp. 245–269, 2014. doi:10.1007/s10817-014-9305-1.

98. Tsarkov, D. and Horrocks, I., "Fact++ Description Logic Reasoner: System Description", *In the Proceedings of Automated Reasoning*, Springer, pp. 292–297, 2006.

99. Grangel-González, I., "A Knowledge Graph Based Integration Approach for Industry 4.0", Doctoral Thesis, University of Bonn, 195pp., 2019.

100. Noy, N., Gao, Y., Jain, A., Narayanan, A., Patterson, A. and Taylor, J., Industry-scale Knowledge Graphs: Lesson and Challenges. *Communications of the ACM*, vol. 62, no. 8, pp. 36–43, 2019.

2 Leveraging Semantic Web Technologies for Veracity Assessment of Big Biodiversity Data

Zaenal Akbar, Yulia A. Kartika, Dadan R. Saleh, Hani F. Mustika, Lindung P. Manik, Foni A. Setiawan, and Ika A. Satya

CONTENTS

2.1 INTRODUCTION

In the last couple of years, we have generated and organized an unprecedented amount of data. The number of data created worldwide is huge and grows exponentially. In 2020, we generated about 64.2 Zettabytes, 30 times more than 10 years ago [1]. This phenomenon, known as "big data", where data is characterized by five Vs

DOI: 10.1201/9781003309420-2

(Volume, Velocity, Variety, Veracity, and Value) [2]. Volume refers to the amount of data, Velocity refers to the speed of data generation, Variety refers to different types of data. Veracity and Value refer to the quality of data and benefits of the data respectively. While the first three Vs are used to characterize the key properties of big data, the 4th V (i.e., Veracity) is important to make big data operational [3]. It has become one of the critical factors for creating value because of big data's inherent uncertainty in the form of biases, ambiguities, and inaccuracies [4]. As the consequence, with a large amount of available data, often from a diversity of sources, it is impossible to assess data veracity manually [5]. Methods, algorithms, as well as tools for assessing data veracity automatically, are highly required.

Data veracity refers to the inconsistency and data quality problems, where poor quality data would affect the results of data analyses [3]. The quality of data influences the extraction of useful and valuable knowledge [6], where the presence of uncertainty in the data may negatively impact the effectiveness and the accuracy of the analyses [7]. Therefore, an assessment method for data veracity that deals with uncertain or imprecise data need to look into multiple factors including data inconsistency, incompleteness, ambiguities, as well as deception [8]. More than that, data from multiple sources introduce data conflicts, making data veracity assessment even more challenging [9]. Veracity is also compromised by the occurrence of intentional deceptions such as fake news, malicious rumors, and fabricated reviews [5].

Due to the wide variety of factors and sources that could affect data veracity, we limited our work to the unintentional factor only, specifically those are originated from data consistency and data uncertainty. There are three main sources for data inconsistency, namely the difference in storage format, semantic expression, and value [10]. Storage format refers to the types of the medium where data is stored, most likely in various formats including structured, semi-structured, and unstructured. The semantic expression refers to the way an object is described, where multiple expressions can be used to describe the same object. A value refers to a measured result of the physical quantity, where various types of inconsistency could happen when measuring an object due to human as well as equipment factors. Sources of data uncertainty are also varied, including data collection variance, concept variance, and multi-modality [7]. Variance in data collection could be introduced by environmental conditions as well as data sampling. The same concept might be used not similarly, introducing concept variance. Data complexity and noise from multiple sensors are examples of multi-modality sources for data uncertainty. Dealing with these multitude sources of data consistency and uncertainty is about filtering out or amending the data through data cleaning and data reduction [11], a necessary step toward successfully big data analytics.

2.1.1 MOTIVATION AND RESEARCH CHALLENGES

Biodiversity research is organized into domains that cover distinct spheres of biodiversity knowledge, e.g., taxonomy, geographical distribution, or functional traits of organisms [12]. In this work, we focus on the studies of the distribution of life across space and time, providing a key link between organisms and their environment, also known as biogeography. To study the link, a biodiversity information system would hold various information about the organisms and their observed environments. Typically, biodiversity data contains observations of the occurrences of specific species that can be identified by a specific taxonomic name at a specific geographic location at a specific time [13]. As an example, the Global Biodiversity Information Facility (GBIF)[1] has recorded 1 billion records of species occurrences in 2018.[2] The record is constantly growing as data are provided by more than 1,600 institutions across the globe. Another example is Pl@ntNet,[3] a citizen science project that helps users to identify plants based on pictures provided by citizens. The project has published more than 6,6 million records of occurrences in two datasets.[4] Another example is eBird,[5] a citizen science project that enables volunteered observers to report bird observations. The project has published an observation dataset with more than 700 million records of occurrences.[6]

In general, biodiversity data can be produced in two methods, data collection and data mobilization [14]. In the first method, data generation will be started with a field survey where researchers would visit (predetermined) locations to observe species occurrences or even to collect specimens. In this step, information that is related to the locations, as well as the observation time, will be also recorded. In the case of citizen science projects, observation can also be performed by amateurs, normally by using instruments such as wireless devices or portable microscopes [15,16]. For specimen collection, the specimens will be preserved in a special room such as a laboratory. If not done yet, each record or specimen will be identified further to determine the correct taxonomic name. Then, all those data will be entered into a biodiversity information system. In the second method, data will be extracted from preserved specimens or works of literature such as checklists and taxonomic monographs.

Our work is motivated by the relatively complex procedures to produce high-quality biodiversity data as described above. Based on multiple factors involved during data production, we hypothesized that biodiversity data tends to be noisy. The noise which will affect data veracity could come from multiple factors as follow:

1. First, location factors. At least, two locations are involved in data generation, namely field observation, and laboratory identification. This situation contributes to data corruption such as missing values. For example, information about the habitat of a species that supposed to be collected from the field was forgotten. In the case of extraction of distributional information from preserved specimens, many important characteristics, e.g., plant growth form, vegetative height, or stem specific density could be missing [14].
2. Second, time factors. the id time differences between field observation to lab identification. Process errors such as data redundancy are highly possible.

3. Third, human factors. error also possible to have happened in the last phase, data entry. While data entry is performed independently it will be performed manually by humans so data entry errors can happen.

We believe that these three factors have a tremendous impact on the data veracity of biodiversity data. To the best of our knowledge, there is no existing work yet to investigate this problem.

2.1.2 RESEARCH OBJECTIVES

In this chapter, we propose an automatic method to assess the veracity of biodiversity data through data consistency analysis. Data consistency analysis can be performed from a variety of perspectives, including database development, computing strategy, and data science [10]. In this work, we use the perspective of data science especially big data management, where data that are scattered in distributed sources is required to be integrated, where data consistency across multiple sources is important to ensure high quality integrated data. The proposed method utilizes a data mapping solution to align data from multiple sources into a predefined structure by using a standardized vocabulary in a way data consistency can be measured and compared. Data mapping solutions bring many advantages, for example, exposing underneath schema of relational databases [17]. Exposing the schema of multiple databases is important in big data analytics. The expose would help users to have a better understanding of the structure of the databases, and at the same time help users to optimize queries to those databases. Data mapping solutions also would enable dataset trustworthiness by exposing the provenance of mapping quality [18]. It is more effective to assess and refine data mapping definitions than to assess and refine the quality of a dataset directly. Furthermore, a data mapping definition can be refined further to improve the quality of data [19]. Based on the assessment of data mapping quality, if a problem (for example data types inconsistency) is detected then a mapping definition refinement can be suggested to automatically improve the mappings.

Our research question is as follows: "How have data inconsistency in structures, value types, and granularities affected the veracity of open biodiversity data?". As reflected in this research question, we would like to investigate the impact of three sources of data inconsistency on the veracity of biodiversity data. First, data structure consistency refers to how elements of data are structured. Second, value types consistency refers to how a similar element data is used across multiple sources. Third, data granularity consistency refers to how specific a similar element data has been described. The rest of the chapter will be organized as follows: relevant and related works including our contributions will be explained in Section 2.2. Our method to measure data consistency will be described in Section 2.3. After presenting our results in Section 2.4, we summarize our chapter in Section 2.5.

2.2 RELATED WORK

We align our work with two prominent research areas, namely big data veracity analysis and data quality analysis of biodiversity data. In this section, we describe several related works from each area and outline our contributions.

Even though the veracity dimension of big data remains under-explored compared to the other dimensions, many works have been done to analyze it in several big data applications. Reference [4] proposes a big data veracity index by defining three main theoretical dimensions of veracity, namely objectivity, truthfulness, and credibility. The index was used to assess systematic variations of textual information across multiple datasets. They found that each dimension might contribute to the overall quality to a different degree, and therefore should be assigned different weights. Since the multi-modality of information sources could amplify the veracity of data, reference [9] proposes cross-modal truth discovery by predicting the truth label of claims through linkage analysis of various events from multiple sources. The approach was able to infer the reliability of sources with no or little prior knowledge. In another work, reference [20] proposes a platform to estimate data veracity by extracting entities, relations, and structures of claims to be combined in a way the veracity label of data and trustworthiness scores of the sources can be determined. Multiple methods can be used to determine the veracity data of electronic medical records, including process mining and using ontology [8]. In the process mining method, data quality will be assessed by mapping the chronological time/date within the records. Ontology can be used to share quality metrics. Standardized terminology can also provide data correction for misspelt words in unstructured text fields. And most recently, the big data should be transformed into smart data where data must be appropriately sorted, structured, and analyzed [6,11]. Smart data aim to filter out or amend imperfect data through data cleaning and data reduction, for example for dealing with data redundancy or contradiction.

Data quality is also becoming a major issue in the field of biodiversity science. Numerous factors could affect data veracity, including observation error, expertise, and reliability of the primary data collectors, possible data corruption during secondary data management and analysis, and any other factor that might increase uncertainty [21]. The integration of biodiversity data deals with the availability, quality, and interoperability of data which are mostly based on disaggregated data types [14]. One of the challenges is missing or inconsistent data items that can be solved with the data imputation method where a value will be estimated (using logical and statistical approaches) to replace the missing data. The increase of "big unstructured data" has highlighted the discrepancy in global data availability between data quantity and data quality [22]. It is necessary to do benchmarking big unstructured data against high-quality structured datasets, as well as developing purpose-specific rankings to assess data quality. A controlled vocabulary and data annotation could improve data quality and fitness for use [23]. Also, the communities which have the necessary expertise to validate, curate, and improve data from diverse sources should be integrated into the data [24]. This approach would enable researchers to engage effectively and efficiently with vast volumes of complex data, to contribute through simple curatorial actions to improve digital knowledge. Furthermore, integrating and transforming biodiversity data into a knowledge graph requires extensive data cleaning and cross-linking [13]. For instance, converting data from multiple sources into a specific format requires multiple steps. Even though a set of declarative mapping rules can be used to align entities from multiple sources to a targeted scheme [25], it remains challenging to reconcile entities across multiple sources.

In line with these two broad research areas explained above, we outline our main contribution as a method to assess data veracity in biodiversity data, such as sources, comparison of veracity types. Based on the definition of sources of noises from [6], our work assesses the "attribute noises" which can be explained as a corruption of data in the values of the input attributes. In this type of veracity source, the factors can be erroneous attribute values, missing values, or incomplete attributes. In contrast with existing works that mainly rely on machine learning approaches, we lay our work on the fundamental approach for data integration, namely data mapping. We map element data from multiple sources into one defined data structure in a way the noise will prevail. Our work is different from the data mapping approach in [8], which was utilized for the correction of misspelt words only. Our method goes beyond that, namely assessment of data consistency in structure, data values, as well as data granularity.

2.3 METHOD

In this section, we introduce our method to assess data veracity of open biodiversity data. We measure and compare data inconsistency across multiple data sources. Three sources of data inconsistency will be investigated, namely data structure inconsistency, data value types inconsistency, and data granularity inconsistency. First, we define our data definitions as an approach to representing data from multiple sources into one generic schema. Second, we describe how to measure three types of inconsistency from the obtained mappings. And finally, our research procedures will be explained at the end of this section.

2.3.1 DATA DEFINITIONS

To be able to measure the three types of data consistency, we introduce several definitions and formalizations as follows:

1. **Vocabulary**: Vocabulary is a collection of data attributes that can be used to describe an object. Each attribute has a name and an expected type of value. As an illustration, to describe a biological specimen, it is necessary to have a few data attributes such as the name of the specimen, where and when the specimen was collected, and so on. Further, an

attribute "name" should have a textual value, an attribute "date" should have a date value, etc. In the field of biodiversity, several existing vocabularies have been used widely. One of them is Darwin Core,[7] a data standard for publishing and integrating biodiversity information [26]. We use this vocabulary due to its wide adoption.

2. **Dataset**: Dataset is a collection of data objects, where each object is described with one or more attributes that are available in the selected vocabulary.

3. **Dataset vocabulary**: Dataset vocabulary is a collection of data attributes, where each object is described with one or more attributes that are available in the selected vocabulary. It is worthy to mention that every dataset could use less or a greater number of attributes available in the vocabulary.

2.3.2 DATA CONSISTENCY ANALYSIS

After introducing the basic definitions of our data mapping solution, we constructed our data consistency measurement as follows:

1. **Data structure consistency**: The first measurement applies to the dataset level, meaning that it can be used to measure consistency across datasets. Two datasets will be called consistent if every related data objects in both datasets utilize similar data attributes. One dataset may have a richer structure than the other.

2. **Data type consistency**: The second measurement is applicable to attribute level, meaning that it will be used to compare attributes across data objects within a dataset or to compare related data objects across datasets. Two data objects will be called consistent if both objects utilize a similar data type for their relevant attributes.

3. **Data granularity consistency**: The third measurement is applicable at the data value level, meaning that it will be used to compare values of related attributes of data objects within a dataset or across multiple datasets. We use a semantic similarity[8] metric to measure the distance between two values. Semantic distance is a metric to measure how far a concept is from other concepts in a knowledge organization system. By identifying concepts in a data value of an attribute, we can map each concept to a knowledge organization system such that the semantic distance between them can be measured. If an attribute is data granularity consistent then all values of this attribute should be mapped to the same concept.

We model the value of an attribute by using a Simple Knowledge Organization System (SKOS),[9] a vocabulary and data model for expressing knowledge organization systems for data referencing and reusing. For example, the concept "location" will be modeled in SKOS as shown in Table 2.1. There are three important semantic relations, namely "broader", "narrower", and "related". The relation "broader" and

TABLE 2.1

Modelling of Concept "Location" Using SKOS

Subject	Predicate	Object
ex:Island	rdf:type	skos:Concept
	skos:prefLabel	"Pulau"@id
	skos:prefLabel	"Island"@en
	skos:related	ex:StateProvince
	skos:related	ex:Country
ex:StateProvince	rdf:type	skos:Concept
	skos:prefLabel	"Provinsi"@id
	skos:prefLabel	"State Province"@en
	skos:altLabel	"Province"@en
	skos:broader	ex:Country
ex:Country	rdf:type	skos:Concept
	skos:prefLabel	"Negara"@id
	skos:prefLabel	"Country"@en
	skos:narrower	ex:StateProvinc

"narrower" are transitive relations to represent if a concept is broader or narrower than others respectively. The relation "related" is a reflexive relation to represent that a concept is related to the other and vice versa. The model can also be visualized in a graph representation as shown in Figure 2.1. This graph depicts that "StateProvince" has a broader concept so-called "Country" and "Country" has a narrower concept so-called "StateProvince". An "Island" is related to both "StateProvince" and "Country".

2.3.3 DATA MAPPING PROCEDURES

After describing our formalization to measure data consistency in the previous sub-sections, now we introduce our data mapping procedures. Figure 2.2 shows our mapping procedures that consist of three main activities as explained in the following sub-sections.

1. **Data crawling and extraction**: In this first activity, data will be crawled and extracted from several sources. Since most of the data are available over the Web, a web-scraping method will be employed to extract the required data. The input of this activity is a list of Uniform Resource Locator (URL). The output will be a collection of files, where each file contains data as a tuple in the form of (key, [values]). In each tuple, a "key" has a list of zero, one, or more "values".

2. **Data mapping**: From every tuple obtained in the previous activity, its "key" will be mapped to a relevant attribute of the selected vocabulary. The relevancy will be determined by users that have a wide variety of

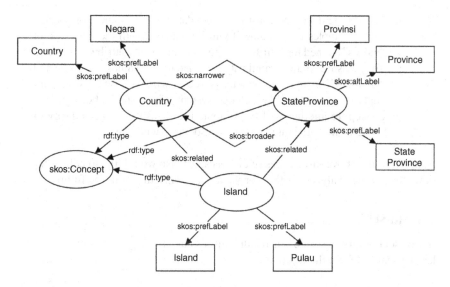

FIGURE 2.1   Visualization of concepts "location" using SKOS.

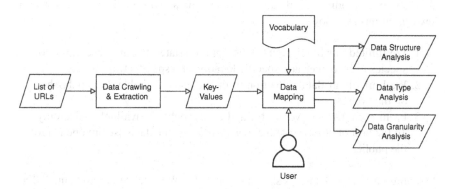

FIGURE 2.2   Research methodology.

expertise in data science. As output, a list of (key, attribute) will be produced. In case of no relevant attribute can be identified, then a tuple (key, Null) will be produced.

3. **Data analysis**: For analysis, we constructed a third collection of tuples based on the defined mapping in the previous activity. Technically, the process will be performed as (key, [values]) + (key, attribute) = (attribute, [values]). To answer our research question, three types of analyses will be performed, explained as follow:

   a. To measure data structure consistency, the tuples (key, attribute) will be compared from one dataset to another. We expected to be able to identify which attributes are widely used and which ones are not.

b. To measure data type consistency, the data type of "values" of selected "key" in tuples (key, [values]) will be matched with the data type of "values" defined for "attribute" in the relevant tuples (attribute, [values]).

c. To measure data granularity consistency, "values" of selected "attribute" in tuples (attribute, [values]) will be aligned with concepts from a defined knowledge system. The distance between one concept to another will be computed to measure the granularity of the relevant "attribute" which is associated with "key".

Apart from the above three types of analyses, we also would like to collect tuples (key, Null) to be analyzed further for vocabulary enrichment in the future.

2.4   RESULT

In this section, we describe our results, discuss our findings, and summarize the lessons learned from the findings.

2.4.1   DATASET

To assess open biodiversity data, we collected data from multiple sources with a few limitations as follows:

1. We limited our biodiversity data that are related to data observation of species occurrences at a specific location at a specific time.
2. We limited our data collection to the publicly available data, published on the Web.
3. From each dataset, we are limited by the publicly available fields only. To the best of our knowledge, not all fields of the database are opened to the public.

The summary of the dataset is shown in Table 2.2. We collected more than 60,000 records of species occurrences from nine sources. It covers botanical as well as zoological data.

2.4.2   DATASET VOCABULARY

As explained in Section 2.3, we need to construct a dataset vocabulary based on the mapped "attributes" of the selected vocabulary i.e. Darwin Core Terms. As result, there are 72 terms were used in our collections of datasets as shown in Figure 2.3. The top three terms belong to class "Taxon", "Location", and "Event" respectively.

2.4.3   DATA STRUCTURE ANALYSIS

To analyze data structure across multiple datasets, we computed how each dataset uses our vocabulary and which attributes are not used in each dataset as shown

TABLE 2.2

Collection of Datasets

No.	URL	# Specimen
1	http://ibis.biologi.lipi.go.id/	10,629
2	http://ibis.biologi.lipi.go.id/mzb/	1,749
3	http://bankbiji.krbogor.lipi.go.id/katalog	114
4	http://sindata.krcibodas.lipi.go.id/	1,969
5	http://ipbiotics.apps.cs.ipb.ac.id/	1,074
	http://indobiosys.org/	14,222
6	http://ipt.biologi.lipi.go.id/	17,250
7	Herbarium of Andalas University	1,128
8	Museum Zoologi Bogor	14,043
9	Tambora Muda Indonesia	
	Total	62,178

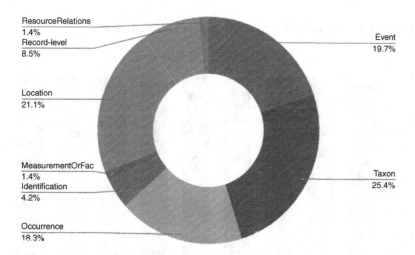

FIGURE 2.3   The proportion of 72 Darwin Core terms that are used as our dataset vocabulary.

in Figures 2.4 and 2.5 respectively. As shown in Figure 2.4, class "Location" was widely used across datasets but has an uneven distribution. Class "Taxon" is also widely used across datasets with better distribution. Other classes were used only partially, for example, class "Event" was used in eight datasets only, class "MeasurementOrFact" was used in two datasets, or class "ResourceRelationship" was used in one dataset only. When identifying which terms are used in each dataset, we obtained results as shown in Figure 2.5. The ratio of terms used that are available in our vocabulary across multiple datasets was very low. Only two datasets were utilized above 50%, two datasets use less than 40%, 1 less than 30%, 2 less than 20%, and even less than 10% for one dataset.

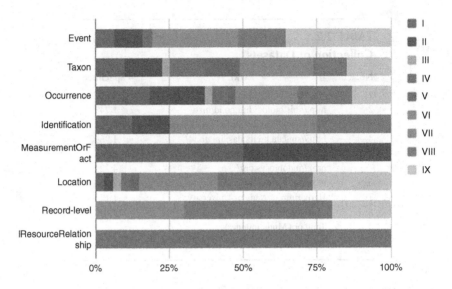

FIGURE 2.4 Proportion of our dataset vocabulary across datasets.

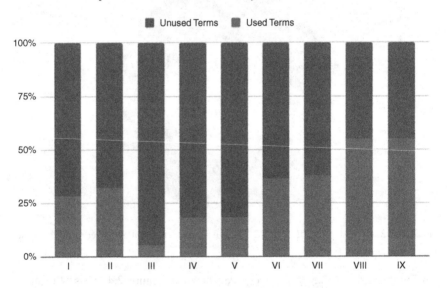

FIGURE 2.5 Proportion of used and un-used attributes across datasets.

From these two results, we explain the situation as follows:

1. Data structure if very heterogeneous
2. High consistency is found in several classes, namely "Taxon" and "Event" due to their relatively equal distribution across multiple data-sets. This indicates that the potential for data integration can be started with these two classes.

Even though we found a relatively low consistency across multiple datasets, we would like to outline potential biases during our data collection and processing as follow:

1. data collection bias refers to how data were collected by different organizations according to the research question that they would like to answer. For example, one organization is probably concerned only about specific species in specific locations and therefore terms that are related to the habitat would not be recorded.
2. data mapping bias refers to a possible error caused by an incorrect alignment between "key" of data to "attribute" in the selected vocabulary.

2.4.4 Data Type Analysis

To analyze data type consistency across multiple datasets, we selected the attribute "eventDate" of class "Event" which according to the definition of Darwin Core can be used to specify the date-time or interval during which an "Event" occurred. It is recommended to use a date that conforms to ISO 8601-1:2019.[10] The standard provides an unambiguous representation of dates and times to avoid misinterpretation of numeric representations of dates and times across different conventions. The attribute was used in seven datasets, but in two datasets the value of the attribute was presented as a combination with values from other fields, and therefore they were discarded. We also disregarded records that have empty values in their corresponding attributes. The result from five datasets is shown in Figure 2.6.

As shown in Figure 2.6, seven formats were employed, where one format represents date intervals, one represents a specific date with time intervals and the other represents a specific date and time. One dataset uses four different formats, while four others use only one format. We also found that only one format was

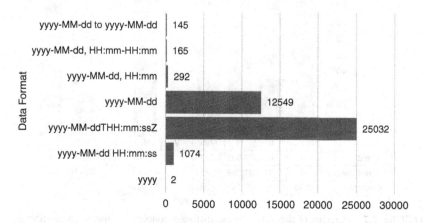

FIGURE 2.6 Number of records that use different data formats in our datasets.

used consistently across two datasets, namely (yyyy-MM-ddTHH:mm:ssZ). Only one format specified a time zone designator (Z) explicitly, which is turn out to be employed by most of the data records.

We explain the situation as follows:

1. Multiple formats can be used across multiple datasets that still conform to a standard (such as ISO 8601).
2. A time zone designator is not adopted wide enough, which could lead to multiple issues when integrating data from multiple sources.

2.4.5 DATA GRANULARITY ANALYSIS

To assess data granularity consistency, we selected three attributes in our dataset vocabulary that refer to a geographical location. The attributes are "country", "locality", and "island" from class "Location". An attribute "country" refers to the name of the country or major administrative unit in which a "Location" occurs, "locality" refers to the specific description of the "Location", and "island" refers to the name of the island on or near which the "Location" occurs. These three attributes were selected because, based on our mapping, they were used across multiple datasets. Three datasets were used attribute "country", five and one datasets were used attributes "locality" and "island" respectively. From every selected attribute, we constructed a knowledge system as depicted in Figure 2.1. The knowledge system consists of three concepts, namely "Country", "StateProvince", and "Island". After that, we analyze how those concepts were utilized in our datasets and the result is shown in Figure 2.7.

As shown in Figure 2.7, the three concepts to represent a geographical location were used differently across multiple datasets. Concepts of "Country" and

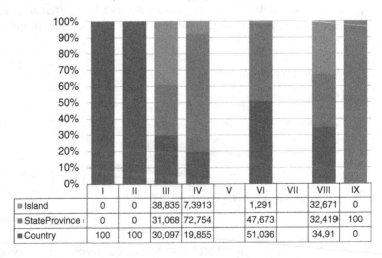

	I	II	III	IV	V	VI	VII	VIII	IX
■ Island	0	0	38,835	7,3913		1,291		32,671	0
■ StateProvince	0	0	31,068	72,754		47,673		32,419	100
■ Country	100	100	30,097	19,855		51,036		34,91	0

FIGURE 2.7 Portions of datasets that use different concepts to specify a geographical location.

"StateProvince" were used consistently in two and one datasets, respectively. Two datasets used a different concept, namely "GeoCoordinate". In the remaining four datasets, those three concepts have been used simultaneously. We describe this situation as follows:

1. The level of granularity is varying across multiple sources. Forcing to follow a specific granularity level is hard, especially in the case of data mobilization. In this case, the information related to a geographical location can be missing due to various factors. For example, the name of a place has been changed.
2. The use of attribute "GeoCoordinate" is becoming popular and should be used as the first option. In the case of data mobilization, it is important to have a global mapping from textual name to this attribute such that the granularity level can be consistent.

2.5 CONCLUSION

In this chapter, we introduced a method to assess data veracity of open biodiversity data. The method performs a consistency analysis on three important sources of data consistency, namely data structure, data type, and data granularity. To the best of our knowledge, our work is the first one that investigated these sources thoroughly. The analysis was performed in the context of the big data ecosystem, where data is distributed across multiple sources, presented in various formats, and maintained by multiple organizations. Our main objective was to assess the veracity of biodiversity data automatically which can be used further to improve the quality of data.

Three sources of data consistency were investigated. First, data structure analysis, applied at the datasets level, was intended to measure how a defined vocabulary was utilized across multiple datasets. Second, data type analysis, applied at the data attribute level, was intended to measure how the data type of an attribute is used within a dataset or across multiple datasets. Third, data granularity analysis, applied at the data value level, was intended to measure how multiple concepts were used in values of selected attributes. As our datasets, we collected publicly available biodiversity data more than 60,000 records of species occurrences that are available in nine distributed data sources. Our analysis was conducted in several systematic steps, namely:

1. Data collection, where biodiversity data from multiple sources were collected. In most cases, web-scraping techniques were used to extract the relevant pair of (key, [values]) for every element data that was presented on a website.
2. Data mapping, where every extracted "key" is mapped to the most suitable attribute in a selected vocabulary to produce a pair of (key, attribute). We used the Darwin Core as our vocabulary due to its wide adoption in the biodiversity area.

3. Based on the mapping, we constructed a final collection of (attribute, [values]). After that, several statistics regarding data consistency were computed.

As a result, we obtained:

1. There is a high number of terms (72 in total) that were utilized across nine datasets. The terms are classified in Taxon, Location, Occurrence, Event, Record-level, Identification, MeasurementOrFact, and ResourceRelations.
2. The terms were utilized imbalance across sources. A class of terms such as Occurrence is widely used but three sources underused it compared to the others. The same case with terms in class Location, widely used but under usage by five sources.
3. There is a high diversity of ways to represent a specimen. Even though we obtained 72 terms in our vocabulary, only two data sources utilized more than 50%.
4. We also identified data inconsistency in data type and data granularity. We identified different ways to define data value for a term with type DateTime. Different granularities were used to specify the value for a location.

In conclusion, due to the high inconsistency found in all three sources, performing analysis of big biodiversity data requires more effort in the step of preprocessing step. Data integration as the first step toward analytics requires the implementation of recent technologies to fill the gap. Those technologies are including:

1. Data value completion, which is including link prediction by using machine learning.
2. Data modelling can be used to suppress the level of inconsistency that can be developed further to extend the existing vocabulary. For example, to have a better representation of location or habitat.

ACKNOWLEDGEMENT

This research was supported by the Research Center for Informatics, National Research and Innovation Agency, Indonesia. We would like to thank all the members of the Knowledge Engineering Research Group for their valuable suggestions and feedback.

REFERENCES

1. Statista.com, "Volume of data/information created, captured, copied, and consumed worldwide from 2010 to 2025," 2021. https://www.statista.com/statistics/871513/worldwide-data-created/.
2. M. Younas, "Research challenges of big data," *Serv. Oriented Comput. Appl.*, vol. 13, pp. 105–107, 2019, doi:10.1007/s11761-019-00265-x.
3. B. Saha and D. Srivastava, "Data quality: The other face of big data," in *2014 IEEE 30th International Conference on Data Engineering*, Mar. 2014, pp. 1294–1297, doi:10.1109/ICDE.2014.6816764.

4. T. Lukoianova and V. L. Rubin, "Veracity roadmap: Is big data objective, truthful and credible?" *Adv. Classif. Res. Online*, vol. 24, no. 1, p. 4, Jan. 2014, doi:10.7152/acro.v24i1.14671.

5. M. García Lozano et al., "Veracity assessment of online data," *Decis. Support Syst.*, vol. 129, p. 113132, Feb. 2020, doi:10.1016/j.dss.2019.113132.

6. D. García-Gil, J. Luengo, S. García, and F. Herrera, "Enabling smart data: Noise filtering in big data classification," *Inf. Sci. (Ny).*, vol. 479, pp. 135–152, Apr. 2019, doi:10.1016/j.ins.2018.12.002.

7. R. H. Hariri, E. M. Fredericks, and K. M. Bowers, "Uncertainty in big data analytics: Survey, opportunities, and challenges," *J. Big Data*, vol. 6, no. 1, p. 44, Dec. 2019, doi:10.1186/s40537-019-0206-3.

8. A. P. Reimer and E. A. Madigan, "Veracity in big data: How good is good enough," *Health Informatics J.*, vol. 25, no. 4, pp. 1290–1298, Dec. 2019, doi:10.1177/1460458217744369.

9. L. Berti-Equille and M. L. Ba, "Veracity of big data: Challenges of cross-modal truth discovery," *J. Data Inf. Qual.*, vol. 7, no. 3, pp. 1–3, Sep. 2016, doi:10.1145/2935753.

10. P. Shi, Y. Cui, K. Xu, M. Zhang, and L. Ding, "Data consistency theory and case study for scientific big data," *Information*, vol. 10, no. 4, p. 137, Apr. 2019, doi:10.3390/info10040137.

11. J. Luengo, D. García-Gil, S. Ramírez-Gallego, S. García, and F. Herrera, *Big Data Preprocessing: Enabling Smart Data*, Cham: Springer International Publishing, 2020, doi:10.1007/978-3-030-39105-8.

12. J. Hortal, F. de Bello, J. A. F. Diniz-Filho, T. M. Lewinsohn, J. M. Lobo, and R. J. Ladle, "Seven shortfalls that beset large-scale knowledge of biodiversity," *Annu. Rev. Ecol. Evol. Syst.*, vol. 46, no. 1, pp. 523–549, Dec. 2015, doi:10.1146/annurev-ecolsys-112414-054400.

13. R. D. M. Page, "Ozymandias: A biodiversity knowledge graph," *PeerJ*, vol. 7, p. e6739, Apr. 2019, doi:10.7717/peerj.6739.

14. C. König, P. Weigelt, J. Schrader, A. Taylor, J. Kattge, and H. Kreft, "Biodiversity data integration—The significance of data resolution and domain," *PLoS Biol.*, vol. 17, no. 3, 2019, doi:10.1371/journal.pbio.3000183.

15. S. Kelling, "eBird: A human/computer learning network to improve biodiversity conservation and research," in *Twenty-Fourth IAAI Conference*, Jul. 2012, p. 11.

16. D. R. Hardison et al., "HABscope: A tool for use by citizen scientists to facilitate early warning of respiratory irritation caused by toxic blooms of Karenia brevis," *PLoS One*, vol. 14, no. 6, p. e0218489, Jun. 2019, doi:10.1371/journal.pone.0218489.

17. M. Rodriguez-Muro and M. Rezk, "Efficient SPARQL-to-SQL with R2RML mappings," *J. Web Semant.*, vol. 33, pp. 141–169, 2015.

18. T. De Nies, A. Dimou, R. Verborgh, E. Mannens, and R. Van de Walle, "Enabling dataset trustworthiness by exposing the provenance of mapping quality assessment and refinement," in *The 4th International Workshop on Methods for Establishing Trust of (Open) Data (METHOD2015)*, 2015, p. 7.

19. A. Dimou et al., "Assessing and Refining Mappings to RDF to Improve Dataset Quality," in *The Semantic Web - ISWC 2015*, vol. 9367, M. Arenas, O. Corcho, E. Simperl, M. Strohmaier, M. D'Aquin, K. Srinivas, P. Groth, M. Dumontier, J. Heflin, K. Thirunarayan, and S. Staab, Eds. Cham: Springer International Publishing, 2015, pp. 133–149.

20. M. L. Ba, L. Berti-Equille, K. Shah, and H. M. Hammady, "VERA: A platform for veracity estimation over web data," in *Proceedings of the 25th International Conference Companion on World Wide Web - WWW '16 Companion*, 2016, pp. 159–162, doi:10.1145/2872518.2890536.

21. S. S. Farley, A. Dawson, S. J. Goring, and J. W. Williams, "Situating ecology as a big-data science: current advances, challenges, and solutions," *Bioscience*, vol. 68, no. 8, pp. 563–576, Aug. 2018, doi:10.1093/biosci/biy068.

22. E. Bayraktarov et al., "Do big unstructured biodiversity data mean more knowledge?" *Front. Ecol. Evol.*, vol. 6, p. 239, Jan. 2019, doi:10.3389/fevo.2018.00239.

23. T. Robertson et al., "The GBIF integrated publishing toolkit: Facilitating the efficient publishing of biodiversity data on the internet," *PLoS One*, vol. 9, no. 8, p. e102623, Aug. 2014, doi:10.1371/journal.pone.0102623.

24. D. Hobern et al., "Connecting data and expertise: A new alliance for biodiversity knowledge," *Biodivers. Data J.*, vol. 7, p. e33679, Mar. 2019, doi:10.3897/BDJ.7.e33679.

25. Z. Akbar, Y. A. Kartika, D. Ridwan Saleh, H. F. Mustika, and L. Parningotan Manik, "On using declarative generation rules to deliver linked biodiversity data," in *2020 International Conference on Radar, Antenna, Microwave, Electronics, and Telecommunications (ICRAMET)*, 2020, pp. 267–272, doi:10.1109/ICRAMET51080.2020.9298573.

26. J. Wieczorek et al., "Darwin Core: An evolving community-developed biodiversity data standard," *PLoS One*, vol. 7, no. 1, p. e29715, Jan. 2012, doi:10.1371/journal.pone.0029715.

NOTES

1 https://www.gbif.org/, Accessed: 30/10/2021.

2 https://www.gbif.org/news/5BesWzmwqQ4U84suqWyOQy/, Accessed: 30/10/2021.

3 https://plantnet.org/, Accessed: 30/10/2021.

4 https://www.gbif.org/publisher/da86174a-a605-43a4-a5e8-53d484152cd3, Accessed: 30/10/2021.

5 https://ebird.org, Accessed: 30/10/2021.

6 https://www.gbif.org/dataset/4fa7b334-ce0d-4e88-aaae-2e0c138d049e, Accessed: 30/10/2021.

7 https://dwc.tdwg.org/, Accessed: 30/10/2021.

8 https://en.wikipedia.org/wiki/Semantic_similarity, Accessed: 30/10/2021.

9 https://www.w3.org/2004/02/skos/, Accessed: 30/10/2021.

10 https://en.wikipedia.org/wiki/ISO_8601, Accessed: 30/10/2021.

3 Semantic Web Technologies
Latest Industrial Applications

Michael DeBellis

CONTENTS

DOI: 10.1201/9781003309420-3

3.1   INTRODUCTION

This chapter is a survey of how Semantic Web Technology (SWT) is being leveraged in industry. This is an exploratory survey rather than a systematic survey. A systematic survey must be repeatable and follow a rigorous methodology based on peer-reviewed sources. An exploratory survey is more flexible [1]. As a new technology transfers from the lab to industry some of the most important sources are not in peer-reviewed journals but also in case studies from vendors and reports by industry analysts. Thus, an exploratory survey is most appropriate. I will focus on a few vendors that are leaders in SWT however, this is by no means an exhaustive review of SWT vendors nor a vendor comparison.

The structure of this chapter is as follows: this introduction provides an overview of the chapter. The next section describes some of the most important business functions that SWT provides. The section on Data Fabrics provides a framework for how these various business functions fit together. The section on applications describes some of the most common types of SWT applications and examples. The conclusion will discuss some of the issues that may impede the adoption of SWT and provide suggestions for the next steps to learn and utilize it.

The core SWT concepts are defined in the text. In addition, there is a summary glossary at the end of this chapter. For detailed overviews of SWT see [2,3]. One of the benefits of SWT is that there are already standards for the technology defined by the World Wide Web Consortium (W3C) [4], the same organization that defines Internet standards such as HTML, HTTP, and TCP/IP. There are also alternative non-standard technologies such as Property Graphs, however, for simplicity I will focus only on the W3C standards in this chapter. See the references above for details on technologies that overlap with these standards but are currently not standardized.

3.2   BUSINESS VALUE OF SEMANTIC
WEB TECHNOLOGY (SWT)

This section describes some of the most essential Semantic Web technologies from the standpoint of business value as well as a discussion of how SWT and Machine Learning relate to each other.

3.2.1 GRAPH DATABASES

The foundation for SWT is the graph database. A graph database is an alternative to a relational database. Relational databases are excellent for data that needs to process large numbers of small transactions. However, for complex highly structured and interconnected data, graph databases provide better performance.

In traditional databases, the foundational data structure is a table. In a graph database, it is a triple. Hence, graph databases are also known as triplestores. A triple has the structure: Subject -> Predicate -> Object. Graph databases are also known as *knowledge graphs*. The language used to define graph databases is the Resource Description Framework (RDF) [5].

Relational databases were primarily designed for systems that focused on one specific domain such as payroll processing. Modern systems need to deal with complex data from multiple domains. For example, a system for drug discovery may require data from clinical trials, drug databases, biomolecular, genetic, and legal data. To perform with large datasets that cross different domains a graph database is more efficient. A database based on tables will tend to degrade when relations span depths of three or more levels. Graph databases on the other hand can scale to many levels of depth because the database is designed in terms of triples rather than tables [6]. Also, the SPARQL query language [7] for graph databases is more powerful, flexible, and scalable for highly connected data than SQL.

3.2.2 SWT AND MACHINE LEARNING

Since both SWT and Machine Learning have their roots in Artificial Intelligence research and work on big data an inevitable question is what the relation between them is. The two technologies are synergistic rather than competitive. Machine Learning (ML) develops an algorithm based on sample data known as training sets. ML is most useful for problems where the goal is to separate a signal from noisy data (supervised learning) or to provide structure to noisy data (unsupervised learning). In supervised learning, we provide a learning algorithm with examples of input data as well as associated output data. The ML system creates a mathematical model that represents the relation between the input and output data. This can be especially useful for problems where the goal is to extract a signal from noisy data such as recognizing speech phonemes from a data stream. Unsupervised learning is typically used to take data and discover collections that have similarities. Essentially to discover new categories. These results can be fed into SWT.

SWT and ML seldom compete with each other. It is rare for a possible application to be a good fit for both technologies. Typically, if a business problem is a good fit for ML, it is not a good fit for SWT and vice versa. What's more, the two technologies complement each other. There are many SWT problems such as the transformation from "strings to things" [8] where ML is an excellent solution. Similarly, as Andrew Ng (one of the leading experts in ML) recently stated, the best way to improve the performance of an ML algorithm is usually not to give it more training sets but rather to improve the data quality of the training sets [9]. The definition and enforcement of constraints on data integrity is one of the major advantages of SWT and hence can greatly benefit ML.

3.2.3 SEMANTIC MODELS

The graph database provides the foundation for SWT. However, to deal with complex highly connected big data it is also essential that the data is understandable to humans as well as machines. This is where the Web Ontology Language (OWL) [10] is utilized. OWL defines the semantics of data in terms of ontologies which combine many different logical formalisms to represent data in an intuitive way. The concept of ontology goes back to the earliest days of artificial intelligence and knowledge representation. OWL ontologies represent knowledge using classes, instances, properties, rules, and axioms. Classes, properties, and instances in OWL are similar to the concepts utilized in OOP. However, there are also significant differences between OWL and OOP. One of the most important differences is that SWT properties are predicates in triples. Also, in OWL properties can be defined in hierarchies from general to specific just as classes can be.

Property values in OWL can be instances of classes or simple data types such as strings or integers. Simple data types are also known as Literals.

Axioms enable the semantic capabilities of SWT to go beyond the class hierarchies in OOP. In OOP, while there are guidelines for when one class should be a sub-class of another these guidelines are only defined in methodologies and are not explicitly represented. OWL enables axioms to describe classes based on number and kind of property values as well as additional concepts from logic and set theory. Axioms enable automated reasoners that can first validate the data model (ensure that it has no contradictory statements) and then perform additional reasoning such as restructuring the class hierarchy and inferring values for properties. For a simple example, in OWL one can specify that two properties are inverses. With this information, the reasoner can automatically infer appropriate values for inverses. E.g., if has_Parent is the inverse of has_Child then if X has_Parent Y the reasoner will infer that Y has_Child X.

Another example of how OWL provides additional semantics is property hierarchies. For example, has_Parent would be a super-property to has_Father. Thus, if X has_Father Y the reasoner will infer that X has_Parent Y as well.

3.2.4 AGILE DATA

One of the advantages of the graph model is that graph-based models such as OWL make adding to and changing the data model much simpler. For example, defining properties as first-class objects means that properties can be added at run time. Traditional OOP properties must be defined as part of a class. Thus, to add a new property one must have access to the code for the class definition, change the class, recompile, and distribute the new version of the code. With relational databases, it is also non-trivial to add new relations to the database. At a minimum, this will slow the performance of the database and it may require the database to be taken offline for the change.

Using SWT, one can add new properties to the graph without changing or even having access to the class definition. A new property can simply be added directly to a graph using SPARQL as easily as one would add new class instances or new property values. This greatly facilitates the rapid evolution of data to meet new business opportunities.

A graph model enables any part of the model to be changed at run time. This is because everything about the data including the semantics such as axioms is represented as part of the graph. Thus, one can dynamically change the structure of the model at runtime as easily as one can add or change data: by simply editing links and nodes in the graph.

3.2.5 DATA INTEGRITY

OWL provides semantic models for defining and reasoning about data. Another SWT standard is the Shapes Constraint Language (SHACL) [11]. SHACL provides a language to define constraints on each property. These constraints include legal datatypes, number of values, ranges for numeric properties, patterns for string properties, logical relations with other properties, etc. These constraints are known as *Shapes* in SHACL. A SHACL Shape provides a collection of constraints that apply to a class and its properties.

SWT include reasoners that analyze SHACL shapes and report when a property value is inconsistent with a SHACL constraint. In addition, SHACL shapes can include code that attempts to rectify constraint violations when possible.

Some of the capabilities in OWL and SHACL such as the ability to define the number of values and datatype for a property overlap. The distinction is in how these definitions are utilized. In OWL definitions are used for *reasoning*. In SHACL definitions are utilized for *data integrity validation*.

3.2.6 W3C STANDARDS

SWT technologies such as OWL, SPARQL, and SHACL are W3C standards. This is an important difference between SWT and previous game-changing technologies such as the first wave of AI expert systems. Standards are a way to "future proof" the technology of an enterprise [12].

In addition to technology standards, there are standards for reusable domain models called vocabularies. A vocabulary in SWT is often an ontology, defined by the W3C or an industry organization. E.g., the W3C standardizes the definition of horizontal domains such as data provenance [13], while industry groups such as ICD and SNOMED define vocabularies for vertical domains such as healthcare. Standards prevent being locked into a specific vendor and facilitate reuse.

3.2.7 DATA GOVERNANCE

In SWT meta-data is defined as part of the model. Models of meta-data, users, permissions, workflows, and provenance are defined with ontologies and stored in knowledge graphs just as domain knowledge is. This makes it easy to reuse standards for meta-data such as Dublin Core [14] as well as to automate the inclusion and editing of meta-data as part of the process of editing data.

Models of data ownership and responsibility are especially important due to emerging laws giving consumers rights over their data such as the General Data Protection Regulation (GDPR) 2016/679 in the European Union. The GDPR is especially significant for social media companies such as Facebook, Instagram, and Twitter. Especially as it has become a model for other legislation in the UK, Brazil, Japan, and the California Consumer Privacy Act [15]. Another requirement that drives the need for data governance is the growing demand for legislation controlling the usage of AI technology [16].

3.2.8 EXPLAINABLE INFERENCES

One of the advantages of the OWL reasoner is that it provides a built-in framework to generate explanations for reasoning. This can also be combined with reasoning engines that go beyond basic OWL such as the Semantic Web Rule Language (SWRL) [17]. SWRL provides a very powerful forward chaining inference engine that leverages the logical foundation of OWL. Any inference of the OWL or SWRL reasoner can be explained by utilizing the inference history. As a trivial example, recall the use case described above of automatically inferring the value of a property based on the value of its inverse. In tools such as the Protégé ontology editor [18], one can select such inferred values and an explanation based on the axioms and rules of the ontology can automatically be generated. This is one of the core goals of SWT: to make the same data understandable both to users and to machines.

Explainable inferences are also one of the important differences between SWT and ML. ML algorithms are based on mathematical models generated from sample data. These mathematical models seldom are directly mappable to domain concepts that users can understand. SWT is based on ontologies that model the world in concepts that are intuitive to users and hence their inferences can be explained with these concepts [19].

3.2.9 LINKED DATA

One of the first major achievements of the Semantic Web was the creation of open crowd-sourced datasets such as Dbpedia [20], Wikidata [21], and Geonames [22]. In addition, there are companies that are leveraging the technology to provide data and other services such as Refinitiv and Crunchbase. Refinitiv is a spin-off company from Reuters. It provides data on financial markets and companies, as well as various financial service products that leverage this data [23]. Crunchbase is a graph database that describes companies, their investors, mergers, acquisitions, etc [24].

These and many other data sources are available on the web as SPARQL endpoints. A SPARQL endpoint is an Internationalized Resource Identifier (IRI) that can receive SPARQL queries, process them, and return the results to the client. An IRI looks similar to a URL but IRIs are not required to be viewable in browsers and often can describe small granularity resources such as a specific class, property, or instance. One of the advantages of the SPARQL query language is that it can access data from many different data sources, both within a corporate firewall and from the public Internet all in the same query. This is known as *Linked Data*. Linked Data allows each user to turn the Internet into one huge, distributed database that can easily be accessed from their PC.

3.2.10   DATA VISUALIZATION

Representing data as a graph allows developers to use several well-understood techniques to visualize the data from graph layout algorithms to matrices [25]. There are efficient algorithms that have existed for decades for laying out tree-based and directed graphs. The layout of network graphs is more complex, but several SWT vendors have developed tools such as the Gruff editor from Franz Inc [26]. That can efficiently layout network graphs and thus visualize data in many intuitive ways. Figure 3.1 shows an example where the Gruff tool generated a visualization from a knowledge graph about actors based on Dbpedia data. Each node and link are color coded according to the key in the left column of the Gruff GUI although the color coding is not displayed in the black and white screen print.

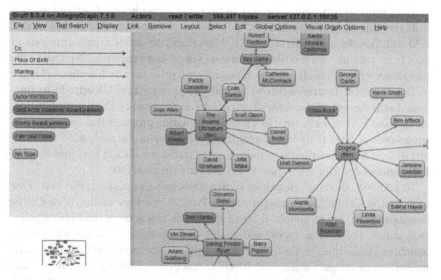

FIGURE 3.1   Example of network graph visualization with gruff.

3.2.11 Data Virtualization

Data virtualization is the definition of a semantic model that integrates all enterprise data regardless of how that data is stored. This allows the advantages of SWT to be applied to structured data (e.g., relational databases, graph databases, Comma Separated Values, XML), binary data (e.g., video, audio, compiled code) and unstructured data (e.g., email, office documents, web pages).

SWT does not require an organization to transform all its data into a triplestore in order to take advantage of the technology. Such a requirement would be unrealistic for any large organization. Also, for many types of data, it is most productive to leave it in their native formats such as relational databases and documents, video, and other types of content. Data virtualization creates a repository of data stored in many different formats. Such a repository is known as a *Data Lake* [27].

One of the most important advantages of data virtualization is the elimination of data silos. Data silos are databases or other data stores that hold one aspect of the business (e.g., customer orders) but are not integrated with other data sources that also have relevant data (e.g., customer trouble tickets). Data silos result in redundant work, redundant data, and inconsistent data. As big data becomes ever more critical to all businesses the elimination of data silos becomes essential to the productive running of the business.

The most common tool for data virtualization is to define pattern matching rules or functions that look for patterns in data and transform them into objects and property values. This approach goes back to Enterprise Application Integration (EAI) middleware products such as Tibco. EAI products integrate systems not via common data models but by having interfaces that send and receive asynchronous messages to a publish and subscribe message bus. Since EAI tools integrate diverse applications, these applications have different data models and different ways of representing data. One system may call a customer a *User* and another a *Buyer*. One system may use an XML model and another a relational model. EAI middleware tools often have GUI interfaces to define transformations of data from various data models and formats via pattern matching [28]. Due to the prevalence of EAI middleware in industry, enterprise data architectures (known as Data Fabrics, described below) typically require EAI middleware to be integrated into the overall framework, especially at the level of Data Virtualization. Also, some EAI vendors such as Tibco are including products that integrate with triplestores or have their own triplestores [29]. Some of the tools used for virtualization in SWT are similar to those used to transform data in EAI middleware. EAI vendors describe what they do as data virtualization [30].

In addition, there are free open source tools that can perform this task. One such tool is Cellfie [31], a plugin for the Protégé ontology editor. With Cellfie developers can define simple rules to transform spreadsheet data into classes, individuals, literal data types, and values of properties. However, the pattern matching capabilities in Cellfie are fairly limited. For more sophisticated transformations developers utilize the REGEX function in SPARQL and Regex and other string manipulation functions in programming languages in combination with APIs to manipulate the triplestore. One example of this type of more complex transformation is the transformations in the CODO system for analyzing epidemiological data for the COVID-19 pandemic [32]. A more sophisticated approach to this problem applies various Machine Learning algorithms. These typically represent the knowledge graph as an n-dimensional vector space and use ML techniques such as deep learning and embedding to enrich and extend the graph [33].

Another type of tool for data virtualization is natural language processing (NLP) tools that use variations of words based on tense, homonyms, homophones, and other grammatical concepts. There are many NLP tools both in vendor products as well as in open source software. An example vendor product is the Free Text Index feature in the AllegroGraph triplestore from Franz Inc [34]. Another example from Top Quadrant is shown in the section on Harmonization below. Two of the most powerful open source NLP data virtualization tools are StanfordNER [35] and IllinoisNER [36].

Another aspect of data virtualization is to connect a large collection of documents known as a corpus to the relevant nodes in a knowledge graph. This is known as *deep text analytics*. It is similar to standard tagging except matching to a knowledge graph provides additional semantics beyond keywords. As a simple example, imagine that we find the word *owl* frequently used in a document. Using standard keywords, we have no way to distinguish if this refers to the species of bird or to the Web Ontology Language. With a knowledge graph, we can find other resources that are closely connected to both concepts in the graph. In the case of the owl species, we would find resources such as "Bird", "Raptor", and "predator". In the case of the ontology language, we would find nearby resources such as "W3C", "SPARQL", and "ontology". We would then search the document for occurrences of both sets of words and would use these occurrences to determine the correct concept [37].

3.2.12 DIGITAL TWINS

The concept of *digital twins* predates knowledge graphs. The idea is to have digital copies (twins) of physical devices. These digital twins can provide information about the status of devices (e.g., a tower in a utility power grid) based on sensors that feed information from the field to a database of the digital twins. This facilitates real-time monitoring in order to anticipate and prevent problems before they occur. In addition, digital twins can be utilized to do simulations to extrapolate what the results of a storm or other events might have on the twin and the infrastructure that it is a component of in order to anticipate and prevent catastrophic failures [38].

Knowledge graphs are ideal for representing digital twins for the same reason that OOP first made a big impact in the domain of stochastic simulation. Representing physical objects as knowledge graph objects is highly intuitive. The real world information about these objects can be represented in class and property hierarchies as well as other axioms that define semantic models. In addition,

Linked Data can be utilized to leverage information about locations in time and space and to download data from remote sensors from the physical objects to their digital twins. Digital Twins are a key enabling technology for the Internet of Things [39].

3.3 THE BIG PICTURE: A DATA FABRIC

These technologies come together in what is known as a Data Fabric. The Gartner Group is a consulting firm that specializes in reports meant for CIOs and CTOs. Unlike other consultants they don't do system integration or software development and are recognized as unbiased evaluators. All they focus on is analyzing the current IT landscape and having their analysts project trends. One of their most well-known concepts is the Gartner technology Hype cycle [40].

Gartner coined the term Data Fabric for an integrated approach to all enterprise data and the term has since been picked up by most of the major vendors. According to Gartner, a data fabric is: "a design concept for attaining reusable and augmented data integration services, data pipelines and semantics for flexible and integrated data delivery... This results in faster, informed and, in some cases, completely automated data access and sharing" [41]. I think a more succinct definition for the essential business reason a Data Fabric is so impactful is that it is a proposed alternative to the standard method of system integration of the past, in most cases EAI message-oriented middleware [42]. If that is true it will completely transform the way system integration is done in future corporate IT departments.

Figure 3.2 shows how the various concepts described in the Business Value section combine together in a data fabric. A data fabric is a strategic architecture goal. There are no commonly agreed upon specifications of the components and

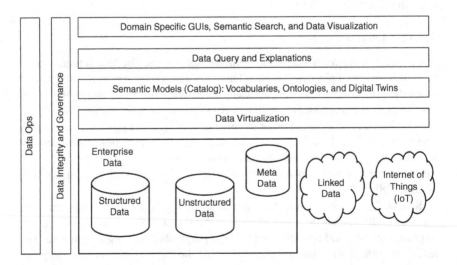

FIGURE 3.2   Data fabric functional architecture.

how they work together. To the extent Data Fabrics currently exist they are very different and customized to the specifics of the organization. Hence, this diagram is meant to provide an integrated picture of the concepts discussed in the Business Value section. It is not intended as a technical architecture diagram.

The Linked Data and IoT clouds represent data that exist on the public Internet. Enterprise data is primarily data that is behind the firewall of an organization. Data virtualization provides an integrated logical view of all data and meta-data which is modeled via semantic models defined in the Web Ontology Language (OWL) and SHACL. This collection of models is typically referred to as a *Data Catalog*. Queries in SPARQL and explanations generated by OWL reasoners provide interfaces between end users and enterprise data. Domain-specific GUIs, visualization, and semantic search tools (described in the next section) provide the interfaces for end users.

There are two types of systems that span all aspects of the data fabric:

1. Data integrity and governance: Data integrity is defined via SHACL shapes. Governance is maintained with workflows and other tools from vendors such as Top Quadrant.
2. Data Ops refers to the management of the infrastructure. This is typically done with technologies such as HADOOP and J2EE application servers that facilitate highly distributed, fault-tolerant infrastructures as well as middleware busses such as Kafka.

3.4   SEMANTIC WEB TECHNOLOGY APPLICATIONS

The semantic web is an enabling technology similar to object-oriented programming (OOP). When OOP first emerged from the lab to the business world there were specific types of applications where it was utilized because it was an especially good fit such as the development of graphical user interfaces and stochastic simulations. However, it was a general-purpose technology and could, and since has been, used for virtually any type of system. The same is true for semantic web technology (SWT). The initial uses for it tend to be focused on a few types of applications where it is especially well suited, but it is a general software paradigm and can be used for most applications. In this section, we will describe some of the most common types of applications that SWT is currently being utilized for.

3.4.1   SEMANTIC SEARCH

One of the first uses for an ontology language that supported a classifier was the BACKBORD system for semantic search developed at the Information Sciences Institute (ISI) in the 1980s. BACKBORD utilized the Loom language which was one of the precursors to OWL. BACKBORD enabled users from the Defense Logistics Agency to semantically search an ontology of electronic parts [43]. Systems such as BACKBORD allow queries to be created by browsing the

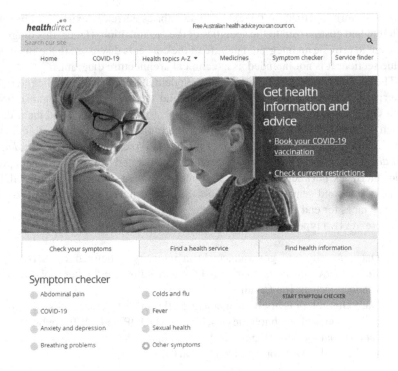

FIGURE 3.3   Pool party semantic search for Australian healthcare.

knowledge graph (e.g., the class hierarchy represented as a tree graph) and filling in forms with parameters that can be utilized by the reasoner.

A modern example of this type of semantic search is the portal created for the Australian ministry of health by Pool Party [44] shown in Figure 3.3. This portal allows consumers to search the Australian Health Ministry knowledge graph to find information on diseases, drugs, providers, etc.

Another example of semantic search is question answering systems. Probably the most famous example of question answering semantic search is the Google Knowledge graph. When users type in a search string in Google the results from the Google knowledge graph are shown in a box at the beginning of the search results before the actual list of relevant links. An example is shown in Figure 3.4 with the query: "What bands has Eric Clapton been in?"

Google calls these Knowledge Graph results as Knowledge Panels. Unfortunately, but understandably, Google has shared little information about the implementation details of their Knowledge Graph as it is an essential component of their flagship service. What we do know is that it utilizes crowd-sourced information such as Dbpedia or Wikidata as well as for-profit linked data sources such as Refinitiv and Crunchbase. These sources are only used as examples of the type of sources as Google doesn't share which specific sources they use. Google does provide documentation for the API to the knowledge graph that identifies schema.org as the

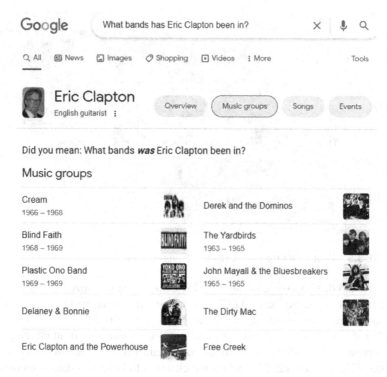

FIGURE 3.4   Example of Google knowledge graph semantic search.

source of their data types as well as links to some of the most important classes [45] and a process for organizations to rectify information in the knowledge graph that is inaccurate [46]. It was Google that first coined the term Knowledge Graph in [8].

3.4.2   INTERNET OF THINGS (IoT)

As computer chips become smaller and more powerful, the trend is to put them into everything from cars to household appliances. This is known as the Internet of Things (IoT). Knowledge graphs utilizing digital twins are ideal for representing and reasoning about the IoT. The semantic knowledge provided in knowledge graphs provides context on the structure, location, time, security, and other critical information to utilize the IoT. Sophisticated models of security are essential to provide users with the trust required to enable the utilization of the IoT. There have already been issues with hackable digital cars and other IoT devices [47].

One example of the power SWT brings to the IoT is the Graph of Things (GoT) project [48]. The GoT provides access to quasi-real-time data from a global network of sensor digital twins displayed as a global map that can display the entire globe or can drill down to 3D models and pictures of city blocks. It also leverages linked data from sources such as Twitter to provide additional contextual information.

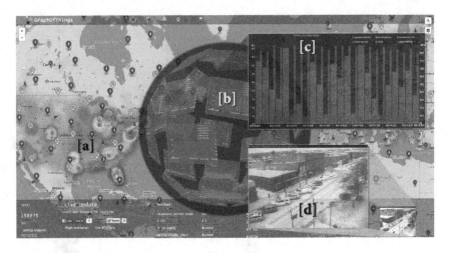

FIGURE 3.5   Four views from the Graph of Things (GoT).

Figure 3.5 displays some of the ways IoT data can be visualized in GoT. All of the data were retrieved via SPARQL queries generated via the GUI. Section [a] displays a heat map aggregated from temperature readings for the last hour for the United States. Section [b] is a 3D layout of live thumbnails from 64 temperature sensors within 200 miles from the coordinate where the mouse was clicked (67.003–178.917). Section [c] displays bar charts with historical data from one of the snapshots in the 3D sphere of section [b]. Section [d] is camera data associated with one of the bar charts in section [c] (i.e., camera data for the time and place associated with one of the bars in [c]).

An example of SWT used for the IoT on a smaller scale is to collect and analyze data from digital twins of a manufacturing production line [49]. This enables the system to provide answers to questions such as:

- Which production line has the minimum/maximum response rate?
- What is the turnaround time for product X?
- Which product has the minimum/maximum turnaround time?
- Are any production lines showing delays from their normal performance past some threshold?

In addition, digital twins for manufacturing can be utilized to do various types of mathematical and stochastic "what if?" scenarios and simulation to find more productive designs for factories.

3.4.3 EXPERT SYSTEMS AND AI

With the emphasis on Big Data, we shouldn't forget that the roots of the Semantic Web go back to the earliest days of research on knowledge representation in Artificial Intelligence (AI) [2] and that ontologies and knowledge graphs provide an ideal platform for expert systems and other types of AI. In the academic community of the present the vast majority of semantic AI research currently conducted utilizes SWT. The traditional capabilities of rule-based and frame-based systems of the past are available in SWT and unlike many previous such environments (e.g., the Lisp language) SWT provides a framework that is both scalable and widely acceptable to industry. Most of the sophisticated AI systems both in the lab and deployed in industry utilize some combination of SWT and ML. Examples include autonomous vehicles and the Watson system from IBM [50].

3.4.4 HARMONIZATION

Harmonization leverages data virtualization to align various vocabularies that an enterprise utilizes both within the enterprise and with business partners in Business to Business (B2B) systems. The problem is most obvious in healthcare where there is an abundance of standards: SNOMED CT, ICD, LOINC, HL7 FYRE, UMLS and countless others. In addition, many organizations use different versions of the standards and/or have customized the models based on requirements in different nations and their own specific requirements. Figure 3.6 shows a simple example of harmonization: two different models, MedDRA and SNOMED CT have two different terms for the symptom of Chest Pain.

While healthcare is the most obvious domain where harmonization is a critical problem it is also important in several other domains.

Harmonization is such a critical issue that there is an SWT startup in Luxembourg called Dynaccurate that is focusing on this specific problem. The Dynaccurate AI is the result of academic research in the field of semantic interoperability and is offered as a tool to maintain harmonization at scale. In a trial with a Life Sciences Drug Pricing Firm, their AI was used to 'remap' the links between SNOMED CT and in-house Disease Identifier codes. Previously,

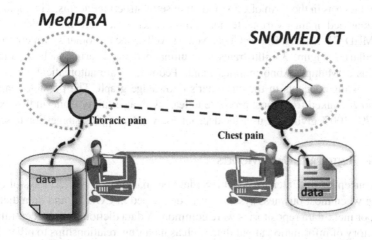

FIGURE 3.6 Simple example of harmonization.

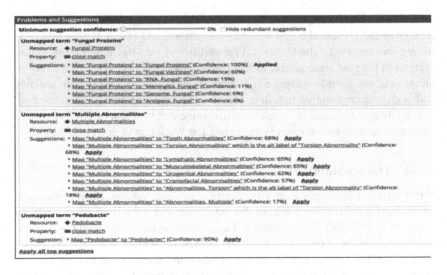

FIGURE 3.7   Automated suggestion of matching terms from TopBraid EDG.

this work had been carried out semi-automatically, using an in-house system of SQL rules. However, this approach was leaving many thousands of mappings for human reconciliation for each new release of SNOMED. By contrast, the AI performed with over 90% precision, and reduced the human workload to only a few hundred reconciliations. In human terms, this meant a reduction from 6 to 10 weeks to only a few days, with a consistent and explainable approach.

Another example of harmonization is shown in Figure 3.7. This is from TopBraid EDG, a product from Top Quadrant. This figure is from a Top Quadrant Life Sciences customer that used their tool to harmonize data coming from many different sources for clinical drug trials. This is also an example of Data Virtualization as the data was in the form of CSV and other semi-structured forms. The input data were matched against various terms from healthcare standard ontologies such as SNOMED CT, MESH, Entrez Gene, Ncit, as well as the customer's own controlled vocabularies. Figure 3.7 illustrates how unmapped input terms such as "Fungal Proteins", "Multiple Abnormalities", and "Pedobacte" are automatically matched with existing concepts in the customer's knowledge graph. The possible matches are shown ranked from most probable to least with a certainty from (in this example) 17%–100% giving the user the decision on which mapping suggestion to select.

3.4.5   ENTERPRISE DATA MODELS

The concept of an Enterprise Data Model goes back to the earliest days of computing when most business systems were developed in COBOL and data dictionaries, or meta-data repositories were common. A data dictionary is a "centralized repository of information about data such as meaning, relationships to other data, origin, usage, and format" [51].

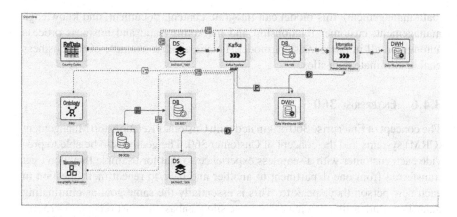

FIGURE 3.8   Information lineage from TopBraid EDG.

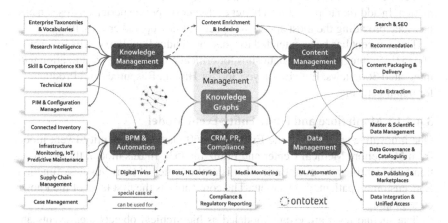

FIGURE 3.9   Semantic web technology as the core of an enterprise data model.

SWT provides the foundation for a much richer enterprise data model than a data dictionary. To be truly useful, an enterprise data model requires more than common definitions for data. It requires the various permissions associated with the data. Who "owns it", who can see it, who can change it, and so on. In addition, the model requires the various workflows that define the processes associated with the lifecycle of the data and the meta-data associated with it such as who created it, who has edited it, when it was last changed, and the domain models (e.g., class and property hierarchies) that define the data. Figure 3.8 shows a diagram generated from TopBraid EDG that illustrates the various systems involved in the Information Lineage (the systems that generate and utilize the data) for part of one customer's enterprise data model.

Figure 3.9 shows a diagram from Ontotext that illustrates how SWT (knowledge graphs and metadata) can form the core of an Enterprise Data Model. In addition to providing a common repository for the development of systems (data management), this model can integrate content, document, and knowledge management; customer relationship and marketing data; and business process automation all in one consistent model. This is another example of the business goal to eliminate data silos.

3.4.6 ENTERPRISE 360

The concept of Enterprise 360 originated with Customer Relationship Management (CRM) systems and the concept of Customer 360. The goal was to be able to provide each customer with a seamless experience so that for example they don't get transferred from one department to another and have to repeat their question to each new person they speak to. This is essentially the same goal as eliminating data silos. Rather than having a disparate silo database for past orders, in-process orders, technical support tickets, and so on there should be integrated data views so that each employee the customer deals with has access to all their relevant information. In addition to providing a better customer support experience, this can also facilitate providing the most appealing special offers, up and cross selling, etc.

This concept has been extended to what the vendor Pool Party calls Enterprise 360 [52]. To integrate all the data silos for critical entities to the organization with integrated 360 views. E.g., Employee 360 for Human Resources and Product 360 for manufacturing.

3.4.6.1 Montefiore and Franz Entity-Event Model

One innovative approach to the Enterprise 360 vision is the Entity-Event model from Montefiore Medical Center and Franz Inc [53]. In this model, the organization first identifies the core Entity that is at the center of their business function. E.g., customer, patient, product, etc. The core entity is the first level of the knowledge graph.

The second level are events modeled as hierarchical objects, i.e., events and sub-events.

As the example in Figure 3.10 illustrates, in a hospital example Patient would be the critical entity and events would be a hospital visit with sub-events such as diagnostics, tests, medication orders, procedures, vital signs, etc. All of these have a start time and end time and other common properties of events such as super-event and sub-event. The third level represents data from diverse sources such as laboratory systems, admitting systems, billing systems, and Electronic Medical Record systems.

Some of the benefits of this approach are:

1. A simple and intuitive model that puts the patient (customer, employee, product, etc.) as the central part of the model. The events that apply to the core entity are then utilized to break down data silos and provide an integrated 360 view of the core entity.
2. Easy and efficient queries. Figure 3.11 shows two examples of a query to find "all the people that had gallbladder calculus in 2010 or later". The

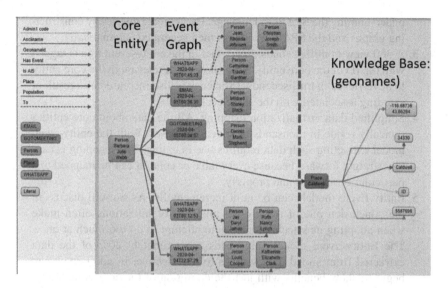

FIGURE 3.10 Montefiore and Franz entity-event model.

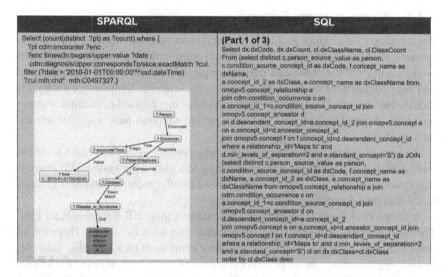

FIGURE 3.11 Comparison of SPARQL and SQL queries with entity-event model.

left half of the figure shows a SPARQL query utilizing the entity-event model. The right half shows part 1 of 3 of the equivalent SQL query to get the same information from a database. This is based on a real healthcare customer. In addition to demonstrating the power of SPARQL and the entity-event model, the figure demonstrates how the GUI in the Gruff tool makes it possible for end users to develop queries without

understanding the SPARQL language. Users create queries by constructing graphs and the SPARQL code is then generated automatically.

3. A 360 view in milliseconds. Every fact in the entity-event tree is tagged
   with the ID of the core entity. This allows data relevant to the core entity
   to be retrieved in milliseconds. Efficiency is also increased as a result of
   sharding associated with the ID of each entity.

4. Simplified data virtualization mapping. All the data about core entities
   typically reside in thousands of siloed databases. With the entity-event
   model extracting such data requires one mapping. The mapping is usually straight forward because every table or column can be mapped to a
   particular event class and property.

5. Entity-Event models can be built incrementally. As we will discuss in
   the conclusion one of the biggest mistakes organizations often make
   when adopting new technologies is to attempt to do too much at once.
   The Entity-Event model only needs approximately 20% of the data
   extracted from siloed data sources to be effective. In addition, it can
   begin to show benefits with just the definition of the core entity and a
   few events [53].

3.4.7 RECOMMENDATION ENGINES

Recommendation engines also have a central class such as Customer and various
connections for each instance of this class created in the knowledge graph based
on the events that the individual initiates or participates in. In [28] the individual
and their connections in the graph are called the semantic footprint for each individual. The semantic footprint for each individual can be used in multiple ways to
recommend products, services, and other people who may have a similar semantic footprint.

As part of the tutorials for the Pool Party product, user can create a simple
wine and cheese pairing recommendation system [54]. This system [55] demonstrates how a knowledge graph can be utilized to offer recommendations tailored
to the semantic footprint of each individual.

Another Pool Party recommendation system is their HR Recommender [28].
The most obvious use is for organizations that find work for clients. However, in
some businesses, such as law and consulting firms such tools are also valuable.
In such organizations, the most important asset is not a product or even a service
but rather the people. Such organizations recruit highly motivated people and
will go out of their way to retain their best people because they realize that in the
long run such a policy benefits both the individual and the firm. Staffing for such
firms can be complex. Staff will often be temporarily assigned to projects far
from their local office due to their specific experience, expertise, and career goals.
Staff are also proactive and will seek out assignments themselves in order to best
further their career goals. Thus, an HR recommendation system could be used
both by project leaders to find the best staff and by staff to find the best projects
by matching the semantic footprints of individuals and projects. The Pool Party

system leverages a reusable vocabulary called the European Skills, Competences, Qualifications and Occupations (ESCO) vocabulary.

3.5 CONCLUSION

In the conclusion we will discuss two issues:

1. Barriers to Industry Adoption of Semantic Web Technology.
2. Suggested next steps.

3.5.1 BARRIERS TO ADOPTION

The work in [56] describes barriers to the adoption of SWT based on the Technology – Organization – Environment framework, a model for analyzing the change in the IT industry. The work in [57] uses a different approach. The author began with common critiques of the Semantic Web he collected from message boards, discussions with colleagues, etc. He then created a questionnaire which he distributed to the Semantic Web mailing list. Many of the issues, although often expressed differently, were similar:

1. The Semantic Web will be eclipsed and made unnecessary due to Machine Learning. This was only an issue in [56].
2. The Semantic Web is too difficult for industry developers to comprehend due to its formal foundations and the complexity of languages such as RDF/XML.
3. Semantic Web technology such as RDF and OWL will not scale to meet the needs of large corporations. This is also discussed in [3]. A detailed discussion of this criticism and an alternative approach to W3C standards for graph data can be found in [58].

Regarding issue 1, for the early years of SWT, ML quite justifiably received a great deal of attention due to it being able to solve problems that semantic AI was unable to solve. However, for problems such as eliminating data silos, providing explainable solutions, semantic search, and others ML is at best a band aid solution [12].

Issue 2 is primarily a question of perception. One doesn't need to be a logician to use SWT. A basic background in logic and set theory are very useful but these concepts are useful for software development in general and quite easy for any competent developer to understand. The issue about RDF/XML is not significant because there are alternative formats such as Turtle that are less verbose and more intuitive. More importantly, with GUI tools such as Protégé, one can do design and even much development without needing to work at the level of these languages.

The question of whether RDF can scale has been addressed by recent large-scale knowledge graphs such as the ones discussed in this chapter and countless others.

The question of whether OWL can scale for the requirements of Big Data is still unresolved. Most of the large knowledge graphs currently in use are based on technologies such as RDF that do not have the strong semantics of OWL. Three possible solutions to this issue are:

1. OWL Profiles. OWL has different profiles [59] with different levels of semantics. This is a classic trade-off of knowledge representation in AI. The closer a language comes to the full power of First Order Logic (FOL) the slower reasoners for that language become [60]. Triplestores that are designed to scale typically use the OWL profiles that are less powerful or design their own profiles that limit the semantics of OWL even further. Even with these scaled down profiles OWL still provides very expressive semantics.
2. Modular knowledge graphs. Knowledge graphs can be divided into different modules [61]. Modules used for complex AI systems requiring sophisticated reasoning can use fuller OWL profiles while modules that have larger data sets can use less powerful profiles.
3. Improved performance. Reasoning algorithms can be made more efficient and can be designed to work incrementally with large data similar to garbage collection in Java and other advanced languages [62]. Distributed computing can be leveraged to also increase performance [63].

3.5.2 SUGGESTED NEXT STEPS

The following are some suggested next steps.

3.5.2.1 Avoid Analysis Paralysis

When I was consulting for clients who were adopting OOP one of the most common pitfalls, I saw was that they would spend months creating models without writing any code. This waterfall approach is based on the concept that designing software is like designing a bridge [64]. Changing the design of a bridge after you have started building it is exponentially more expensive than getting the design right before you build. This analogy was valid in the earliest days of software development but hasn't been valid for most systems for a very long time. There are several reasons:

1. Humans have been designing bridges for centuries. The principles and alternatives are well understood. Software development is a much newer process. Software development is a knowledge-intensive process. The best software architectures emerge over time as developers understand the limits and opportunities of their software.
2. The most impactful software redefines how an organization works. The understanding of how to reengineer the business emerges as teams gain experience with new technologies. As a result, software requirements are far more fluid than requirements for bridges [64].

3. Modern software makes it easier to create simple architectures and then to refactor them as requirements dictate rather than to design the perfect system all at once [64].

I often see the same analysis paralysis in semantic web projects. If anything, due to the analytical background of many semantic data scientists, the tendency to spend too much time on design is even more common in my experience than with OOP systems. This is ironic because one of the benefits of semantic technology is that it supports Agile design and development. As described above, it is far easier to make design changes at run time for SWT than for OOP or relational systems since the design itself is part of the graph and the design can be modified at run time as easily as the data.

3.5.2.2 Utilize Agile Methods

Agile methods are very appropriate for developing SWT systems. Agile is a proven technique to manage risk, especially with new technologies [63]. One of the most important Agile principles for SWT is to have test data defined from day one. This is another very common error I see in new users of SWT. They spend months designing a model but never create any test data. Then the first time they create test data the reasoner detects many inconsistencies. The OWL reasoner is a powerful tool but to get the best out of it one needs to run it frequently when designing an ontology and one needs sample test data to find design errors from day one.

3.5.2.3 Start with an "Easy Win" First Project with Business Value

The first reaction for many IT leaders is to begin designing a data fabric infrastructure. However, focusing on infrastructure delivers little immediate business value and may encourage the management perception that the new technology is just the latest fad rather than something that can help the business create competitive advantage. Before focusing on infrastructure, I suggest clients first find small projects that are a good fit for SWT, can be done with relatively little risk, and will deliver tangible business value.

For example, when the Internet was still considered leading edge technology one project that my consulting team would sell to clients was to replace the printed documents from their HR department with electronic documents available on the company's Intranet. The amount of money spent every quarter to print and distribute new versions of these documents was staggering. Replacing print documents with electronic documents was an easy project with little risk that could pay for itself in a short time while at the same time growing the team's skills.

3.5.2.4 Expand Your Knowledge of SWT

For technical people who want to be hands on with the technology, the best place to start is with the Protégé ontology editor [65]. Protégé is a free tool, but it is robust, has a very intuitive user interface, and an active support community. In addition, there is an active community of developers who have created many plugins that extend the functionality of Protégé. The Protégé editor is for editing OWL ontologies. However, plugins exist for virtually all of the most important Semantic Web technologies such as SHACL and SPARQL. The Protégé editor is robust enough that it is often used for ontology design in industry. Some of the plugins are primarily for learning purposes and not powerful enough for industry use yet but they are a good starting point to understand the technologies and then to advance to other vendor or open source tools. The latest Protégé tutorial [66] includes detail on all SWT concepts as well as step by step instructions on how to create and query a tutorial ontology.

After learning Protégé Java developers are encouraged to become familiar with the Apache Jena toolkit [67]. Jena provides several Java libraries for manipulating OWL and RDF knowledge graphs. OWLReady2 [68] provides similar functionality for Python.

The references [2,3] provide more detailed descriptions of SWT. Reference [12] is an excellent discussion of the business case for SWT.

Each of the major SWT vendors highlighted in this chapter have demonstration and/or free community versions available for download. Each vendor has different strengths, and they are all worth the time to work through their demos.

3.6   GLOSSARY

In the following any word in *italics* refers to another term defined in this glossary.

- **Axiom**: A logical definition in the *Web Ontology Language* that describes a model. Examples of axioms include definitions of super and sub-*classes*, *property* hierarchies, restrictions on the number and type of *property* values, etc.
- **Class**: A Class is already something that most developers are familiar with via Object-Oriented Programming (OOP). In SWT classes are defined with the *Web Ontology Language (OWL)*. *OWL* classes are similar to OOP classes in some ways but there are also important differences such as that OWL classes have formal definitions based on set theory that enable sophisticated validation and reasoning. An *OWL* class is semantically a set.
- **Concept**: In this chapter, we use the term concept to refer to objects that can be either *classes, properties, axioms, literals*, and/or *individuals*.
- **Individual**: An individual is similar to an instance in OOP. Each individual is an instance of at least one *class*. An individual that is an instance of a *class* has the same semantics as an element of a set.
- **Knowledge graph**: A graph model of data typically stored in a *triplestore* database. The core concept for a knowledge graph is the Subject -> Predicate -> Object *triple*. I.e., a triple is to a knowledge graph as a table is to a relational database. Triples result in non-directed graphs because the subject of one *triple* can be the object of another and vice versa.
- **IRI**: An Internationalized Resource Identifier (IRI) looks like a URL, but it is not required to be viewable in a web browser and often is a link to very small-grained objects such as individual *classes* or *instances*.

- **Literal**: Literals are common data types such as integers and strings. SWT typically reuses datatypes from XML, especially the XML Schema Definition (XSD).
- **Model**: An *ontology* or *knowledge graph*.
- **Ontology**: The term ontology goes back to the earliest days of knowledge representation in Artificial Intelligence (AI) and Information Science. An ontology models a domain usually represented as one or more taxonomies (tree structured graphs). In SWT ontologies are defined using the *Web Ontology Language (OWL)*.
- **Property**: A property in SWT is similar to properties in OOP or relations and attributes in relational databases. The main difference is that in *OWL*, properties are equivalent to relations in set theory. Thus, common mathematical characteristics that apply to relations in set theory such as property hierarchies, inverses, transitive properties, etc. can be defined and used for inferencing by *OWL reasoners*.
- **Reasoner**: A reasoner is similar to an inference engine in a rule-based system except that whereas rule-based inference engines are constrained to if-then rules with forward and/or backward chaining, *OWL* reasoners can take advantage of the richer set theoretic semantics of *OWL* and can provide many more types of reasoning. A reasoner first ensures that an *ontology* is consistent (has no contradictory *axioms*) and then can redefine the structure of the *ontology* and assert new data values based on the *axioms* in the *model*. An *OWL* reasoner is a type of automated theorem prover.
- **Resource**: Any concept with an *IRI*, such as a *class*, *property*, or *individual*.
- **Resource Description Framework (RDF)**: The W3C standard language for defining knowledge graphs as collections of *triples* where each item in the triple is a *resource*. *OWL* is built on top of RDF. *SHACL* is an RDF *vocabulary*.
- **Shapes Constraint Language (SHACL)**: The language standard for defining data integrity constraints on SWT data. A Shape defines a set of constraints on *properties* for a *class*. Examples of constraints include cardinality of values, patterns for strings, value constraints for numbers, and logical relations between two or more properties. SHACL includes a reasoner that checks the data integrity of a model, reports integrity violations and may attempt to repair such violations. To some extent *OWL* and SHACL have overlapping capabilities. The difference is in the way the definitions are used. *OWL axioms* are used for reasoning whereas SHACL shapes are used to define and enforce data integrity constraints.
- **SPARQL**: SPARQL is the query language for SWT. SPARQL is to *triplestores* as SQL is to relational databases. SPARQL is powerful because it can match the Subject, Predicate, and/or Object of a *triple*. A query that matches all 3 of the Subject, Predicate, and Object will return every *triple* in the graph. SPARQL can also perform federated queries

on public data sources known as SPARQL *endpoints*. These types of federated queries are known as Linked Data. SPARQL can also transform *triples* (change, add, or delete them).

- **SPARQL endpoint**: An Internationalized Resource Identifier (*IRI*) that can receive and process *SPARQL* queries.
- **Triple**: The fundamental data structure of a *knowledge graph*. All triples have a Subject -> Predicate -> Object structure.
- **Triplestore**: A database that utilizes *triples* as its core representation rather than tables. Although *models* can be stored in files, large industry scale *models* are stored in triplestores.
- **Vocabulary**: A *model* designed to be reusable for a horizontal or vertical domain. Vocabularies are usually public. Examples of vocabularies are the Friend of a Friend (FOAF) vocabulary for describing social networks, the Dublin Core vocabulary for defining meta-data and the Prov vocabulary for defining provenance information about data.
- **Web Ontology Language (OWL)**: The language used to define *models* in SWT. OWL is an implementation of Description Logic which is a decidable subset of First Order Logic

REFERENCES

1. Chatterjee, A., Nardi, C., Oberije, C., and Lambin, P. Knowledge Graphs for COVID-19: An Exploratory Review of the Current Landscape. *J. Pers. Med.* 2021, 11, 300. doi:10.3390/jpm11040300.
2. DeBellis, Michael. Semantic Technology Pillars: The Story so Far. With Jans Aasman. In *Data Science with Semantic Technologies: Theory, Practice, and Applications*. Edited by Archana Patel, Narayan Debnath, and Bharat Bhushan. Wiley. To be published March 2022.
3. Uschold, Michael. *Demystifying OWL for the Enterprise*. Morgan & Claypool Publishers, San Rafael, CA; 1st edition (May 29, 2018).
4. World Wide Web Consortium web site. https://www.w3.org/.
5. RDF Specification. https://www.w3.org/RDF/.
6. Traiaia, Karim. Why Your Business Should Use a Graph Database. July 8, 2020. https://memgraph.com/blog/why-your-business-should-use-a-graph-database.
7. SPARQL Specification. https://www.w3.org/TR/sparql11-query/.
8. Singhal, A. Introducing the Knowledge Graph: Things, Not Strings. *Google White Paper*, 2012. https://www.blog.google/products/search/introducing-knowledge-graph-things-not/.
9. Ng, Andrew. A Chat with Andrew on MLOps: From Model-centric to Data-centric AI. March 24, 2021. https://www.youtube.com/watch?v=06-AZXmwHjo.
10. https://www.w3.org/TR/owl2-overview/.
11. SHACL Specification. https://www.w3.org/TR/shacl/.
12. Booth, David. Principles of the Semantic Web: Enabling Global Data Reuse. W3C Presentation. 2002. https://www.w3.org/2002/Talks/1218-semweb-dbooth/Overview.html.
13. The Prov-O Vocabulary. https://www.w3.org/TR/prov-o/.
14. Dublin Core Web site. https://dublincore.org/.

15. Francesca Lucarini, The Differences between the California Consumer Privacy Act and the GDPR. April 13, 2020. https://advisera.com/eugdpracademy/blog/2020/04/13/gdpr-vs-ccpa-what-are-the-main-differences/.

16. Lander, Eric. Americans Need a Bill of Rights for an AI-Powered World. With Alondra Nelson. Wired Magazine. 2021. https://www.wired.com/story/opinion-bill-of-rights-artificial-intelligence/.

17. SWRL Specification. https://www.w3.org/Submission/SWRL/.

18. Musen, Mark. The Protégé Project: A Look Back and a Look Forward. *A.I. Matters* 1(4), 2015. doi:10.1145/2757001.2757003.

19. McComb, Dave. *Software Wasteland: How the Application-Centric Mindset is Hobbling our Enterprises.* Technics Publications. February 6, 2018. ISBN-13: 978-1634623162.

20. https://www.dbpedia.org/.

21. https://www.wikidata.org/.

22. http://www.geonames.org/.

23. Craig, David. Meet Refinitiv. 2018. https://www.refinitiv.com/perspectives/financial-crime/meet-refinitiv/.

24. Mannes, John. Crunchbase Pro Brings New Search and Analysis Features for Power Users. September 12, 2016. https://techcrunch.com/2016/09/12/crunchbase-pro-brings-new-search-and-analysis-features-for-power-users/.

25. Antoniazzi, Francesco, et al. RDF Graph Visualization Tools: A Survey. In *2018 23rd Conference of Open Innovations Association (FRUCT).* IEEE. https://ieeexplore.ieee.org/abstract/document/8588069.

26. AAsman, J. and Cheatham, K. RDF Browser for Data Discovery and Visual Query Building. Workshop on Visual Interfaces to the Social and Semantic Web (VISSW2011) Co-located with ACM IUI 2011. Palo Alto, US. 2011.

27. Denodo. Data Virtualization, An Overview. https://www.denodo.com/en/data-virtualization/overview.

28. Hohpe, Gregor. Enterprise Application Integration Patterns: Designing, Building and Deploying Messaging Solutions. With Bobby Wolf. Addison-Wesley. December 2003. pp. 85–95 and 327–361.

29. Tibco. TIBCO® Graph Database - Enterprise Edition 3.0.0. October 2020. https://docs.tibco.com/products/tibco-graph-database-enterprise-edition-3-0-0.

30. Tibco White Paper. Ten Things You Need to Know About Data Virtualization. https://www.tibco.com/sites/tibco/files/resources/WP-ten-things-about-DV.pdf.

31. O'Connor, M. J., Halaschek-Wiener, C., and Musen, M. A.. Mapping Master: A Flexible Approach for Mapping Spreadsheets to OWL. *9th International Semantic Web Conference (ISWC)*, Shanghai, China, 2010.

32. DeBellis, Michael. The Covid-19 CODO Development Process: An Agile Approach to Knowledge Graph Development. *With Biswanath Dutta, Proceedings of the KGSCW 2021 Conference.* Nov. 22–24. 2021. https://tinyurl.com/CODO-KGSWC.

33. Shaoxiong, Ji. A Survey on Knowledge Graphs: Representation, Acquisition and Applications. With Shirui Pan, Erik Cambria, Pekka Marttinen, and Philip S. Yu. IEEE Transactions on Neural Networks and Learning Systems, 2021. https://arxiv.org/pdf/2002.00388.pdf.

34. AllegroGraph Freetext Indexing. AllegroGraph documentation. Franz Inc. https://franz.com/agraph/support/documentation/current/text-index.html.

35. Krishnan, Vijay and Manning, Christopher D. An Effective Two-Stage Model for Exploiting Non-local Dependencies in Named Entity Recognition. *In Proceedings of the 21st ACL-COLING*, pp. 1121–1128, 2006.

36. Ratinov, Lev and Roth, Dan. Design Challenges and Misconceptions in Named Entity Recognition. In *Proceedings of the Thirteenth CoNLL*, pp. 147–155, 2009.
37. Auto-Classification 101: Optimizing CMS Search with Automated Tagging. PoolParty On Demand Webinar. https://www.poolparty.biz/events/webinar-auto-classification-101-optimizing-cms-search-with-automated-tagging.
38. Pureweb. Smart Assets: The Role of Digital Twins in the Utilities Industry. https://www.pureweb.com/resources.
39. Blumauer, Andreas. *The Knowledge Graph Cookbook: Recipes that Work. With Helmut Nagy*. 1 edition, 2020. Monochrom Publishing. ISBN: 978-3-902796-70-7.
40. Gartner Technology Hype Cycle. https://www.gartner.com/en/research/methodologies/gartner-hype-cycle.
41. Zaidi, Ehitsham. Data Fabrics Add Augmented Intelligence to Modernize Your Data Integration. With Eric Thoo. Gartner Group. 17 December 2019. ID G00450706.
42. Singh, Bipin. Top 5 Data Fabric Takeaways from 2021 Gartner D & A Summit. Data Operations/Nexla News. https://www.nexla.com/top-5-data-fabric-takeaways/.
43. Yen, John. BACKBORD: Beyond Retrieval by Reformulation. With R Neches and M DeBellis. *ACM SIGCHI Bulletin* 20(1), 77.
44. PoolParty. Healthdirect Australia Uses Poolparty Semantic Suite Technology as The Basis. For Innovative Health Services. White Paper. https://www.poolparty.biz/healthcare-information-system/.
45. Google Inc. Knowledge Graph Search API. Jul 28, 2021. https://developers.google.com/knowledge-graph.
46. How Google's Knowledge Graph works. 2021. Google, Inc. https://support.google.com/knowledgepanel/answer/9787176?hl=en&ref_topic=9803953.
47. Dibaei, Ho Mahdi, Zheng, Xi, Jiang, Kun, Abbas, Robert, Liu, Shigang, Zhang, Yuexin, Xiang, Yang, and Yu, Shui. Attacks and Defences on Intelligent Connected Vehicles: A Survey. *Digital Communications and Networks*, 6(4), 2020, pp. 399–421, ISSN 2352-8648. doi:10.1016/j.dcan.2020.04.007.
48. Le Phuoc, Danh et al. The Graph of Things: A Step towards the Live Knowledge Graph of Connected Things. 2016. https://www.researchgate.net/publication/303789187.
49. Banerjee, Agniva et al. Generating Digital Twin Models Using Knowledge Graphs for Industrial Production Lines. *9th International ACM Web Science Conference*. June 2017. https://www.researchgate.net/publication/319356723.
50. Ferrucci, David et al. Building Watson: An Overview of the DeepQA Project. *AI Magazine*. 31(3), Fall 2010. https://ojs.aaai.org/index.php/aimagazine/article/view/2303.
51. *IBM Dictionary of Computing*. September 1993. McGraw-Hill, Inc. ISBN:978-0-07-031488-7.
52. Blumauer, Andreas. Benefit from Enterprise 360. *Pool Party White Paper*. https://www.poolparty.biz/wp-content/uploads/2021/05/White-Paper-Benefit-from-Enterprise-360-.
53. Aasman, Jans. Entity Event Knowledge Graphs for Data Centric Organizations. *Franz Inc. White Paper*. https://allegrograph.com/wp-content/uploads/2020/06/-Entity-Event-Knowledge-Graphs-White-Paper-v692020.pdf.
54. Burg, Thomas. The Fusion of Search and Recommendation Functionalities. With Christian Blaschke. Pool Party Use Case White Paper. https://www.poolparty.biz/-wp-content/uploads/2017/09/Recommender-Engine-Wine-Cheese-Pairing.pdf.
55. To see a live version of the demo go to: http://vocabulary.semantic-web.at/GraphSearch/ If prompted for a name and password use: User: demouser Password: poolparty.

56. Kim, Dan J. Exploring Determinants of Semantic Web Technology Adoption from IT Professionals' Perspective. With John Hebeler, Victoria Yoon, Fred Davis. *Computers in Human Behavior*, 86, 2018, pp. 18–33, ISSN 0747-5632. https://www.sciencedirect.com/science/article/abs/pii/S074756321830178X.

57 Hogan, Aidan. The Semantic Web: Two Decades On. 10/01/2019. Semantic Web Journal. http://semantic-web-journal.net/content/semantic-web-two-decades-0.

58. Chalupsky, Hans. Creating and Querying Personalized Versions of Wikidata on a Laptop. With Pedro Szekely, Filip Ilievski, Daniel Garijo, and Kartik Shenoy. 2021. Information Sciences Institute (ISI) research paper. https://arxiv.org/abs/2108.07119.

59. OWL Profiles Specifications. https://www.w3.org/TR/owl2-profiles/.

60. Levesque, Hector. A Fundamental Tradeoff in Knowledge Representation and Reasoning. With Ronald Brachman. *In Ronald Brachman and Hector J. Levesque. Readings in Knowledge Representation*. Morgan Kaufmann. 1985.

61. Shimizu, C., Hammar, K., and Hitzler, P. (2020) Modular Graphical Ontology Engineering Evaluated. In: Harth A. et al. (eds) *The Semantic Web. ESWC 2020. Lecture Notes in Computer Science*, vol 12123. Springer, Cham. doi:10.1007/978-3-030-49461-2_2.

62. Kolovski, V., Wu, Z., and Eadon, G. Optimizing Enterprise-Scale OWL 2 RL Reasoning in a Relational Database System. In: Patel-Schneider P. F. et al. (eds) *The Semantic Web – ISWC 2010. ISWC 2010. Lecture Notes in Computer Science*, vol 6496. Springer, Berlin, Heidelberg, 2010. doi:10.1007/978-3-642-17746-0_28.

63. Mutharaju, R. Very Large Scale OWL Reasoning through Distributed Computation. In: Cudré-Mauroux P. et al. (eds) *The Semantic Web – ISWC 2012. ISWC 2012. Lecture Notes in Computer Science*, vol 7650. Springer, Berlin, Heidelberg, 2012. doi:10.1007/978-3-642-35173-0_30.

64. Beck, Kent. *Extreme Programming Explained*. Addison Wesley, Boston, MA, 2000.

65. Protégé Web Site. https://protege.stanford.edu/.

66. New Protégé Tutorial. https://www.michaeldebellis.com/post/new-protege-pizza-tutorial.

67. Apache Jena Open Source Site. https://jena.apache.org/.

68. OWLReady Open Source Site. https://pypi.org/project/Owlready2/.

4 Latest Applications of Semantic Web Technologies for Service Industry

Godspower O. Ekuobase and
Esingbemi P. Ebietomere

CONTENTS

DOI: 10.1201/9781003309420-4

4.1   INTRODUCTION

Semantic web technologies (SWTs) are collections of encoded knowledge dedicated to making possibly large heterogeneous and distributed data or information, seamlessly comprehensible to machines. This knowledge manifests as competencies, standards, methods, and tools. Of these technological components, the most valuable but rarely acknowledged are competencies. Competencies, generally, are encoded in humans through intuition, learning, and experience; and in machines or their software components through computation, learning, and experience. Encoding competencies into machines or any of its software components is a different but complementary vision to Semantic Web (SW) called artificial intelligence.[1] SW standards are computational norms set toward the actualization of the SW vision of seamless comprehension of heterogeneous and distributed data by machines. SW methods are the "how" – procedures or techniques employable to realize the SW vision or develop and maintain any of its tools. SW tools are abstract or intangible entities that can be used to develop or maintain intangible entities or artifacts that enable the actualization of the SW vision, or any such entity so developed. In general, however, it is the collection of SW standards and tools that are usually termed SWTs [1,2].

The essence of the SW vision is to enable machines' efficient and effective performance of cognitive or decision-oriented tasks either as human surrogates or support humans in the performance of such tasks [3]; using available and accessible data which are necessarily heterogeneous and distributed. The heterogeneous and distributed nature of these data (web of data) demands that these data be logically linked or interconnected (linked data) and decorated (knowledge graph) to enable seamless integration, exchange, and processing of these data for machine comprehension and subsequent performance of the role of cognitive surrogate or decision support by the machine as if it were human. In particular, the semantic heterogeneity of data within and across domains forces a formal and explicit specification of shared conceptualization of the data in a given domain (ontology) for the specialized and effective performance of cognitive or decision support tasks in that domain; by machines. These possible cognitive or decision support tasks have found varying economic or industrial applications. The essence of this chapter is to explicate the latest applications of SWTs in the industries.

An industry is a set of organizations that carry out similar activities of value production [4]. The value production process of an organization should be fair to the environment and persons (natural or juristic) involved in the production chain with the value so produced satisfying to all that subscribed to the organization. These values are encapsulated as goods or services – tangible or intangible value products respectively. We can categorize industries into primary, secondary, and tertiary industries [5]. Simply put, primary industries produce raw goods; secondary industries produce processed goods while tertiary industries (co-) produce services. From primary through secondary to tertiary industries is a radical shift from labor-intense activities to knowledge-intense activities with increasing value addition [5]. While the primary and secondary industries are dwindling in value production, the service industry is increasing in value production. The service industry accounts for about 75% of developed economies', and 50% of the developing economies', gross domestic product (GDP), trade, and employment and as well, drives other industrial value production activities [6]. In particular, the service industry is predominantly driven by information and communications technologies (ICTs) [7,8]; SWTs inclusive. Thus, the industrial applications of SWTs exposed in this chapter are those of the service industry.

Since the chapter's focus is on the service industry – the industry that produces services; a quick one at the concept of service will not be out of place. The author [9] defined service as,

> an interactive mechanism that adds value to participating entities: individuals, people, organisations, society, service or service objects (goods), with the consent or cooperation of the entities or their surrogates which could also be entities, procedures or technologies with specialized competencies (p. 2).

The output of this mechanism is what is called services – a time perishable, intangible experience that satisfies the needs and wants of participating entities. The key participating entities are the service providers and service consumers. The service consumers are usually the customers while the service providers are those thought to produce services but services are co-produced by both the service providers and service consumers. Goods are services carriers i.e. they have services embedded in them. It is these services that are actually of interest to consumers of goods. This is the beautiful future of the service industry where "Everything is a Service" (XaaS) [10–12] and autonomous SWTs applications will drive its economic production and growth.

Also not out of place, is the concept of SWTs industrial applications. Technologies are products of human innovations and can only receive industrial attraction and use when it assumes a sufficient level of economic viability, reliability, efficiency, convenience, trust, and safety (e-VRECTS) for use in one or more value production activities. The author [9] posited that sustained commercialization is the hallmark of (service) innovation and emphasized mutual benefits (tactical or strategic) to both the service providers and service consumers. In this chapter, therefore, SWTs industrial applications are commercialized value production tasks compelled by SWTs for enhanced value addition. Only the recent industrial applications in each selected service industry are of interest to this work, and good attempts were made to evade promoting brands particularly due to the unobtrusive nature of SWTs [13].

Already, SWTs are invaluable to the service industry because of the growing burden of information overload, rapidly changing heterogeneous business models, technology, and data, and the need for sustained collaboration and shared intelligence [14]. Machines being superfast and reliable [15,16] can seamlessly rescue humans from these stifling burdens with the exploitation of SWTs. How this is being exploited in the service industry for enhanced productivity is the crux of this chapter. Although [4] identified 21 broad industry types with 15 of them being service industry types, this chapter for concision, the similarity of service mechanisms and cognitive skill sets required by their service providers, identified 12 dominant service industry types; slightly different but consistent with those of [4]. The identified service industry types are: business and finance, law, health, security, education, research and development, communication, hospitality, utility, governance, logistics and transportation, and meteorology. Each of the following 12 sections is dedicated to the latest SWTs applications in the identified service industries. The last section holds the chapter's summary and conclusion.

4.2   APPLICATIONS OF SWTs IN BUSINESS AND FINANCE

Business[2] is any legal mechanism that enables the exchange of goods and services for social or economic gains [17]. When the economic value is exchanged for social or economic gains, the enabling legal mechanism is a special business called finance.[3] In recent times, SWTs have found notable industrial applications in business intelligence (BI), customer relationship management, collaboration, and content discovery. These tasks, though preceded the SW vision, are being revolutionized by SWTs for enhanced productivity and competitiveness of the business and finance industries in particular and the entire service industry in general.

4.2.1   BUSINESS INTELLIGENCE APPLICATIONS

BI applications enable prompt evidence (data) based on cutting-edge decision-making by business stakeholders, customers inclusive, toward efficient, competitive, optimal and sustainable value production for the host organization. BI applications support large, heterogeneous and distributed data extraction, integration, summarisation, organization, processing, analysis and presentation of a specialized set of information to users or machines in real time. SWTs currently drive critical functionalities of BI systems such as data extraction, integration, annotation, analytics, and publishing [18–22] in many business and financial enterprises. These activities may precede the SW consciousness [23] but would have been antiquated without the SWTs such as knowledge graph (KG), ontology, RDF, RDFS, OWL, and SPARQL considering the sustained reality of exhilarating business competitiveness and unprecedented data volume, velocity, and variety.

Common Industrial BI applications driven by SWTs are question and answering (Q&A) and chatbot systems. SWT such as KG is particularly used as self-updating knowledge source, to factor in personalization and context into these BI applications [24]. Q&A and chatbot applications are easily visible on organizational websites and decision support systems (DSSs). KGs have become an upstream industrial SWTs application [25–28] that drives modern downstream (business) applications with excellent industrial maturity. The works of [26] and [27] discussed some successful upstream BI applications of KGs to known industrial business products. KGs are ontologies with real-time updating capabilities and, possibly, domain-independent. Unlike RDF, OWL, and SPARQL which are primitive SW tools, ontology and KG are engineered and reusable knowledge repositories built on the primitive SW tools. Ontologies and KGs are complete software components publishable and consumable as service. Unlike preceding knowledge bases, ontologies and KGs are embedded with cognitive intelligence and are machine comprehensible. The authors [29] description of how to build KG is a good place to begin learning the craft of engineering KGs for BI.

Business forecasting (BF) is another industrial aspect of BI that has been strengthened by SWTs. BF is analytical BI that predicts future possibilities based on present and past trends in existing data, and is critical for planning and investment decisions. The volatile nature of the business and finance industries makes such look-ahead priceless. SWTs like KGs and ontologies help to enhance the reliability and precision of business forecasts because they guarantee data freshness, and can conveniently handle large data volume and data variety. This is already gaining traction in the business and finance industries [30–33].

4.2.2 CUSTOMER RELATIONSHIP MANAGEMENT APPLICATIONS

The greatest asset of any business is its customers. Customer relationship management (CRM) is a business strategy for enhanced value production, customer experience, and customer loyalty. The term social CRM (SCRM) is used to describe CRM enriched with social media channels. A broader CRM is the e-CRM that engages available online channels to bring about BI enriched CRM. Already, KG-mediated e-CRMs for personalized and contextual customer engagements and value production are gaining traction in e-business platforms [24,34,35].

4.2.3 CONTENT DISCOVERY

Content discovery is the identification of value (patterns, facts, objects, or services) from possibly large, heterogeneous, distributed, and fast changing data repository, and it usually involves automatic or machine search. The deployment of SWTs in content discovery can bring about exactness and completeness of discovery in context, intent, purpose, and preference. Search and content discovery is one notable application area of SWTs in the service industry [36]. Ontology, KG, and SPARQL are veritable tools for semantic search and content discovery. Typical end user semantic search and content discovery application is the content-based recommender system [37–40]. Search and recommender systems built on SWTs abound with varying intermediate or end usage across the service industry, and some of these industrial usages will be discussed under their respective industrial application section.

4.2.4 COLLABORATION

Knowledge exchange is a critical component of the business world. Interactive business knowledge exchange among stakeholders including customers toward enhanced value production is termed collaboration. Specifically, CRM is a collaboration strategy involving business entities and customers. Collaboration can be synchronous or asynchronous. SWTs enabled collaboration systems for businesses to abound. In particular, SWTs enable machine-machine or human-machine collaboration in addition to human–human collaboration. With SWTs, the machine can better get involved in knowledge exchange for improved business competitiveness and value production; and has been successful at it [41,42]. For intelligent and effective machine-* collaboration, the machine must rely on a knowledge management system that is sound and fresh. SWTs as KGs or ontologies interfaced with SPARQL may be sufficient for such system or any higher level system built on the SWTs. The authors [41,43,44] discussed how collaboration system driven by SWTs can be built using SWTs and other associated technologies.

It is important to note that despite the successes of SWTs in the business and finance industry; a lot still needs to be done for optimal and sustained value production in the areas of privacy, availability, security, and integrity of engaged or resultant systems and data sources; because of the volatility of the business and finance industry.

4.3 APPLICATIONS OF SWTs IN LAW

Law as an industry is any legal mechanism that enables the exchange of judicial services for social or economic gains. Judicial services consist of administration of justice, interpretation of law, trial of offenders and adjudication. The authors [45] opined that "law has remained an indispensable part of man and every society, and organization has not only become a product of law but is now also intrinsically and intricately operated by it" (p. 1). Law is a strongly antecedent-based service with sustained non-volatile indigenous information growth. The industry's core activities are information access, comprehension, integration, analysis, synthesis, and publishing; driven by efficiency, accuracy, strict rules, and discretion. The law profession due to its strong dependence on information and reasoning and the huge amount of such information in the domain has long adopted information and communication technology (ICT) and AI as assistive tools. Lately, SWTs with their notable strength in information integration, personalization, and contextualization are being embraced by the law industry, particularly for retrieval, Q&A, summarization and storage of law information (e.g. case law) by the machine for efficient and accurate adjudication and judgment by the law professionals. These activities are unique in the law profession because the law is case-based with unique vocabulary.

4.3.1 STORAGE OF INFORMATION IN LAW

It has been noted that how legal information is organized in memory is critical to the efficiency and effectiveness of legal reasoning and judicial verdicts [46]. Ontology, a SWT, is a common legal knowledge base and is the most excellent means of preserving legal information [45]. The semi-automatic means of building ontology for legal information and systems is recommended [47–49]. The integration capability across heterogeneous and distributed data sources has gotten the industrial attraction of KG in law. KG is currently being used to develop legal compliance system for cross-lingua and cross-jurisdictional Europe [50].

4.3.2 INFORMATION RETRIEVAL IN LAW

SWTs-driven information retrieval systems (IRSs) for legal information are gradually replacing other syntactic and semantic IRSs due to the increasing desire for excellent precision and the antecedent nature of most legal information. The author [47] built a semantic retrieval system for case law based on ontology. The IRS called Law Torch was designed to displace the syntactic IRS for case law presently in their country's law industry and its clear potential in this regard has been established [45]. Syntactic IRSs are a nuisance to law and a call is hereby made for partnership with the authors toward this displacement. Similarly, IRSs are already in use across the globe. Still, on IRSs, the authors [51] built an IR for case law based on ontology and RDF with a particular nick on recommending similar court rulings. These search systems are geared toward enhancing the productivity of law professionals.

4.3.3 QUESTION AND ANSWERING IN LAW

The Q&A systems in law are deeper than their IRS counterpart and are able to give contextual answers (information), not documents like IRS, with possible sources (documents) as available to the system. It serves the legal inquisitiveness of users and it is useful to none law professionals as well. As exposed by [52], Q&A systems are less common, and more demanding to build than IRS. The authors [48] and [53] discussed some successful Q&A systems in law realized using ontology and KG.

4.3.4 SUMMARIZATION IN LAW

Summarization presents the source document(s), data or information as a significantly shorter version with the source semantics preserved. A key advantage of using a summary instead of the source document(s), data or information is that it promotes reading, comprehension or processing efficiency [54,55]. The importance of summarization in law is not only to promote the efficiency of legal reasoning or adjudication among the law professionals but also to help none law professionals particularly consumers of (online) goods and services give informed consent as most of their terms of service agreements are too lengthy. Industrial applications of SWTs to the summarization of (legal) information or document are scarce. However, automatic summarization of this information or documents is of high industrial applicability [56–58]. Nonetheless, the application of automatic semantic summarization techniques on SWTs-driven law knowledge bases (KBs) such as RDF/S KBs or semantic RDF graphs are common industrial practices [54,59,60].

4.4 APPLICATIONS OF SWTs IN HEALTH

The world health organization (WHO) defines health as "a state of complete physical, mental and social wellbeing and not merely the absence of disease or infirmity".[4] Thus, we define the health industry as any legal mechanism that enables the exchange of health services for social or economic gains. Broadly, these services include nursing, medical, pharmaceutical, laboratory/diagnostic, and auxiliary health services. The health industry being ubiquitous and cross-domain with growing heterogeneous and distributed data has found a safe haven in SWTs for enhanced wellbeing of humans. SWTs have been used in the health industry for knowledge representation, access, exchange, analytics, and integration [61–64]. Knowledge representation is the foundation of other SWTs applications to health. Reference [65] details the various aspect of knowledge representation for health care (KR4HC) and exposed the dominance of SWTs, ontology in particular, in KR4HC. Ontology is being used in the health industry to achieve consistency in terminology, and interoperability across various health units and facilities; and for clinical history and decisions. Irrespective of the type of health services, three basic applications of SWTs in health have been identified: standardization/interoperability, clinical information management, and precision medicine; and are discussed in the following subsections.

4.4.1 SWTs AND STANDARDIZATION/INTEROPERABILITY IN HEALTH

A current application of SWTs to health is interoperability usually based on standard reference ontologies – upstream SWTs applications for the health industry. Hundreds of referential health ontologies are being built and updated: a visit to a popular health ontology repository[5] in mid-October 2021 puts the growing list of ontologies at 930 with a total of 13,473,548 classes, 36,286 properties, and 55,648,584 mappings. The three most visited of these ontologies were PW, MedDRA, and SNOMEDCT. Mapping and aligning health application ontologies with appropriate referential health ontologies guarantee the resultant ontologies' interoperability and seamless data integration. The work [66] exposed how interoperable application ontology for clinical information management can be realized by mapping and aligning it with appropriate referential health ontologies. Intra- and inter-health facilities interoperability is an essential attribute of health services system due to patient mobility, disease comorbidity, the ubiquitous and cross-domain nature of the health industry.

4.4.2 SWTs and Clinical Information Management

Health services must be timely, precise, and holistic for lasting well-being of patients. This underscores the deployment of computers and ICTs in health facilities to acquire, organize, store, analyze and access patient clinical information seamlessly for prompt and accurate clinical decision by relevant health professionals without compromising patients' rights and privacy. Such service mechanism is termed clinical information system (CIS) and currently drives health care, particularly in the developed economies. Basic functions of CISs include clinical management, patient decision support, diagnostic support, improved documentation, patient safety, universality, efficiency, and convenience [67,68]. In recent times, CISs are increasingly dependent on SWTs such as ontology [63–65,69–72] for improved health services. Ontology is used to enforce medical compliance, universality, improved search, as well as enhanced diagnostic and prescription support. The works [62,64,73] exposed tools and techniques successful at building health ontology.

4.4.3 SWTs and Precision Medicine

Precision medicine (PM) is a contextual and personalized health service based on patient clinical information and biomedical data. It is a modern health service that extends traditional medicine practice by synchronizing clinical and biomedical data using SWTs (ontology, linked data, and KG) and big data to realize data-based clinical decisions – highly effective and low-risk health care for patients [74–77]. PM enabled by SWTs is particularly useful for diseases such as cancer, thrombosis, cystic fibrosis, HIV/AIDS, alcohol-use disorder, and covid-19 among others [74,77,78]. The workings of PM are akin to that of BI in the business and finance industry – while BI uses business data, PM uses clinical and biomedical data for decision support.

4.5 APPLICATIONS OF SWTs IN SECURITY

The authors [79] stated that "Security is not the absence of threat but the ability to respond to security breaches and threats with expediency and expertise" (p. 1). Security as an industry is any legal mechanism that enables the exchange of security services for social or economic gains. Security therefore has to do with the protection of life, dignity of life, liberty, and anything (tangible or intangible) of value from harm, appearances of harm or fear. Any act or intent that is capable of compromising security is often tagged crime by a people, organization or state and usually discouraged. Security services can be offered in the physical- or cyber -space. Irrespective of the space, security service is strongly dependent on information or security intelligence (SI). SI must be timely, correct, reliable, complete, specific, accurate and sometimes acquired covertly. SWTs particularly ontology and KG have since found upstream applications as security knowledge repository in SI because of their capability to extract and integrate information or data from multiple sources in real time. SWTs also drive many downstream security decision support systems (SDSS) because of their flair for personalization and context-awareness in the sea of heterogeneous information. These downstream and upstream applications of SWTs to SI will be discussed under cyber-security, counter-terrorism, and policing.

4.5.1 APPLICATION OF SWTs TO CYBER-SECURITY

Cyber-security is the protection of the cyber space – programs and data – from false or forced failure, mal-function, acquisition, deletion, alteration, destruction, comprehension, access or use as well as the protection of their environment – computers and its peripheral devices – from forced failure or mal-function. The protection of programs and data's environment from false or forced acquisition, destruction, access and use is not within the boundary of cyber-security but that of physical security. SWTs particularly RDF, ontology, and KG are actively involved in the generation and storage of the SI of our cyber space [80–87] including insider threat [88,89] and access control [27,90–92]. The work [93] is a good take-off point for building ontology and KG for cyber-security intelligence.

4.5.2 APPLICATION OF SWTs TO COUNTER-TERRORISM

Terrorism as a security threat is devoid of global definition – its meaning varies across nations. We however see terrorism as an organized and sophisticated (anti-) security mechanism capable of sustained massive destruction, dehumanisation, and forced or false sack or control of a people or government(s). Any entity (natural or juristic) that perpetuates this mechanism is a terrorist. SWTs have also found huge applications in counter-terrorism in the form of attack prediction or construction of knowledge bases for SI in counter-terrorism [94,95]. As expected, most of these works are usually classified and thus remain unpublished.

4.5.3 APPLICATION OF SWTs TO POLICING

Policing is a security mechanism that enforces safety, law, and order in an organization or region or state. Policing usually involves crime (threat) prevention, detection, or investigation – complex tasks almost dependent on information.

SWTs are already supporting surveillance and automobile safety systems [96,97]. Again, such systems and their documentation are usually classified. Generally, however, data integrity and analytic precision are vital issues that require urgent attention in SI.

4.6 APPLICATIONS OF SWTs IN EDUCATION

Education can be defined as any legal mechanism that enables the exchange of instructional services for social or economic gains. Instructional services are client (student) acquired competency facilitated by the service provider (school); and evaluated and certified also by the service provider and/or recognized third party. The service provider delivers the required technology – competencies, methods, standards, and tools – that enables the client acquisition of specified degree of competency; and may also evaluate the level of the client's new competency and so declare. The mode of delivery of instruction determines whether it is face-to-face (traditional learning), online (e-learning) or a blend of both. Whether the instructional services are delivered via traditional learning, e-learning or hybrid learning modes, SWTs such as RDF, ontology, SWRL, KG, and SPARQL have attracted industrial usage in a digital library or e-learning platforms for improved resource retrieval, personalized resource recommendation, and adaptive learning. Digital libraries (DLs) and learning management systems (LMSs) are two notable instructional technologies that drive the modern education industry.

4.6.1 SWTs AND RESOURCE RETRIEVAL

DLs and LMSs are, in the least, educational content repositories with search and retrieval capabilities. In recent times, the growing increase in educational content and pursuit of quality teaching and learning experience has necessitated DLs and LMSs integration of SWTs into education content representation for personalized, complete, and precise retrieval of content [98–101]. RDF, ontology, KG, and SPARQL are common SWTs used for educational content representation (RDF, ontology) and integration (KG) for effective search (SPARQL) and retrieval experience for both students and teachers. Ontology helps to guarantee the precision and completeness of search results, KG helps with the real-time update particularly with the open and standard educational contents while SPARQL handles the query tasks.

4.6.2 SWTs AND RESOURCE RECOMMENDATION

Personalized resource recommendation (PRR) is an important attribute of recent DL, LMS, and other e-learning systems where different contents are recommended depending on prior or ongoing learning performance, knowledge structure, and learners' preferences. PRR has been shown to increase search satisfaction, teaching/learning efficiency, and performance. PRR is more effectively and efficiently realized using SWTs such as RDF/S, SPARQL, and ontology [102–104].

4.6.3   SWT AND ADAPTIVE E-LEARNING

A common and serious mistake in instructional services is to assume all learners to be on the same cognitive and behavioral footing for every course or topic and at all times. Adaptive learning assumes otherwise and makes frantic efforts to personalize learning in the context of individual competence and changing behavior [105]. Adaptive learning may not be possible in the traditional mode, particularly with large classes but sure possible with SWTs-driven LMS and other e-learning platforms. The works [102,106,107] exposed some considerations, techniques, and tools for integrating personalized adaptive learning into learning management platforms in practice. Ontology is a key SWT enabling adaptive learning in digital learning platforms. Adaptive e-learning systems can be made more effective by complementing the system's Q&A diagnostic component used in gauging learners' changing behavior with real-time sensor data intelligence-driven capability.

4.7   APPLICATIONS OF SWTs IN RESEARCH AND DEVELOPMENT

Research and development (R&D) is any legal mechanism that enables the exchange of intellectual services for social or economic gains. Intellectual services involve the production, accumulation, and application of "new knowledge"[6] for the development of human potential and enhanced value production[7] [108]. Knowledge is therefore the main resource, and the value production of R&D. The three basic components of R&D: production, accumulation, and application of "new knowledge" are being facilitated by SWTs in some aspects. The following subsections discuss how these components of R&D are facilitated by SWTs.

4.7.1   SWTs AND KNOWLEDGE PRODUCTION

The production of knowledge commences with access to, comprehension, and analysis of scholarly knowledge through knowledge opportunities to experimentation, theorization or observation, and subsequent notes or documentations and unbiased scrutiny or peer review. It is at the point of acceptance of research contributions as new knowledge by competent authority that scholarly knowledge is said to be produced. SWTs have found particular industrial engagements in efficient knowledge access – digital search systems. The search system has proved invaluable in the search and retrieval of data or information/document in this era of digitization and internet. Semantic search systems for varying nature of information now abound that take advantage of SWTs for excellent precision and recall. Key SWTs used for this purpose are ontology and KG [109–119]. The use of ontology undisputedly yields better semantic search results. The work [120] described the developing procedure of a basic ontology-driven search engine.

4.7.2 SWTs AND KNOWLEDGE ACCUMULATION

Knowledge accumulation has to do with the documentation, presentation, aggregation, preservation or storage in natural or juristic persons, exposure or publishing, and dissemination or transfer of new knowledge perpetually. Knowledge accumulation is currently enjoying the potential of SWTs such as Knowledge graph and ontology in the aggregation and storage of new knowledge [26,121–124]. It should be noted that education is a means of accumulating tacit or explicit knowledge specifically in natural persons. Thus, applications of SWTs in education are also germane to knowledge accumulation.

4.7.3 SWTs AND KNOWLEDGE APPLICATION

The logical use of sets of accumulated knowledge, labor, and technological or material resources to address man's need or want is termed knowledge application. When a knowledge application enjoys sustained commercialization, the term innovation is preferred. The SWTs-driven industrial applications discussed in this chapter and more are evidence of how SWTs have enabled knowledge applications and commercialization.

4.8 APPLICATIONS OF SWTs IN COMMUNICATION

Communication as an industry is any legal mechanism that enables the exchange of information among persons or entities for social or economic gains. Notable enablers of communication in contemporary times are telecommunication service providers and social media platform providers. These communication service providers are already enjoying SWTs as ontology and KG for enhanced service delivery. This work treats a one-way communication mechanism like broadcasting, not as communication, and hence their exclusion from the communication industry in this chapter.

4.8.1 SWTs AND TELECOMMUNICATION SERVICE DELIVERY

Service delivery by telecommunication systems requires real-time monitoring and management of its interacting components (i.e. communicating entities) which is often hampered by issues of interoperability. Ontology and KG are being exploited in the monitoring and management of service delivery in telecommunications [125–128].

4.8.2 SWTs AND SOCIAL MEDIA PLATFORMS

Social media platform or site is web 2.0 where "the user is also the developer and publisher usually with no known publishing or hosting obligation" (p. 230) [129]. Several social media platforms (SMPs) abound with about half of every population using one or more of them [129]. The industrial application of SWTs to SMPs is a symmetric one: SWTs are used to enhance SMPs communication and interactive effectiveness while SMPs serve as fresh data sources used to enhance SWTs intelligence as an upstream application in business, security, governance, and so on. Popular SMPs and SWTs are already exploiting each other to further: (i) enhance user experience with SMPs [26,130] and (ii) enrich SWTs for improved decision support [95,131–133]. These exploitations are non-trivial but sufficiently exposed and discussed in the associated literature.

4.9 APPLICATIONS OF SWTs IN HOSPITALITY

The hospitality industry is concerned with the social and emotional well-being of humans. It is any legal mechanism that enables the exchange of hospitality services for social or economic gains. Hospitality is the welcome, safety, empathy, servitude, acknowledgment, autonomy, surprise, entertainment, and efficiency experienced by an individual or group on a social exposure usually in a service environment [134]. Any industry dedicated solely to giving persons on social exposure to these human experiences is a hospitality industry. SWTs have helped the hospitality industry improve their guests' visit experience, particularly in the area of guest attraction and retention.

4.9.1 SWTs AND GUEST ATTRACTION

Guest attraction is critical to hospitality providers because however excellent their services, guest need to be aware first of their existence and the experiences in the offering in their service environment. It is a fact that over 60% of tourists make their tourist choices purely based on online information and recommendations of tourist sites or environments [135]. SWTs are already helping hospitality providers to compete for attention and attraction in the industry: taking advantage of SWTs as ontology and KG capability to aggregate distributed and heterogeneous open data and produce meaningful (hospitality) intelligence "as programmed" for recommender systems, online advertisement and search engines to help and encourage possible guests make their choices [34,135–142]. Take note of the clause, "as programmed": hospitality providers are already providing unique content and data, influencing these programs to attract new guests to their service environment.

4.9.2 SWTs AND GUEST RETENTION

It is one thing to attract a guest and another thing to engage the guest into a repeat, and possibly loyal, guest. Although this is more on character, business integrity, and professionalism, SWTs as ontology and KG are already helping hospitality providers give guests personalized context-aware services, surprises in particular [34,140,143]. Although much more work still needs to be done in this regard to realize what we term "precision hospitality": say, personalized meals, drinks temperature and type, favorite music or video, room temperature, etc. depending on preference, situation, demographic and biomedical data. The basic difference between the KG in guest retention and that of the guest attraction is the kind of content extracted and aggregated for use.

4.10 APPLICATIONS OF SWTs IN UTILITY

Utilities are the essential daily consumables or disposables of man for survival or living. They include water, gas, sewerage, electricity, etc. The utility industry is any legal mechanism that enables the exchange of utility services for social or economic gains. Utility service is the safe and consistent provision of utilities in an environment while utility services are varying comfort or satisfaction gained from the utility service provided by utility service providers to the customers. SWTs are already been engaged to control, monitor, estimate, and adapt utility consumption systems in response to the dynamic situations and circumstances that influence them; thus addressing the problem of inadequate and unsustainable utility consumption or supply [144–148]. Typical SWTs used in the utility industry are ontology and KG. The work of [149] is a good place to begin the craft of constructing and evaluating utility ontologies.

4.11 APPLICATIONS OF SWTs IN GOVERNANCE

This chapter appreciates the multiple definitions of governance that "governance encompasses the system by which an organization is controlled and operates, and the mechanism by which it, and its people, are held to account".[8] Governance as an industry therefore is any legal mechanism that enables the exchange of public service for social or economic gains. Public service is the mechanism of state control, operations, and accountability; usually not for profit but specifically for the peace, prosperity, and progress of the people of the state and the general good of humanity. Of interest to this work is e-government – the utilization of ICT for enhanced public service. E-government facilitates transparency, collaboration, and citizen participation in governance which has been strengthened by SWTs: SWTs have proved to be the most enabling technology for these goals of e-government [150].

Ontologies, linked open (government) data, and KGs have been or are being developed by some economies not only for citizen information, participation, or collaboration but also for reuse by inter- and intra- national agencies to ensure service compliance, cooperation, and trade facilitation [151–155]. Governments of nations have successfully deployed SWTs-enabled e-government systems on the web for citizen information [156–159]; citizen participation [153,160] and collaboration with citizens [161]. More economies of the world particularly the developing economies are encouraged to fully adopt SWTs enabled e-government systems for enhanced reportage, citizen participation, and collaboration. More work is needed on the security of government open data to avoid temporary or permanent misinformation which can tense the polity and cause irreparable social-economic damage. Also, feedback from citizens to appropriate authorities through the aggregation of citizens' feelings or popular positions on government policies and projects calls for further involvement of SWTs in e-government systems.

4.12 APPLICATIONS OF SWTs IN LOGISTICS AND TRANSPORTATION

Transportation is the movement of persons or goods from place A in the universe to another place B in the universe with no change to the goods or persons except their locations. Logistics is "the time-related positioning of resources to meet user requirements" (p. 147) [162]. While logistics consists of transportation, logistics in addition to location change of goods may also lead to a change of form and nature of goods usually with value addition. Besides, the term logistics usually excludes conveying persons. Logistics and Transportation were however grouped as one industry in this chapter not just for convenience but because both involve the safe conveying of goods and persons from one location to another on schedule. Location is any point or a specific place on land, water, air, and space.[9] We therefore define the logistics and transportation (L&T) industry as any legal mechanism that enables the exchange of conveying services[10] for social or economic gains. The L&T mechanism will be discussed under the three popular location arenas viz land, air, and water with the exclusion of space. The L&T industry has since embraced SWTs for improved safety, reliability, and efficiency in service delivery.

4.12.1 SWTs AND LAND L&T INDUSTRY

L&T in land consists of road and railway transportation. Linked open data (LOD) on road accidents have been integrated with street maps to help road users identify road sections prone to accidents and guide against further accidents in those sections of the road [163]. Ontologies already exist for semantics-aware Q&A transportation systems that provide travelers on query with transportation utility information like nearest parking slots or bus stops [164]. General upstream ontologies and KG for road transport services including traffic control are in use [165,166]. In the railway transportation also, upstream ontologies and KG have been developed for seamless data exchange among the heterogeneous railway sub-systems for railway asset monitoring and decision support [167,168].

4.12.2 SWTs AND AIR L&T INDUSTRY

In the air L&T industry, upstream ontologies and KG have been developed for seamless data exchange among the heterogeneous air carriers, air traffic control, and secondary aviation systems for flight safety, and efficient and cost-effective air traffic management [169–172]. Ontologies are now also being used to build knowledge-based systems, and integrate incident databases and training organizations for effective airline pilots' training [173].

4.12.3   SWTs AND WATER L&T INDUSTRY

SWTs notably ontologies have found use in maritime surveillance and incident management toward mitigating disasters, particularly at sea [174,175]. Seaport (risk) management is now also enjoying the data integration capabilities of SWTs as linked data and ontology to synchronize and coordinate the activities of port stakeholders [176].

4.13   APPLICATIONS OF SWTs IN METEOROLOGY

Meteorology as an industry is a legal mechanism that enables the exchange of meteorological services for social or economic gains. Meteorological service culminates in the scientific and reliable prediction of weather (sunny, cloudy, windy, rainy, snowy, etc.) and natural disasters (hurricanes, storms, floods, tornados, etc.) usually in real time. Weather forecast is critical to human settlements, primary industries, and the L&T industry among others for safety: unfavorable weather can hardly be prevented but an accurate weather forecast gives humans some time lag to avoid or mitigate its impact. The constantly growing meteorological data are heterogeneous and distributed and thus the meteorology industry has since embraced SWTs such as KG, ontologies, and LOD for seamless and efficient data integration, interoperability, global accessibility and collaboration, and decision support [177–180]. Meteorology is one industry that requires seamless and global data exchange, interoperability, and unhindered human and machine comprehension and collaboration which SWTs naturally provide.

4.14   SUMMARY AND CONCLUSION

The vision of making the computer seamlessly comprehend the data or information it is made to preserve, process, and exchange particularly in the prevalent distributed and heterogeneous environment coupled with the traditional computer attributes: superfast, reliable, limitless storage and processing has engendered to mankind the applications of SWTs – the technologies that exclusively support the SW vision. The SW vision has been shown to enable machines' performance of varying cognitive or decision-oriented tasks most efficiently and effectively across the information and communication obsessed activities of humans – the service industry. In particular, 12 specific such industrial sectors have been exposed to be currently driving their value co-production mechanisms using the SWTs. This chapter exposed how the business and finance industry is enjoying the application of SWTs for enhanced BI, CRM, collaboration and content discovery. Typical BI applications strengthened by SWTs are knowledge bases, Q&A, chatbot, and BF systems. A recurrent SWT in these BI applications is the KG, and how to build KG was guided in the chapter. For CRM, e-CRM for e-business was shown to be a typical CRM application enjoying the SWTs, KG in particular. Under content discovery, the content-based recommender system built on ontology, KG, and SPARQL is also holding sway in the business and finance industry. The industry was exposed to be presently boasting of robust knowledge management systems built on ontologies/KG and interfaced with SPARQL. Inherent in these business and finance SWTs applications however is the issue of privacy/security and the integrity of data sources.

Law has been exposed to be presently profiting from SWTs in the area of information storage and retrieval, Q&A, and summarization systems. Key SWT in these applications is ontology and the semi-automatic means of building ontology has been recommended. How this can be realized was also properly guided in the chapter. The health sector was also shown to be in this league in terms of standardization, interoperability, CIS, and precision medicine: ontology, linked data, and KG were exposed as useful SWTs maintaining the health industry's position in the league. In the area of security, SWTs were shown to be potent cyber-security, counter-terrorism, and policing systems driving components. It was made clear that this most covert area of mankind is enjoying the beauty of ontology and KG and how this is done was properly directed in the chapter; not without stating the stifling issues of data integrity and analytic precision. Education did not also escape the radiant of SWTs. DLs, LMSs, and adaptive e-learning were shown to be employing SWTs such RDF/S, ontology, KG, and SPARQL to realize excellent precision and recall, personalization, and contextualization of fresh educational resource. Incorporating real-time sensor data intelligence into LMSs and e-learning platforms for enhanced adaptive learning was advocated. R&D activities of knowledge production, accumulation, and application being facilitated by SWTs as ontology, LOD and KG were also interesting expositions in the chapter.

In the social area of the economy – communication, hospitality, utility, and governance; the SWTs have also been of immense benefit. SMPs and SWTs enjoy a mutually beneficial stand for enhanced usage experiences. Telecommunication service delivery are better monitored and managed by the use of ontology and KG. How this works were properly directed in the chapter. Also directed is how these SWTs have been used to improve guests' attraction and retention in the hospitality industry; in the same way, they have been used to improve control, monitor, estimate, and adapt utility consumption systems. E-government is also benefiting from the capabilities of SWTs the chapter indicated. SWTs were exposed as the most enabling technology for facilitating e-government's tripod goals of transparency, collaboration, and citizen participation. However, the security of government open data is an issue that deserves more attention, the chapter noted. L&T is another critical sector of the economy that has been exposed to have embraced the SWTs such as ontologies, LOD, and KG for improved safety, reliability, and efficiency in service delivery. Finally, the chapter exposed how the same set of SWTs is involved in meteorology for timely and accurate weather forecast.

Broadly, the SWTs are of both upstream and downstream industrial usage in the service industry. It is hoped that this navigation of some of the latest industrial applications of SWTs is a sufficient introduction enough to captivate developers, entrepreneurs, and policy makers toward taking advantage of the SWTs capabilities of cooperatively engaging man and machine for prompt, reliable, and accurate information cognition and decision support tasks despite the unyielding burden of fast changing information overload from a distributed and heterogeneous environment – a burden whose attenuation is not in sight.

REFERENCES

1. Janev, V. and Vranes, S., "Maturity and Applicability Assessment of Semantic Web Technologies", In *Proceedings of I-KNOW '09 and I-SEMANTICS '09*, Graz, Austria, pp. 530–541, 2009.
2. Patel, A. and Jain, S., "Present and Future of Semantic Web Technologies: A Research Statement", *International Journal of Computers and Applications*, 2019. doi:10.1080/1206212X.2019.1570666.
3. Jain, S., *Understanding Semantics-Based Decision Support*, CRC Press, Taylor & Francis Group, LLC, Boca Raton, FL, 2021.
4. United Nations, "International Standard Industrial Classification of All Economic Activities, Rev.4.", In *Statistical Papers*, Series M No. 4/Rev.4, Economic and Social Affairs Statistics Division, United Nations, 299pp, 2008.
5. Jalava, J., "Production, Primary, Secondary, and Tertiary: Finnish Growth and Structural Change, 1860–2004", In *Pellervo Economic Research Institute Working Papers*, No. 80, Pellervo Economic Research Institute PTT, 28pp, 2006.
6. United Nations, "The Role of the Services Economy and Trade in Structural Transformation and Inclusive Development", In *United Nations Conference on Trade and Development, Multi-year Expert Meeting on Trade, Services and Development*, Fifth session, 24pp, 2017.
7. Ekuobase, G. O. and Olutayo, V. A., "Study of Information and Communication Technology (ICT) Maturity and Value: The Relationship", *Egyptian Informatics Journal*, Vol. 17, No. 3, pp. 239–249, 2016.
8. Olutayo, V. A. and Ekuobase, G. O., "Exploring the Correlation Between Information and Communication Technology Maturity and Value of Listed Companies in the Nigerian Stock Exchange", *The African Journal of Information Systems*, Vol. 13, No. 3, Article 4, pp. 346–369, 2021.
9. Ekuobase, G. O., "Service Innovation Computing – Nigeria's Pilot to El-Dorado", In *232nd Inaugural Lecture Series, University of Benin Press*, 58pp, 2020.
10. Aspin, G., Collins, G. Krumkachev, P., Metzger, M., Radeztsky, S. and Srinivasan, S., "Everything-as-a-Service: Modernising the Core through a Services Lens", In *Proceedings of Technology Trends 2017: The Kinetic Enterprise*, Deloitte University Press, pp. 78–91, 2017.
11. Classen, M., Blum, C., Osterrieder, P. and Friedlia, T., "Everything as a Service? Introducing the St.Gallen IGaaS Management Model", In *2nd Smart Services Summit*, Zurich, pp. 61–65, 2019.
12. De Bosschere, K. and Duranton, M., "Everything as a Service", In M. Duranton et al., editors, HiPEAC Vision 2021, pp. 206–211, 2021. doi:10.5281/zenodo.4719715.
13. Feigenbaum, L., Herman, I., Hongsermeier, T., Neumann, E. and Stephens, S., "Semantic Web in Action", In *Information Technology, Scientific American*, pp. 90–97, 2007.
14. Sheth, A. P., "Semantic Web: Technologies and Applications for the Real-World", In *CORE Scholar*, Wright State University, 2007. https://corescholar.libraries.wright.edu/knoesis/640.
15. Ekuobase, G. O. and Fajuyigbe, O., "Approaches to Computer Assisted Objective Question Examination", *Benin Journal of Advances in Computer Science*, Vol. 2, No. 1 & 2, pp. 74–86, 2001.

16. Fajuyigbe, O. and Ekuobase, G. O., "Hardware for Computerised Accounting Systems (CAS)", *Benin Journal of Advances in Computer Science*, Vol. 2, No. 1 & 2, pp. 42–62, 2001.

17. Timms, J., *Introduction to Business and Management*, University of London, London, 66pp, 2012.

18. Bozic, K. and Dimovski, V., "Business Intelligence and Analytics for Value Creation: The Role of Absorptive Capacity", *International Journal of Information Management*, Vol. 46, pp. 93–103, 2019.

19. Gacitua, R., Mazon, J. N. and Cravero, A., "Using SemanticWeb Technologies in the Development of DataWarehouses: A Systematic Mapping", *Wiley Interdisciplinary Reviews: Data Mining and Knowledge Discovery*, Vol. 9, No. 3, 40pp, 2019. doi:10.1002/widm.1293.

20. Sayah, Z., Kazar, O. and Ghenabzia, A., "Semantic Integration in Big Data: State-of-the-Art", *Journal of Mobile Multimedia*, Vol. 15, No. 3, pp. 191–238, 2020. doi:10.13052/jmm1550-4646.1533.

21. Hitzler, P., "A Review of the Semantic Web Field", *Communications of the ACM*, vol. 64, No. 2, pp. 76–83, 2021. doi:10.1145/3397512.

22. Quboa, Q. and Mehandjiev, N., "Creating Intelligent Business Systems by Utilising Big Data and Semantics", *Proceedings of IEEE 19th Conference on Business Informatics (CBI)*, Vol. 2, pp. 39–46, 2021. doi:10.1109/CBI.2017.71.

23. Berners-Lee, T., Hendler, J. and Lassila, O., "The Semantic Web: A New Form of Web Content That Is Meaningful to Computers will Unleash a Revolution of New Possibilities", In *Scientific American*, 3pp, 2001.

24. McCrae, J. P. Mohanty, P., Narayanan, S., Pereira, B., Buitelaar, P., Karmakar, S. and Sarkar, R., "Conversation Concepts: Understanding Topics and Building Taxonomies for Financial Services", *Information*, Vol. 12, Article 160, 2021. doi:10.3390/info12040160.

25. Elnagar, S. and Weistroffer, H. R., "Introducing Knowledge Graphs to Decision Support Systems Design", In S. Wrycza, and J. Maślankowski (Eds.), *Information Systems: Research, Development, Applications, Education, SIGSAND/PLAIS 2019, Lecture Notes in Business Information Processing*, Springer, Cham, vol 359, 2019. doi:10.1007/978-3-030-29608-7_1.

26. Noy, N., Gao, Y., Jain, A., Narayanan, A., Patterson, A. and Taylor, J., "Industry-scale Knowledge Graphs: Lesson and Challenges", In *ACMQUEUE*, 28pp, 2019.

27. Xiao, G., Ding, L., Cogrel, B. and Calvanese, D., "Virtual Knowledge Graphs: An Overview of Systems and Use Cases", *Data Intelligence*, Vol. 1, pp. 201–223, 2019. doi:10.1162/dint_a_00011.

28. Elhammadi, S., "Financial Knowledge Graph Construction", Doctoral Thesis, University of British Columbia, Vancouver, Canada, 79pp, 2020.

29. Repke, T. and Krestel, R., "Extraction and Representation of Financial Entities from Text", In S. Consoli et al. (eds.), *Data Science for Economics and Finance*, pp. 241–263, 2021. doi:10.1007/978-3-030-66891-4_11.

30. Qu, H., Sardelich, M., Qomariyah, N. N. and Kazakov, D., "Integrating Time Series with Social Media Data in an Ontology for the Modelling of Extreme Financial Events", In *Proceedings of International Conference on Language Resources and Evaluation (LREC 2016)*, European Language Resources Association (ELRA), SVN, pp. 57–63, 2016.

31. Liu, Y., Zeng, Q., Mere, J. O. and Yang, H., "Anticipating Stock Market of the Renowned Companies: A Knowledge Graph Approach", *Complexity*, Article ID 9202457, 15pp, 2019. doi:10.1155/2019/9202457.

32. Zhang, W., Paudel, B., Zhang, W., Bernstein, A. and Chen, H., "Interaction Embeddings for Prediction and Explanation in Knowledge Graphs", In *Proceedings of Twelfth ACM International Conference on Web Search and Data Mining (WSDM '19)*, ACM, pp. 96–104, 2019. doi:10.1145/3289600.3291014.

33. Deng, S., Rangwala, H. and Ning, Y., "Dynamic Knowledge Graph based Multi-Event Forecasting", In *Proceedings of the 26th ACM SIGKDD Conference on Knowledge Discovery and Data Mining (KDD '20)*, ACM, pp. 1585–1595, 2020. doi:10.1145/3394486.3403209.

34. Fensel, A., Akbar, Z, Kärle, E., Blank, C., Pixner, P. and Gruber, A., "Knowledge Graphs for Online Marketing and Sales of Touristic Services", *Information*, Vol. 11, Article 253, 15pp, 2020. https://doi.org/10.3390/info11050253.

35. Meier, S., Gebel-Sauer, B. and Schubert, P., "Knowledge Graph for the Visualisation of CRM Objects in a Social Network of Business Objects (SoNBO): Development of the SoNBO Visualise", *Procedia Computer Science*, Vol. 181, pp. 448–456, 2021.

36. Janev, V. and Vranes, S., "Applicability Assessment of Semantic Web Technologies", *Information Processing and Management*, Vol. 4, pp. 507–517, 2011. doi:10.1016/j.ipm.2010.11.002.

37. Ristoski, P. and Paulheim, H., "Semantic Web in Data Mining and Knowledge Discovery: A Comprehensive Survey", *Web Semantics: Science, Services and Agents on the World Wide Web*, Vol. 36, pp. 1–22, 2016.

38. Kanza, S. and Frey, J. G., "A New Wave of Innovation in Semantic Web Tools for Drug Discovery", *Expert Opinion on Drug Discovery*, 12pp, 2019. doi:10.1080/17460441.2019.1586880.

39. Alaa, R., Gawich, M. and Fernández-Veiga, M., "Personalized Recommendation for Online Retail Applications Based on Ontology Evolution", In *ICCTA, Turkey, Association of Computer Machinery*, 5pp, 2020. doi:10.1145/3397125.3397134.

40. Chicaiza, J. and Valdiviezo-Diaz, P., "A Comprehensive Survey of Knowledge Graph-Based Recommender Systems: Technologies, Development, and Contributions", *Information*, Vol. 12, No. 6, Article 232, 23pp, 2021. doi:10.3390/info12060232.

41. Mezghani, E., Exposito, E. and Drira, K., "A Collaborative Methodology for Tacit Knowledge Management: Application to Scientific Research", *Future Generation Computer Systems*, Vol. 54, pp. 450–455, 2016. doi:10.1016/j.future.2015.05.007.

42. Mathrani, S. and Edwards, B., "Knowledge-Sharing Strategies in Distributed Collaborative Product Development", *Journal Open Innovation: Technology, Market, and Complexity*, Vol. 6, Article 194, 2020. doi:10.3390/joitmc6040194.

43. Hasan, B., "An Ontological Approach to support Knowledge Sharing between Product Design and Assembly Process Planning (APP)", Doctoral Thesis in Machine Design, Kth Royal Institute of Technology, Stockholm, 2017.

44. Kiptoo, C. C., "An Ontology and Crowd Computing Model for Expert-Citizen Knowledge Transfer in Biodiversity Management", Doctoral Thesis, University of Pretoria, 267pp, 2017.

45. Ebietomere, E. P., and Ekuobase, G. O., "A Semantic Retrieval System for Case Law", *Applied Computer Systems, De-Gruyter (Sciendo)*, Vol. 24, No. 1, pp. 38–48, 2019. doi:10.2478/acss-2019-0006.

46. Ekuobase, G. O., and Ebietomere, E. P., "Ontology for Nigerian Case Laws", *African Journal of Computing and ICTs*, Vol. 6, No. 2, pp. 177–194, 2013.

47. Ebietomere, E. P., "A Semantic Retrieval System for Nigerian Case Law", Doctoral Thesis, University of Benin, 260pp, 2018.

48. Fawei, B., Pan, J., Kollingbaum, M. and Wyner, A. Z., "A Semi-automated Ontology Construction for Legal Question Answering", *New Generation Computing*, Vol. 37, pp. 453–478, 2019. doi:10.1007/s00354-019-00070-2.

49. Humphreys, L., Boella, G., Caro, L. D., Robaldo, L., Torre, L., Vati, S. G. and Muthuri, R., "Populating Legal Ontologies using Semantic Role Labeling", In *Proceedings of the 12th Conference on Language Resources and Evaluation (LREC 2020)*, ELRA, pp. 2157–2166, 2020.

50. Rodríguez-Doncel, V. and Montiel-Ponsoda, E., "LYNX: Towards a Legal Knowledge Graph for Multilingual Europe", *Law in Context*, Vol. 37, No. 1, pp. 175–178, 2021. doi:10.26826/law-in-context.v37i1.129.

51. Kant, G., Singh, V. K. and Darbari, M., "Legal Semantic Web – A Recommendation System", *International Journal of Applied Information Systems*, Vol. 7, No. 3, 2014. doi:10.5120/ijais14-451165.

52. Arbaaeen, A. and Shah, A., "Ontology-Based Approach to Semantically Enhanced Question Answering for Closed Domain: A Review", *Information*, Vol. 12, Article 200, 21pp, 2021. doi:10.3390/info12050200.

53. Sovrano, F., Palmirani, M. and Vitali, F., "Legal Knowledge Extraction for Knowledge Graph Based Question-Answering", In *Legal Knowledge and Information Systems*, IOS, pp. 143–153, 2020. doi:10.3233/FAIA200858.

54. Cebiric, S., Goasdoué, F., Kondylakis, H., Kotzinos, D., Manolescu, I., et al., "Summarizing Semantic Graphs: A Survey", *The VLDB Journal*, Vol. 28, No. 3, 45pp, 2018.

55. Aghaunor, C. T. and Ekuobase, G. O., "Automatic Text Summarisation of Case Law using GATE with ANNIE and SUMMA Plug-ins", *Nigerian Journal of Technology*, Vol. 38, No. 4, pp. 987–996, 2019. doi:10.4314/njt.v38i4.23.

56. Galgani, F., Compton, P. and Hoffmann, A., "Combining Different Summarization Techniques for Legal Text", In *Proceedings of the Workshop on Innovative Hybrid Approaches to the Processing of Textual Data (Hybrid2012)*, EACL 2012, pp. 115–123, 2012.

57. Manor, L. and Li, J. J., "Plain English Summarization of Contracts", In *Proceedings of the Natural Legal Language Processing Workshop*, ACL, pp. 1–11, 2019.

58. Bui, N. D. Q., Yu, Y. and Jiang, L., "Self-Supervised Contrastive Learning for Code Retrieval and Summarization via Semantic-Preserving Transformations", In *Proceedings of the 44th International ACM SIGIR Conference on Research and Development in Information Retrieval (SIGIR'21)*, ACM, pp. 511–521, 2021. doi:10.1145/3404835.3462840.

59. Zneika, M., "Querying Semantic Web/Linked Data Graphs Using Summarization", Doctoral Thesis, University of Paris Seine, University of Cergy Pontoise, 147pp, 2019.

60. Vassiliou, G., Troullinou, G., Papadakis, N. and Kondylakis, H., "WBSum: Workload-based Summaries for RDF/S KBs", In *Proceedings of 33rd International Conference on Scientific and Statistical Database Management (SSDBM 2021)*, ACM, pp. 248–258, 2021. doi:10.1145/3468791.3468815.

61. Dogdu, E., "Semantic Web in eHealth", In *Proceedings of ACMSE'09*, ACM, 4pp, 2009.

62. Zenunia, X., Raufia, B., Ismailia, F., Ajdaria, J., "State of the Art of Semantic Web for Healthcare", In *World Conference on Technology, Innovation and Entrepreneurship, Procedia – Social and Behavioral Sciences*, Vol. 195, pp. 1990–1998, 2015.

63. Pal, M., Ray, R., Maji, P. and Panja, A., "Remote Patient Monitoring during Pandemic Caused by COVID-19 Using Semantic Web Technologies", *Journal of Physics: Conference Series*, Vol. 1797, 10pp, 2021. doi:10.1088/1742-6596/1797/1/012023.

64. Karami, M. and Rahimi, A., "Semantic Web Technologies for Sharing Clinical Information in Health Care Systems", *Acta Informatica Medica*, Vol. 27, No. 1, pp. 4–7, 2019.

65. Riano, D., Pelegb, M. and Teije, A. T., "Ten Years of Knowledge Representation for Health Care (2009–2018): Topics, Trends, and Challenges", *Artificial Intelligence in Medicine*, Vol. 100, 13pp, 2019. doi:10.1016/j.artmed.2019.101713.

66. Ebietomere, E. P., Nse, U., Ekuobase, B. U. and Ekuobase, G. O., "Crafting Electronic Medical Record Ontology for Interoperability", *The African Journal of Information Systems*, Vol. 13, No. 3, pp. 296–315, 2021.

67. Islam, M. M., Nasrin, T. and Li, Y., "Recent Advancement of Clinical Information Systems: Opportunities and Challenges", In *IMIA Yearbook of Medical Informatics*, pp. 83–90, 2018.

68. Sutton, R. T., Pincock, D., Baumgart, D. C., Sadowski1, D. C., Fedorak, R. N. and Kroeker, K. I., "An Overview of Clinical Decision Support Systems: Benefits, Risks, and Strategies for Success", *NPJ Digital Medicine*, Vol. 17, 10pp, 2020.

69. Iftikhar, S., Khan, W. A., Ahmad, F. and Fatima, K., "Semantic Interoperability in E-Health for Improved Healthcare", In *Semantics in Action – Applications and Scenarios*, IntechOpen, pp. 108–138, 2012.

70. Chantrapornchai, C. and Choksuchat, C., "Ontology Construction and Application in Practice Case Study of Health Tourism in Thailand", *SpringerPlus*, Vol. 5, Article 2106, 31pp, 2016. doi:10.1186/s40064-016-3747-3.

71. Peng, C., Goswami, P. and Bai, G., "A Literature Review of Current Technologies on Health Data Integration for Patient-Centered Health Management", *Health Informatics Journal*, Vol. 26, No. 3, pp. 1926–1951, 2020. doi:10.1177/1460458219892387.

72. Adel, E., El-Sappagh, S., Barakat, S., Hu, J. and Elmogy, M., "An Extended Semantic Interoperability Model for Distributed Electronic Health Record Based on Fuzzy Ontology Semantics", *Electronics*, Vol. 10, Article 1733, 27pp, 2021. doi:10.3390/electronics10141733.

73. Liyanage H, Krause P, de Lusignan S., "Using Ontologies to Improve Semantic Interoperability in Health Data", *Journal of Innovation in Health Informatics*, Vol. 22, No. 2, pp. 309–315, 2015. doi:10.14236/jhi.v22i2.159.

74. Wang, Z., Zhang, L. and Zhao, W., "Definition and Application of Precision Medicine", *Chinese Journal of Traumatology*, Vol. 19, pp. 249–250, 2016. doi:10.1016/j.cjtee.2016.04.005.

75. Kamdar, M. R., Fernández, J. D., Polleres, A., Tudorache, T. and Musen, M. A., "Enabling Web-Scale Data Integration in Biomedicine through Linked Open Data", *NPJ Digital Medicine*, Vol. 2, Article 90, 14pp, 2019. doi:10.1038/s41746-019-0162-5.

76. Mukwaya, J. N., "An Investigation of Semantic Interoperability with EHR Systems for Precision Dosing", Master Thesis, Kth Royal Institute of Technology, 49pp, 2020.

77. Hou, L., Wu, M., Kang, H., Zheng, S., Shen, L. Qian, Q. and Li, J., "PMO: A Knowledge Representation Model Towards Precision Medicine", *Mathematical Biosciences and Engineering*, Vol. 17, No. 4, pp. 4098–4114, 2020. doi:10.3934/mbe.2020227.

78. Chatterjee, A., Nardi, C., Oberije, C. and Lambin, P., "Knowledge Graphs for COVID-19: An Exploratory Review of the Current Landscape", *Journal of Personalized Medicine*, Vol. 11, Article 300, 21pp, 2021. doi:10.3390/jpm11040300.

79. Ekuobase, G. O. and Anyaorah, I. E., "A Ubiquitous Technology Framework for Curbing BOKO HARAM Menace in Nigeria", *Computing, Information Systems, Development Informatics and Allied Research Journal*, Vol. 5, No. 1, pp. 17–26, 2014.

80. Lin, Z. and Tripunitara, M., "Graph Automorphism-Based, Semantics-Preserving Security for the Resource Description Framework (RDF)", In *Proceedings of CODASPY'17*, ACM, pp. 337–348, 2017. doi:10.1145/3029806.3029827.

81. Mittal, S., Joshi, A. and Finin, T., "Cyber-All-Intel: An AI for Security related Threat Intelligence", *arXiv:1905.02895v1 [cs.AI]*, 13pp, 2019.

82. Mittal, S., "Knowledge for Cyber- Threat Intelligence", Doctoral Thesis, University of Maryland, 117pp, 2019.

83. Rastogi, N., Dutta, S., Zaki, M. J., Gittens, A. and Aggarwal, C., "MALOnt: An Ontology for Malware Threat Intelligence", In *Proceedings of MLHat: The First International Workshop on Deployable Machine Learning for Security Defense*, ACM, 8pp, 2020.

84. Liu, Z., Sun, Z., Chen, J., Zhou, Y., Yang, T., Yang, H., Liu, J., "STIX-based Network Security Knowledge Graph Ontology Modeling Method", In *Proceedings ICGDA'20*, ACM, pp. 152–157, 2020. doi:10.1145/3397056.3397083.

85. Wang, Y., Zhou, Y., Zou, X., Miao, Q. and Wang, W., "The Analysis Method of Security Vulnerability Based on the Knowledge Graph", In *Proceedings of the 10th International Conference on Communication and Network Security (ICCNS 2020)*, ACM, pp. 135–145, 2020. doi:10.1145/3442520.3442535.

86. Kotis, K. I., Zachila, K. and Paparidis, E., "Machine Learning Meets the Semantic Web", *Artificial Intelligence Advances*, Vol. 3, No. 1, pp. 71–78, 2021. doi:10.30564/aia.v3i1.3178.

87. Kaloroumakis, P. E. and Smith, M. J., "Toward a Knowledge Graph of Cybersecurity Countermeasures", In *The MITRE Corporation*, 11pp, 2021.

88. Kul, G. and Upadhyaya, S., "A Preliminary Cyber Ontology for Insider Threats in the Financial Sector", In *Proceedings of MIST'15*, ACM, pp. 75–78, 2015. doi:10.1145/2808783.2808793.

89. Mavroeidis, V., Vishi, K. and Jøsang, A., "A Framework for Data-Driven Physical Security and Insider Threat Detection", In *IEEE/ACM International Conference on Advances in Social Networks Analysis and Mining (ASONAM)*, IEEE, pp. 1108–1115, 2018.

90. Hosseinzadeh, S., Virtanen, S., Díaz-Rodríguez, N. and Lilius, J., "A Semantic Security Framework and Context-Aware Role-Based Access Control Ontology for Smart Spaces", In *Proceedings SBD'16*, ACM, pp. 1–6, 2016. doi:10.1145/2928294.2928300.

91. Zhang, Y., Saberi, M. and Chang, E., "Semantic-based Lightweight Ontology Learning Framework: A Case Study of Intrusion Detection Ontology", In *Proceedings of WI '17*, ACM, pp. 1171–1177, 2017. doi:10.1145/3106426.3109053.

92. Brewster, C., Nouwt, B., Raaijmakers, S. and Verhoosel, J., "Ontology-Based Access Control for FAIR Data", *Data Intelligence*, Vol. 2, pp. 66–77, 2020. doi:10.1162/dint_a_00029.

93. Aboubacar, M.S., Castelltort, A. and Laurent, A., "Knowledge Graph on Cybersecurity: A Survey", 4pp, 2020. https://upvdoc.univ-perp.fr/medias/fichier/aboubacar-maman-sani-article_1604172827312-pdf?ID_FICHE=770&INLINE=FALSE.

94. Spiliotopoulos, D., Vassilakis, C., Margaris, D., "Data-Driven Country Safety Monitoring Terrorist Attack Prediction", In *Proceedings of IEEE/ACM International Conference on Advances in Social Networks Analysis and Mining*, ACM, pp. 1128–1135, 2019. doi:10.1145/3341161.3343527.

95. Mitzias, P., Kontopoulos, E., Staite, J., Day, T., Kalpakis, G., Tsikrika, T., Gibson, H., Vrochidis, S., Akhgar, B. and Kompatsiaris. I., "Deploying Semantic Web Technologies for Information Fusion of Terrorism-Related Content and Threat Detection on the Web: The Case of the TENSOR EU funded Project", In *Proceedings of International Conference on Web Intelligence (WI '19 Companion)*, ACM, pp. 193–199, 2019. doi:10.1145/3358695.3360896.

96. Pahal, N., Mallik, A. and Chaudhury, S., "An Ontology based Context-aware IoT Framework for Smart Surveillance", In *Proceedings of 3rd International Conference on Smart City Applications (SCA '18)*, ACM, 7pp, 2018. doi:10.1145/3286606. 3286846.

97. Shaaban, A. M., Schmittner, C., Gruber, T., Mohamed, A. B., Quirchmayr, G. and Schikuta. E., "Ontology-Based Model for Automotive Security Verification and Validation", In *Proceedings of the 21st International Conference on Information Integration and Web-based Applications & Services*, 10pp, 2019. doi:10.1145/3366030.3366070.

98. Tang, L. and Chen, X., "Ontology-Based Semantic Retrieval for Education Management Systems", *Journal of Computing and Information Technology*, Vol. 23, No. 3, pp. 255–267, 2015. doi:10.2498/cit.1002493.

99. Kar, S. and Das, R., "Publishing E-resources of Digital Institutional Repository as Linked Open Data: An Experimental Study", *Library Philosophy and Practice (e-journal)*, 20pp, 2020. https://digitalcommons.unl.edu/libphilprac/4699.

100. Hussein, M. and Hassan, M., "Information Retrieval Improvement in E-Learning Systems Using Semantic Web", *Journal of Al-Azhar University Engineering Sector*, Vol. 15, No. 57, pp. 1003–1015, 2020.

101. Alrehaili, N. A., Aslam, M. A., Alahmadi, D, H., Alrehaili, D. A., Asif, M. and Malik, M. S. A., "Ontology-Based Smart System to Automate Higher Education Activities", *Complexity, Hindawi (Wiley)*, Vol. 2021, 20pp, 2021. doi:10.1155/2021/5588381.

102. Pelap, G., Zucker, C. and Gandon, F., "Semantic Models in Web Based Educational System Integration", In *Proceedings of the 14th International Conference on Web Information Systems and Technologies (WEBIST 2018)*, SCITEPRESS, pp. 78–89, 2018. doi:10.5220/0006940000780089.

103. Wu, L., Liu, Q., Zhou, W., Mao, G., Huang, J and Huang, H., "A Semantic Web-Based Recommendation Framework of Educational Resources in e-Learning", In *Technology, Knowledge and Learning*, Springer, 23pp, 2018. doi:10.1007/s10758-018-9395-7.

104. Joy, J., Ray, N. S. and Renumol, V. G., "An Ontology Model for Content Recommendation in Personalized Learning Environment", In *Proceedings of DATA'19*, ACM, 6pp, 2019. doi:10.1145/3368691.3368700.

105. Peng, H., Ma, S. and Spector, J. M., "Personalized Adaptive Learning: An Emerging Pedagogical Approach Enabled by a Smart Learning Environment", *Smart Learning Environments*, Vol. 6, No. 9, 14pp, 2019. doi:10.1186/s40561-019-0089-y.

106. Rani, M., Nayak, R. and Vyas, O. P., "An Ontology-Based Adaptive Personalized e-Learning System, Assisted by Software Agents on Cloud Storage", *Knowledge-Based Systems*, Vol. 90, pp. 33–48, 2015.

107. Iatrellis, O., Kameas, A. and Fitsilis, P., "Personalized Learning Pathways Using Semantic Web Rules", In *21st PCI*, ACM, 6pp, 2017. doi:10.1145/3139367.3139404.

108. Zakharova, E. N., Chistova, M. V., Abesalashvili, M. Z. and Gonenko, D. V., "The Role of the Intellectual Services Sector in the Development of Innovative Processes of the Modern Knowledge Economy", *Revista ESPACIOS*, Vol. 40, No. 40, 11pp, 2019.

109. Jain, R., Duhan, N. and Sharma, A. K., "Comparative Study on Semantic Search Engines", *International Journal of Computer Applications*, Vol. 131, No. 14, 8pp, 2015. doi:10.5120/ijca2015907370.

110. Rashid, J. and Nisar, M. W., "A Study on Semantic Searching, Semantic Search Engines and Technologies Used for Semantic Search Engines", *International Journal of Information Technology and Computer Science*, Vol. 10, pp. 82–89, 2016. doi:10.5815/ijitcs.2016.10.10.

111. Besbes, G. and Baazaoui-Zghal, H., "Fuzzy Ontologies for Search Results Diversification: Application to Medical Data", In *Proceedings of Symposium on Applied Computing (SAC'2018)*, ACM, 8pp, 2018. doi:10.1145/3167132.3167343.

112. Stefanoni, G., Motik, B. and Kostylev, E. V., "Estimating the Cardinality of Conjunctive Queries over RDF Data Using Graph Summarisation", In *Proceedings of the 2018 Web Conference (WWW 2018)*, ACM, 10pp, 2018. doi:10.1145/3178876.3186003.

113. Roy, S., Modak, A., Barik, D. and Goon, S., "An Overview of Semantic Search Engines", *International Journal of Research & Review*, Vol. 6, No. 10, pp. 73–85, 2019.

114. Sreeja, M. U. and Kovoor, B. C., "Object Driven Semantic Multi-Video Summarisation Based on Ontology", In *Proceedings of DATA'19*, ACM, 6pp, 2019. doi:10.1145/3368691.3368697.

115. MacAvaney, S., Sotudeh, S., Cohan, A., Goharian, N., Talati, I. and Filice, R. W., "Ontology-Aware Clinical Abstractive Summarization", In *Proceedings of the 42nd International ACM SIGIR Conference on Research and Development in Information Retrieval (SIGIR '19)*, ACM, 4pp, 2019. doi:10.1145/3331184.3331319.

116. Galhotra, S. and Khurana, U., "Semantic Search over Structured Data", In *Proceedings of the 29th International Conference on Information and Knowledge Management (CIKM '20)*, ACM, 4pp, 2020. doi:10.1145/3340531.3417426.

117. Brutzman, D. and Flotyński, J., "X3D Ontology for Querying 3D Models on the Semantic Web", In *The 25th International Conference on 3D Web Technology (Web3D '20)*, ACM, 6pp, 2020. doi:10.1145/3424616.3424715.

118. Ran, L. and Xinbang, H., "An Clustering-based Ontology Summarization Method with Structural and Semantic Information Integration", In *Proceedings of ICCIR'21*, ACM, 2021. doi:10.1145/3473714.3473743.

119. Alam, N., Graham, Y. and Gurrin, C., "Memento: A Prototype Lifelog Search Engine for LSC'21", In *Proceedings of the 4th Annual Lifelog Search Challenge (LSC '21)*, ACM, 6pp, 2021. doi:10.1145/3463948.3469069.

120. Vanjulavalli, N. and Kovalan, A., "Ontology based Semantic Search Engine", *IJCSET*, Vol. 2, No. 8, pp. 1349–1353, 2012.

121. Bonatti, P. A., Decker, S., Polleres, A. and Presutti, V., "Knowledge Graphs: New Directions for Knowledge Representation on the Semantic Web", In *Report from Dagstuhl Seminar 18371*, ACM, 2018. doi:10.4230/DagRep.8.9.29.

122. Penev, L., Dimitrova, M., Senderov, V., Zhelezov, G., Georgiev, T., Stoev P. and Simov, K., "OpenBiodiv: A Knowledge Graph for Literature-Extracted Linked Open Data in Biodiversity Science", *Publications*, Vol. 7, Article 38, 16pp, 2019. doi:10.3390/publications7020038.

123. Auer, S., Oelen, A., Haris, M., Stocker, M., D'Souza, J., Farfar, K. E., Vogt, L., Prinz, M., Wiens, V. and Jaradeh, M. Y., "Improving Access to Scientific Literature with Knowledge Graphs", *Journal of Library Research and Practice, De Gruyter* [Preprint], 17pp, 2020. doi:10.18452/22049.

124. Garifo, G., Futia, G., Vetrò, A., De Martin, J. C., "The Geranium. Platform: A KG-Based System for Academic Publications", *Information*, Vol. 12, No. 366, 21pp, 2021. doi:10.3390/info12090366.

125. Vergara, J. E. L., Guerrero, A., Villagrá, V. A. and Berrocal, J., "Ontology-Based Network Management: Study Cases and Lessons Learned", *Journal of Network and Systems Management*, Vol. 17, No. 3, pp. 234–254, 2009. doi:10.1007/s10922-009-9129-1.

126. Schneider, J. R., "Resolving Tactical Network Management Interoperability by Using Ontology", *Johns Hopkins APL Technical Digest*, Vol. 33, No. 1, pp. 68–80, 2015.

127. Zhou, Q., "Ontology-Driven Knowledge Based, Autonomic Management for Telecommunication Networks: Theory, Implementation, and Applications" Doctoral Thesis, Heriot-Watt University, 180pp, 2018.

128. Krinkin, K., Kulikov, I., Vodyaho, A. and Zhukova, N., "Architecture of a Telecommunications Network Monitoring System Based on a Knowledge Graph", In *Proceedings of 26th Conference of Open Innovations Association*, 9pp, 2020. doi:10.23919/FRUCT48808.2020.9087429.

129. Ekuobase, G. O., "The Use of Social Media in Business", In V. O. Igbineweka, A. E. Uwubanmwen, H. Alonge, D. Enowoghomwenma, P. Igenegbai and D. Aideyan (Eds.), *Entrepreneurship Development: The Uniben Perspective*, Centre for Entrepreneurship Development, University of Benin, Benin City, pp. 227–249, 2020.

130. He, Q., Yang, J. and Shi, B., "Constructing Knowledge Graph for Social Networks in A Deep and Holistic Way", In *Companion Proceedings of the Web Conference 2020 (WWW '20 Companion)*, ACM, pp. 307–308, 2020. doi:10.1145/3366424.3383112.

131. Brambilla, M., Ceri, S. and Valle E. D., "Extracting Emerging Knowledge from Social Media", In *International World Wide Web Conference (WWW'17)*, ACM, pp. 795–804, 2018. doi:10.1145/3038912.3052697.

132. Junior, R. V. S., Oliveira, G. M. C., Neto, F. M. M., Silva, S. D., Santos, I. F. and Neto, A. F. S., "An Ontology Used to Support Learning in the Field of Heritage Education", In *Proceedings of Euro American Conference on Telematics and Information Systems, (EATIS '20)*, ACM, 5pp, 2020. doi:10.1145/3401895.3402080.

133. García, R. and Gil, R., "Social Media Copyright Management Using Semantic Web and Blockchain", In *Proceedings of the 21st International Conference on Information Integration and Web-based Applications & Services (iiWAS2019)*, ACM, 5pp, 2019. doi:10.1145/3366030.3366128.

134. Pijls-Hoekstra, R., Groen, B. H., Galetzka, M. and Pruyn, A. T. H., "Experiencing Hospitality: An Exploratory Study on the Experiential Dimensions of Hospitality", 9pp, 2015. http://EuroCHRIE+2015+-+experiencing+hospitality+FINAL.pdf.

135. Lu, C., Stankovic, M. and Laublet, P., "Leveraging Semantic Web Technologies for More Relevant E-tourism Behavioral Retargeting", In *Proceedings of International Conference of the World wide web (WWW'15) Companion*, ACM, pp. 287–1292, 2015. doi:10.1145/2740908.2742001.

136. Niemann, M., Mochol, M. and Tolksdorf, R., "Enhancing Hotel Search with Semantic Web Technologies", *Journal of Theoretical and Applied Electronic Commerce Research*, Vol. 3, No. 2, pp. 82–96, 2008. doi:10.4067/S0718-18762008000100008.

137. Soualah-Alila, F., Faucher, C., Bertrand, F., Coustaty, M. and Antoine Doucet, A., In *Proceedings of ESAIR'15*, ACM, 2015. doi:10.1145/2810133.2810137.

138. Zhou K., Zhao, W. X., Bian, S., Zhou, Y., Wen, J. and Yu, J., "Improving Conversational Recommender Systems via Knowledge Graph based Semantic Fusion", In *Proceedings of 26th ACM SIGKDD Conference on Knowledge Discovery and Data Mining (KDD'20)*, ACM, pp. 1006–1014, 2020. doi:10.1145/3394486.3403143.

139. She, K., Haw, S., Chew, L., "Tourism Recommender System Utilising Property Graph Ontology as Knowledge Base", In *Proceedings of ICCMS'20*, ACM, pp. 14–18, 2020. doi:10.1145/3408066.3408102.

140. Gala, A. A., Cardinale, Y., Dongo, I. and Ticona-Herrera, R., "Towards an Ontology for Urban Tourism", In *The 36th ACM/SIGAPP Symposium on Applied Computing (SAC '21)*, ACM, pp. 1887–1890, 2021. doi:10.1145/3412841.3442142.

141. Do, P., Phan, T. H. V. and Gupta, B. B., "Developing a Vietnamese Tourism Question Answering System Using Knowledge Graph and Deep Learning", *ACM Transactions on Asian and Low-Resource Language Information Processing*, Vol. 20, No. 5, 18pp, 2021. doi:10.1145/3453651.

142. Hananto, V. R. Serdült, U. and Kryssanov, V. V., "A Tourism Knowledge Model through Topic Modeling from Online Reviews", In *7th International Conference on Computing and Data Engineering (ICCDE 2021)*, ACM, pp. 87–93, 2021. doi:10.1145/3456172.3456211.

143. Khallouki, H., Abatal, A. and Bahaj, M., "An Ontology-based Context Awareness for Smart Tourism Recommendation System", In *Proceedings of LOPAL'18*, ACM, 5pp, 2018. doi:10.1145/3230905.3230935.

144. Birov, S., Robinson, S., Villalón, M. P., Suárez-Figueroa, M. C. RaúlGar-cía Castro, R. G., Euzenat, J., Fies, B., Cavallaro, A., Peters-Anders, J., Tryferidis, T., et al., "Ontologies and Datasets for Energy Measurement and Validation Interoperability", [Contract], Ready4SmartCities.2015, 135pp, 2015. https://hal.inria.fr/hal-01247616v2.

145. Lefrançois, M., Kalaoja, J., Ghariani, T. and Zimmermann, A., "Smart Energy Aware Systems ITEA2-12004: D2.2 SEAS Knowledge Model", In *Information Technology for European Advancement*, 76pp, 2016.

146. Fossatti, F., "The Operation and Maintenance Domain Ontology for Utility Networks Database Encoding", Master Thesis, University of Twente, 104pp, 2020.

147. Aryan, P. R., Ekaputra, F. J., Sabou, M., Hauer, D., Mosshammer, R., Einfalt, A., Miksa, T. and Rauber, A., "Explainable Cyber-Physical Energy Systems based on Knowledge Graph", In *9th Workshop on Modeling and Simulation of Cyber-Physical Energy Systems (MCPES'21)*, ACM, 6pp, 2021. doi:10.1145/3470481.3472704.

148. Nicola, A. D. and Villani, M. L., "Smart City Ontologies and Their Applications: A Systematic Literature Review", *Sustainability*, Vol. 13, Article 5578, 40pp, 2021. doi:10.3390/su13105578.

149. Gharibi, M., "Building a Knowledge Graph for Food, Energy, and Water Systems", Master Thesis, University of Missouri, Kansas City, 53pp, 2017.

150. Tibau, M., Siqueira, S. W. M. and Nunes, B. P., "Semantic Data Structures for Knowledge Generation in Open World Information System", In *Proceedings of XVI Brazilian Symposium on Information Systems (SBSI'20)*, ACM, 7pp, 2020. doi:10.1145/3411564.3411611.

151. Zeleti, F. A. and Ojo, A., "An Ontology for Open Government Data Business Model", In *Proceedings of ICEGOV'17*, ACM, pp. 195–203, 2017. doi:10.1145/3047273.3047327.

152. Hasan, M. M., Aganostopoulos, D., Loucopoulos, P. and Nikolaidou, M., "Regulatory Requirements Compliance in e-Government System Development: An Ontology Framework", In *Proceedings of ICEGOV'17*, ACM, pp. 441–449, 2017. doi:10.1145/3047273.3047341.

153. Fedorova, S. and Malceva, D., "New Trends in E-Government: Comparative Analysis of E-Participation systems in the United Kingdom and the Russian Federation", In *Proceedings of IMS2017*, ACM, pp. 200–205, 2017. doi:10.1145/3143699.3143703.

154. Sinif, L. and Bounabat, B., "Approaching an Optimizing Open Linked Government Data Portal", In *Proceedings of ICSDE'18*, ACM, 5pp, 2018. doi:10.1145/3289100.3289122.

155. Gerontas, A., "Towards an e-Government Semantic Interoperability Assessment Framework", In *Proceedings of the 13th International Conference on Theory and Practice of Electronic Governance (ICEGOV 2020)*, ACM, pp. 767–774, 2020. doi:10.1145/3428502.3428617.

156. Araújo, L. S. O., Santos, M. T. and Silva, D. A., "The Brazilian Federal Budget Ontology – A Semantic Web Case of Public Open Data", In *Proceedings of MEDES'15*, ACM, pp. 85–89, 2015. doi:10.1145/2857218.285723.

157. Petrušić, D., Segedinac, M. and Konjović, Z., "Semantic Modelling and Ontology Integration of the Open Government Systems", *Tehnički vjesnik*, Vol. 23, No. 6, pp. 1631–1641, 2016. doi:10.17559/TV-20150514115428.

158. Stavropoulou, S., Romas, I., Tsekeridou, S., Loutsaris, M. A., Lampoltshammer, T., Thurnay, L., Virkar, S., Schefbeck, G., Kyriakou, N., Lachana, Z., Alexopoulos, C. and Charalabidis, Y., "Architecting an Innovative Big Open Legal Data Analytics, Search and Retrieval Platform", In *Proceedings of the 13th International Conference on Theory and Practice of Electronic Governance (ICEGOV'20)*, ACM, pp. 723–730, 2020. doi:10.1145/3428502.3428610.

159. Kalogirou, V., Dooren, S. V., Dimopoulos, I., Charalabidis, Y., De-Baets, J. and Lobo, G., "Linked Government Data Hub, An Ontology Agnostic Data Harvester and API", In *Proceedings of the 13th International Conference on Theory and Practice of Electronic Governance (ICEGOV 2020)*, pp. 779–782, 2020. doi:10.1145/3428502.3428619.

160. Kumar, A. and Joshi, A., "Ontology Driven Sentiment Analysis on Social Web for Government Intelligence", In *Proceedings of ICEGOV'17*, ACM, pp. 134–139, 2017. doi:10.1145/3055219.3055229.

161. Sanchez-Nielsen, E. and Chavez-Gutiérrez, F., "Using Semantic Annotations on Political Debate Videos for Building Open Government Based Lawmaking", *Expert Systems*, Wiley, 24pp, 2021. doi:10.1111/exsy.12748.

162. Lowe, D., "Dictionary of Transport and Logistics", In *Institute of Logistics and Transportation, British Library Cataloguing in Publication Data*, 297pp, 2002.

163. Colacino, V. G. and Po, L., "Managing Road Safety through the Use of Linked Data and Heat Maps", In *Proceedings of WIMS '17*, ACM, 8pp, 2017. doi:10.1145/3102254.3102273.

164. Nandini, D. and Shahi, G. K., "An Ontology for Transportation System", *Kalpa Publications in Computing*, Vol. 10, pp. 32–37, 2019.

165. Albiston, G. L., "Semantic-based Framework for the Generation of Travel Demand", Doctoral Thesis, Nottingham Trent University, 433pp, 2019.

166. Chaves-Fraga, D., Priyatna, F., Cimmino, A., Toledo, J., Ruckhaus, E. and Corcho, O., "GTFS-Madrid-Bench: A Benchmark for Virtual Knowledge Graph Access in the Transport Domain", *Web Semantics: Science, Services and Agents on the World Wide Web*, Vol. 65, 28pp, 2020.

167. Lewis, R., "A Semantic Approach to Railway Data Integration and Decision Support", Doctoral Thesis, University of Birmingham, 300pp, 2012.

168. Tutcher, J., "Development of Semantic Data Models to Support Data Interoperability in the Rail Industry", Doctoral Thesis, University of Birmingham, 355pp, 2015.

169. Keller, R. M., Ranjan¢n, S., Wei, M. Y. and Eshow, M. M., "Semantic Representation and Scale-up of Integrated Air Traffic Management Data", In *Proceedings of SBD'16*, ACM, 6pp, 2016. doi:10.1145/2928294.2928296.

170. Keller, R. M., "Building a Knowledge Graph for Air Traffic Management Community", In *Proceedings of WWW'19*, ACM, pp. 700–704, 2019. doi:10.1145/3308560.3317706.

171. Stefanidis, D., Christodoulou, C., Symeonidis, M., Pallis, G., Dikaiakos, M., Pouis, L., Orphanou, K., Lampathaki, F. and Alexandrou, D., "The ICARUS Ontology: A general Aviation Ontology Developed Using a Multi-Layer Approach", In *Proceedings of the 10th International Conference on Web Intelligence, Mining and Semantics (WIMS 2020)*, ACM, pp. 21–32, 2020. doi:10.1145/3405962.3405983.

172. Aghdam, M. Y., Tabbakh, S. R. K., Chabok, S. J. M. and Kheyrabadi, M., "Ontology Generation for Flight Safety Messages in Air Traffic Management", *Journal of Big Data*, Vol. 8, Article 61, 22pp, 2021. doi:10.1186/s40537-021-00449-3.

173. Dapicaa, R. and Peinado, F., "Towards a Semantic Knowledge Base for Competency-Based Training of Airline Pilots", *Procedia Computer Science*, Vol. 192, pp. 1208–1217, 2021. doi:10.1016/j.procs.2021.08.124.

174. Santipantakis, G., Kotis, K. I. and Vouros, G. A., "Ontology-Based Data Integration for Event Recognition in the Maritime Domain", In *Proceedings of WIMS '15*, ACM, 11pp, 2015. doi:10.1145/2797115.2797133.

175. Wen, Y., Zhang, Y., Huang, L., Zhou, C., Xiao, C., Zhang, F., Peng, X., Zhan, W. and Sui, Z., "Semantic Modelling of Ship Behavior in Harbor Based on Ontology and Dynamic Bayesian Network", *International Journal of Geo-Information*, Vol. 8, Article 107, 24pp, 2019. doi:10.3390/ijgi8030107.

176. Pileggi, S. F., Indorf, M., Nagi, A. and Kersten, W., "CoRiMaS—An Ontological Approach to Cooperative Risk Management in Seaports", *Sustainability*, Vol. 12, No. 4767, 23pp, 2020. doi:10.3390/su12114767.

177. Atemezing, G., Corcho, O., Garijo, D., Mora, J., Poveda-Villalón, M., Rozas, P., Vila-Suero, D. and Villazón-Terrazas, B., "Transforming Meteorological Data into Linked Data", *Semantic Web Journal*, Vol. 1, pp. 1–5, 2011.

178. Ramar, K. and Mirnalinee, T. T., "A Semantic Web for Weather Forecasting Systems", In *Proceedings of International Conference on Recent Trends in Information Technology*, IEEE, 6pp, 2014.

179. Barrott, J., Bharwani, S. and Brandon, K., "Transforming Knowledge Management for Climate Action: A Road Map for Accelerated Discovery and Learning", In *PLACARD*, 62pp, 2020.

180. Xiao, Z. and Zhang, C., "Construction of Meteorological Simulation Knowledge Graph Based on Deep Learning Method", *Sustainability*, Vol. 13, Article 1311, 20pp, 2021. doi:10.3390/su13031311.

NOTES

1 For a good understanding of artificial intelligence, readers may begin with the resource: https://doi.org/10.2760/382730.

2 The term business in this section is strictly about the exchange of goods for economic gains.

3 Finance was grouped with business because money (economic value) is still largely seen as tangible i.e. goods.

4 For a global view of health visit: https://www.who.int/governance/eb/who_constitution_en.pdf.

5 This repository consists of active health and biomedical ontologies: https://bioportal.bioontology.org/.

6 New knowledge is a novel or modified old knowledge

7 The authors see innovation as part of R&D; as innovation is the commercial application of new knowledge.
8 This definition appears holistic and for better appreciation visit: www.governance-institute.com.au.
9 Space here refers to outside the earth or its atmosphere.
10 These services excludes conveying the intangibles.

5 Semantic Web Ontology Centred University Course Recommendation Scheme

Gina George and Anisha M. Lal

CONTENTS

5.1 E-LEARNING

Learning is a way of life. It takes place whenever something new is learned than what is already known. In every aspect of our life learning takes place and learning never ends. Learning has never been limited to the academic group alone, rather learning has and remains open to each human being with no age limit. This is exactly why the Internet became a huge need for humans to function. The World Wide Web became available to the public in the early 1990s. There has been no turning back ever since. From the initial uses of email communication, data sharing to social media uses, electronic business to e-learning, the Internet provides it all. A brief history of e-learning goes back to 1997 when Elliot Massie stated

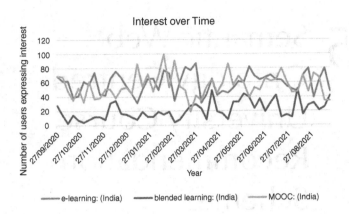

FIGURE 5.1   Interest vs time.

online learning involves network technology. Later in 1998 Jay Cross, founder of the "e-learning" jargon, said, "e-learning is learning on Internet Time." However, in 1999 when he collaborated with Cisco, the company reframed e-learning to Internet-enabled learning [1]. Figure 5.1 shows the level of interest expressed by people in India on topics related to learning for the past 1 year. E-learning, blended learning, Massive open online course (MOCC), online resources among others are some of the member topics people search under the umbrella term of learning.

E-learning takes place when learning happens over an electronic medium but not necessarily using the Internet. The medium for e-learning can be intranet too. Online learning specifically focuses on learning using the Internet. There are various terms related to e-learning with a very fine line of distinction between each of them. Blended learning, flipped classroom, ICT- supported traditional classroom learning, synchronous learning, and asynchronous learning [2]. Blended learning occurs when there is a combination of offline and online classes. The flipped classroom takes place when the students are given the content via the Internet to learn much ahead of the actual class. ICT-supported traditional classroom learning occurs in a majority of classes these days. Synchronous learning is learning through distance mode but with live lectures. Asynchronous learning takes place through distance mode but with pre-recorded lectures. Figure 5.1 also shows statistics on people searching for online learning purpose-related resources along with the massive open online course platforms that provide online courses. The graph clearly shows an increased interest in e-learning as time passes. Along with that the current global pandemic with online classes only leads to more reasons for people switching to e-learning.

E-learning has gone through several version changes from E-learning 1.0 to E-learning 2.0, and finally E-learning 3.0. E-learning 1.0 saw more interaction between the learner and computer via simulation, multimedia, and animation to engage the learners [3]. With the emergence of Web 2.0, the next generation of E-learning 2.0 came into existence. E-learning 2.0 combines the learning available through several social network tools that include wikis, blogs, podcasts, and other applications available on the Internet [4]. Further on, with the development of Web 3.0, new technologies that include semantic web, big data, and artificial intelligence play an even greater role in engaging learners actively in the learning process [5].

5.1.1 E-LEARNING DURING COVID19

With the onset of covid19, all the educational institutions around the world from the primary level to the higher education level had made a shift from traditional classroom learning to online learning. Zoom, Microsoft Teams, and Google Classroom were some of the go-to options for these educational institutions. However, students in several places could not access the Internet making this transition to e-learning a challenge. This section discusses some of the impacts, both positive and negative, of the challenges faced in this shift from the traditional classroom to e-learning during the covid19.

Favale et al. [6] made an intensive study on the effects covid19 has concerning Internet traffic from their university campus. The authors took their sample from Politecnico di Torino University. The authors explored whether students using the e-learning option located at different places and whether students using different network operators affect the Internet traffic. The study concluded that more online teaching content was uploaded, i.e. there was more outgoing traffic was observed. Students made use of online learning to the full extent. Smart working became the norm for the faculty and students of Politecnico di Torino University. Though heavy load was experienced on the teaching server, the bandwidth did not get affected.

Almaiah et al. [7] did a study to understand the challenges and factors for use of the e-learning systems during the pandemic time. The authors covered the study across six universities. The authors identified some of the reasons that the e-learning system failed were lack of technical support, technological challenges, low quality of lecture content, difficulty in making teaching faculty more tech-savvy, and the lack of security while using such systems. The authors next identified some of the challenges faced while using the e-learning systems: financial, technical, and resistance to accepting change were on top of the list. The trust factor, culture factor, and self-efficacy factor were some of the factors that need to be addressed for improving the usage of e-learning.

Radha et al. [8] researched the trends, interests of students across the world towards embracing e-learning. The authors find that majority of the respondent students were: willing to study the e-learning way and found that their self-study skills improved while accessing as much online material. Though the respondents showed an interest in e-learning, the majority of them still preferred the traditional form of learning to e-learning.

5.2 RECOMMENDER SYSTEMS

In today's digital world where everything is available online, one can access movies, do their shopping, study their learning materials, earn an online course

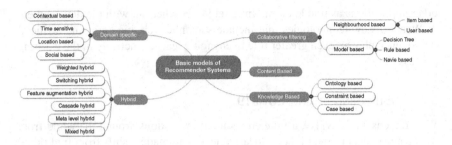

FIGURE 5.2   Taxonomy.

certificate, etc. with a click of their finger. Amazon, Flipkart, Coursera, Udacity, and Netflix among others are some of the websites offering their shopping, e-learning, and movie experiences online. Though such a huge convenience is offered, it comes at a cost of information overload. For example, if a user expresses interest in the fiction category of books on amazon.in, the website offers 60,000 results. That is truly an overload of options. Hence, it is in such situations that recommender systems become useful. The recommender systems filter the options based on the user's previous purchases or based on ratings given by the user or also based on similar genre books rated by other users. Recommender systems work based on two types of data that the system gets mainly based on the user-item centric activity and based on ratings [9]. The recommender system filters the relevant items and brings them to the attention of the user.

Figure 5.2 [10] represents a pictorial view of the different types of recommender systems. The traditional recommender systems used collaborative filtering and content-based filtering. Collaborative filtering is of two types: Neighbourhood-based and Model-based. Neighbourhood-based have two types based on neighbourhood of items and neighbourhood of users. Neighbourhood-based collaborative filtering works mainly on neighbourhood-related user-item centric activity while the model-based is centred on using machine learning models. For instance, say, a user named "Bob" likes mystery genre books and thereby gave a 4 rating to a mystery book "X" he purchased on an e-commerce site. Similarly, say, a user named "Alice" also likes mystery genre books and rated book "X" 5. Then, the recommender system learns that Bob and Alice are similar users. Model-based on the other hand learns similarities using Naïve Bayes, decision trees among other machine learning models. Content-based recommender systems use the attributes of items purchased preciously or rated previously. Attributes like the price of the book, particular authors that a user likes are some of the attributes the recommender system looks into.

As time passed domain-related information became helpful in the recommender systems. Constraint-based recommender systems work by applying certain criteria or filters expressed by the user. For instance, if a user wants to purchase real estate property. The user would apply constraints in terms of the preferred geographical location, area of the property, etc. Case-based recommender system uses the knowledge gained by a similar case and recommends accordingly.

With the advance of the Semantic Web, the use of ontology became more popular. The semantic ontology-based recommender system is knowledge-based as domain knowledge plays a key role in such systems. Though many methodologies are available, each has its limitations. Hybridizing different methods help to overcome individual methods' drawbacks thereby suggesting better and more effective recommendations. Hybridization can be done in a weighted, switched, cascaded, or mixed manner. Also, feature augmentation and meat level ways of hybridization exist.

Recommender systems are extensively and successfully used in many domains like e-commerce, movies, e-learning, and social media platforms [10].

5.2.1 Recommendation Systems in the E-learning Domain

Today's most used jargon is "E-learning" and that is because of the several advantages it has like the lack of rigidness the system provides for learners so that they get to learn at their comfortable pace, availability of quality resources, accessibility to learning anytime from anywhere, ability to provide teaching from renowned faculty which otherwise may be difficult for students to access [11]. As mentioned in the previous section, recommender systems are used across a variety of domains. However, the exact methodology applied in the conventional recommender systems cannot be applied to this area due to a learner's dynamic nature. The learning style, learning goals, knowledge level, a learner's preference, etc. keeps changing and evolving. Thereby, this leads to e-learning-based recommender systems' requirement to give more personalized recommendations.

For instance, a first-time user on a MOOC platform is required to fill in certain details like the user's highest educational qualification, area of interest, user's year of birth, gender, etc. Once the user enters the details and says the user enters Artificial Intelligence as a subject to learn. The user is given suggestions over approximately 1,384 courses. This becomes a massive challenge for users. Also, a survey by Financial Times in 2019 shows a disproportion in the number of users enrolling to the number of users finishing the courses. The article goes on to say certain MOOC platform has a course completion rate of 34%. It is in such cases, that a recommendation system is very much needed to help users make a more informed decision. The recommender system helps to filter and give relevant suggestions. Similarly, another study was done to understand the habits of professional learners who successfully completed a 14-course program in Data Science offered by Microsoft on the MOOC platform edX [12]. The authors compared the completion rate of those enrolling for that program with others enrolling for several other courses on edX. An average completion rate of 6%–31% was only seen across the different courses with the Microsoft offered course reaching a completion rate of 12%. The authors also observed the highly successful learners attempted more assignments and did more courses than the mandatory required for the program.

A system was designed to recommend university courses in [13]. Their system utilizes three models. The first model extracts a sequence of courses that students enrolled for each semester. The next model learns from available features on courses that students enrolled for in different semesters. The authors employed the bag-of-words methodology over the course catalog description. The authors evaluated the models alongside benchmarking recurrent neural network-based recommendations that were already part of their system.

Polyzou et al. [14] created a Markov chain based on the assumption that students will take the next semester courses depending on the courses already taken in the previous semester. The authors introduced a Scholar's Walk founded on a random walk approach to recommend the next possible courses a student can take.

Another work in an e-learning-based recommender system done by [15] is designed to recommend courses based on recurrent neural networks and three assumptions. The authors suggest three assumptions for their model: there is a zone of proximal development for each student, performance in a course helps to identify prerequisite courses for a target course the student could be interested in, and the recommendations generated by the model will be followed by students who succeed in the recommended courses.

Parameswaran et al. [16] proposed an e-learning-based recommender system by applying flow-based technique and heuristic technique. Their work aimed to suggest courses for students at Stanford University that would satisfy university constraints and that would be of interest to students.

A glimpse of some of the different methodologies applied for e-learning-based recommender systems is tabulated in Table 5.1.

However, some limitations are present in recommender systems. These systems work mainly based on ratings. The system looks for similar users based on ratings given by other users of the system. Similarly, the system looks for similar items based on ratings. An issue like cold-start occurs when there is a new user or new item in the system. The recommender system is unable to find ratings for the new user or for the new item based on which the system can give suggestions to the new user or of the new item. Such websites that offer recommendations have a huge record of users who browse their site and of items that are available on the website. However, the ratio of ratings to users is not proportional, i.e., the quantity of ratings is less compared to the users of the system. Additionally, users do not always rate the items available in the system particularly, for those recommender systems in higher education institutes. It is in such scenarios that semantic web-based ontology is useful.

5.3 ONTOLOGY

The concept of the Semantic Web came into existence when the need, to improve integration data across the Web, came so that smart data is available for the smart applications across the Web to perform their full potential [24]. The semantic web standards employ the concept of classes and hierarchy between them. The main essence of modelling on the semantic web is that anyone can give their opinion on any topic and modelling helps to work through the resulting chaos. Ontology is one such tool used for modelling. The semantic web is built using ontologies

TABLE 5.1
Prevue of Some E-Learning Based Recommender System

S. No.	Author	Methodology Applied	Purpose
1.	[13]	Machine learning	Recommends university courses
2.	[14]	Random walk methodology	Recommends next semester course at the university level
3.	[15]	Recurrent neural network	Recommends prerequisite course
4.	[17]	Semantic web rule language and SPARQL protocol and RDF query language	Recommends learning materials and learning sequence
5.	[18]	Browser extension parameters; ontology; Felder Silverman learning style model-based algorithm	To identify the learning style of a learner on MOOC platform
6.	[19]	Pre-filtering based on student's goal, knowledge level and sequential pattern mining	Recommends learning objects
7.	[20]	Web mining; Felder Silverman learning style model; fuzzy C-means clustering	Predicts learning style
8.	[21]	Collaborative filtering and data mining	Recommends learning objects
9.	[22]	Collaborative tagging	Recommends personalized learning material
10.	[23]	Attribute-based collaborative filtering; sequential pattern mining	Recommends learning material

that are inherently based on concepts and relationships between them [25,26]. For the Semantic Web to work with everyone collaborating, it becomes even more necessary that standards for how to share across the Internet and the meaning behind the representations are understood as they should be [27]. Again, ontology serves exactly just that purpose. An ontology represents domain knowledge in a model form. Ontology is one of the major cornerstones of the semantic web [28]. Ontology had its origin in philosophy, but its variant started to be used in computer science. Humans and computers can understand models. Ontology provides a basis for sharing meaning or information across humans and computers. Ontology is a model that defines terms, concepts, rules, and relationships that allows the representation of domains of knowledge. Machines can interpret ontology since it uses knowledge representation techniques. According to [29], "Ontology is a formal, explicit specification of a shared conceptualization." Ontology is machine-readable (formal), showing a precise representation (explicit specification) derived from consensus (shared) on a subject matter (conceptualization). In other words, when the ontology is part of computer science, it is a conceptual and executable model of a particular application domain. Ontologies can be queried for domain knowledge [28]. There are different types of ontologies: Workplace ontology, domain ontology, task ontology, activity-related ontology, and common ontology. Ontologies can be used on different levels [30]:

- as a common vocabulary for agents to communicate
- as conceptual schema
- as an information provider in a knowledge base
- provides answers for understanding
- provides means of standardization across the Internet
- for the semantic transformation of database
- to reuse knowledge available in knowledge base
- to reorganize knowledge in a knowledge base

Ontologies can be expressed using various languages based on the different levels of expressivity. Some of the ontology languages are Resource Description Framework Schema (RDFS) and Web Ontology Language (OWL), Web Service Modeling Language (WSML). Protege, Swoop, SemanticWorks are some of the available ontology editors. Ontology has conceptual entities used for knowledge representation separated from the actual statements that use these elements to express knowledge about the domain in the form of axioms [28]. These entities are classes, instances, relations, and properties within an ontology.

'Class groups' have common characteristics, such as "a class of students belonging to a specific branch of study or class of all vehicles." Class in ontology is a collection of instances. Figure 5.3 shows the OWLViz view of class hierarchy for a sample ontology in the e-learning domain. Person and Course are the different classes under the main class Thing. There are subclasses Faculty, Student, and Researcher for the Person class. Associate Professor and Full Professor are sibling

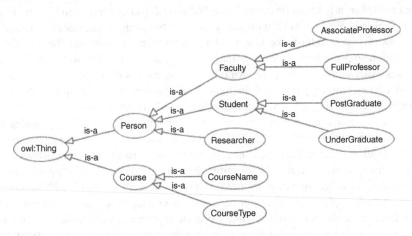

FIGURE 5.3   OWLViz.

classes for the superclass Faculty. Similarly, UnderGraduate and PostGraduate are sibling classes for the superclass Student.

Instances are the individual or single occurrence of the node occurring in the semantic network, such as a particular person or car. The "property" helps to express the relationship between instances of classes [28]. "Property" is further divided into two types: object property and data property. Object property associates one class of individuals with another class of individuals. In other words, object property discusses how classes relate to each other based on their instances.

An example can be student A of a student class "studied" course X of the course class. "studied" becomes the object property or the relation between the student and course classes. Data properties, on the other hand, allow for data values to be assigned to them. If there is a data property assertion hasGrade, it can be assigned a value, say 8.

Figure 5.4 displays the different object property assertions and the data property assertions for a sample ontology containing instances for the student class. For each student instance, there are some object property assertions and some data property assertions. Object property relates instances of one class to another. Here, the object property "studied" makes a relation between the student class instances with the course class instances. For example, the student (instance) "ABC" studied (i.e., object property) courses (i.e., instances) like "Mathematics I, Image Processing, Human-Computer Interaction, and Cryptography." Similarly, the personality and profile attributes are assigned as data property assertions. For example, in the created ontology, "ABC" who is a student (i.e., instance) has data property assertions "hasRegID and hasCgpa". These are assigned data values in string and integer form respectively.

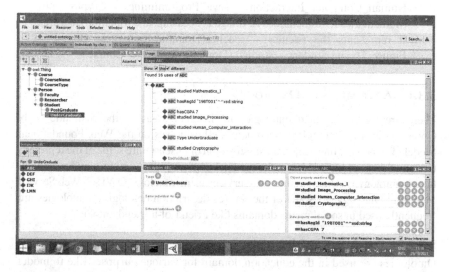

FIGURE 5.4   Assertions view.

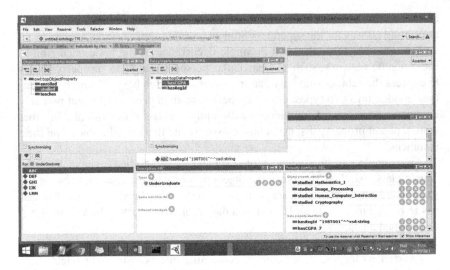

FIGURE 5.5   Object property.

Figure 5.5 displays the different object properties defined for this sample ontology in the e-learning domain. "studied", "enrolled", and "teaches" are some of the object properties that will relate Student class to Course class and Faculty class to Course class. Figure 5.5 also shows the different data property assertions defined, i.e., the Register ID for each student and the CGPA scored by each student.

Figures 5.6 and 5.7 represent two aspects of the sample ontology. They represent:

i. Spring view of the Course class with its different course instances like Human_Computer_Interaction, Java_Programming, Cryptography, Mathematics I, Mathematics II, Image Processing, and Machine Learning.
ii. It represents the Faculty -- teaches(Domain>Range) -> Course in Figure 5.6 and Student -- studied (Domain>Range) -> Course.

5.3.1   Applications of Ontology

There are well-established ontologies in different areas of the Semantic Web. GoodRelations is an ontology used across e-businesses on the Web. Foundational Model of Anatomy, International Classification of Diseases are some of the representations of domain knowledge in the medical domain [28]. Another popular application of ontology is for annotating web services using ontology. OWL-S, Web Service Modeling Ontology are some of the service description ontologies. Ontologies are frequently used in social media domains like Friend-of-a-friend ontology.

5.3.1.1   Ontologies in E-learning

Ontologies are used in the education domain for various purposes like to model a learner, learning resources [31], for instructional design [32], to model the

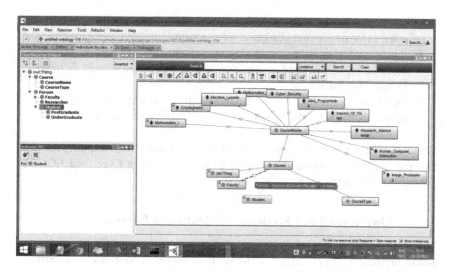

FIGURE 5.6 Spring view 1.

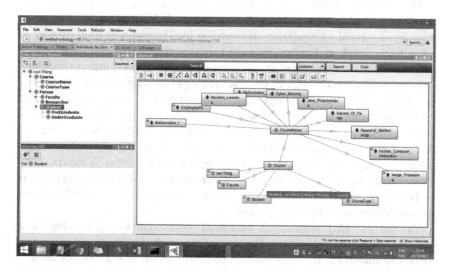

FIGURE 5.7 Spring view 2.

learning sequence [17] among many other reasons [27] reviewed previous literature in the area of ontologies being used in E-learning. The authors' findings are:

- Ontologies are used to model curriculum elements and to link the learning objectives and learning outcomes to the courses.
- Ontologies are used to model subjects and to model various learning
  activities like assignments, feedback.

- Ontologies are used to model learner-related information like their personal information and academic information.
- Since ontologies allow for reusing the information, the authors observe that ontologies are used to share vocabulary and data among different E-learning systems.
- Ontologies can also be used for searching and retrieving learning resources based on the metadata of those learning resources.

There are defined standards ontology follows when used for E-learning services like the IEEE LOM (Learning Object Metadata), Dublin Core, and SCORM (Sharable Content Object Reference Model).

5.3.2 ONTOLOGY-BASED RECOMMENDER SYSTEMS IN E-LEARNING

This section gives a brief overview of different ways ontology is applied in e-learning-based recommender systems. Recommender systems in e-learning aim at recommending learning materials, choice of university, courses that can be of interest to the student, etc.

The authors [33] proposed a system that applies three phases. Initially, a file based on the web log file is created. Next, a knowledge base in the form of an ontology is created by analyzing the web log file. The authors used Protégé to construct the ontology. Further, the users using the system also give some explicit information. The second phase groups users employing fuzzy c-means clustering. Data got from these two phases becomes the input to the proposed system. The final phase makes use of ontology mapping based on item collaborative filtering.

A recommender system was designed based on ontology and dimensionality reduction. Here again, the authors apply two phases. In the first phase, ratings are clustered using an expectation-maximization algorithm. Dimensionality reduction using singular value decomposition is applied to each cluster. A matrix showing similarity between users and items is thereby generated. The second phase is for predicting recommendations. The authors employ ontology to find relations semantically between the concepts [34].

The authors [35] proposed a coalesced system employing ontology and sequence prediction algorithms. The goal of the work is to recommend elective courses to students at the higher education level. Details of students are used to construct an ontology and semantic similarity is found between students to find the neighbourhood of similar students. Elective courses taken by these semantically similar students are taken as input to find a recommendation, by applying a sequence prediction algorithm, for a target student.

A framework was suggested that recommends learning materials based on a learner's request and the system also suggests a learning sequence over those materials. The authors focus specifically on learning topics that occur in the programming domain. The authors use the learner's knowledge level as the input to filter further any of the suggested learning topics. An ontology is built to model the learner and the learning materials in programming. Based on Semantic Web

Rule Language (SWRL) and SPARQL, prerequisite topics for the requested learning material are extracted from the ontology [17].

The authors [31] employ a hybridized method to recommend learning materials. The authors use otology to model a learner and learning resource. Thereafter, the authors calculate cosine similarity between learners to form a neighbourhood of similar users. The top N materials are extracted based on the neighbourhood. Finally, the authors apply sequence pattern mining to filter the suggestions and thereby recommend refined top k materials.

An ontology-based course recommender system (OPCR) was designed in [36]. OPCR requires students to fill in their personal and academic-related data. Data is gathered related to courses, jobs that are possible with respect to courses, and factors that students consider most important while deciding on a university course. Three ontologies are constructed to model knowledge about the course, student, and job. OPCR calculates content-based recommendations and collaborative-based recommendations.

Table 5.2 is a summary of the various recommender systems in the e-learning domain applying ontology. As seen from the table it can be understood that such recommender systems are useful to suggest learning materials, sequence to study those learning materials, courses that a student would be interested in, the university that a student can opt for based on his interests, preferences, etc.

5.3.3 CASE STUDY

Since this chapter discusses semantic web ontology-based recommender systems in the e-learning domain, a case study is discussed here. Unlike an e-commerce product, learners have several attributes with which recommender systems can be designed. The knowledge level of a learner, the learning goals a learner has in mind, the different learning styles a learner has with his peers, and the learner's preferences, interests, cognitive ability, and learning attitude are just some of the list of attributes attributed to a learner. Knowledge level has different categories of basic/moderate/advanced. Learning attitude has characteristics of a learner being active/passive.

Consider the architecture as shown in Figure 5.8. The architecture shows the usage of ontology and sequence prediction mining algorithm, particularly, compact prediction tree being used. Generally, students at higher education level have to earn credits during their degree program. However, students are given mandatory core courses and elective courses across the university that a student can choose from. The challenge in this scenario is the wide choice of electives available. In other words, we see a situation of information overload. Again, it is a common observation to see students taking courses that either their peers chose or that their seniors in college chose. This kind of choice, in the long run, does not necessarily fully satisfy a student's line of interest. The work shown in Figure 5.8 aims to help students with their choice of electives at the university level. To support the ground reality, such as the lack of ratings over courses at the university level, and to support personalization, the work does not rely on ratings to find

TABLE 5.2

E-learning Recommendation System Applying Ontology

Type	Means by Which Ontology Is Applied	Steps Involved	Recommendation Result
Domain ontology [17]	Web Ontology Language (OWL) Protégé 4	Learner submits learning topic request Knowledge level is evaluated Ontology is created Needed prerequisites found using SWRL Querying using SPARQL	Proposed system recommends learning materials and learning sequence a student can follow
Domain ontology [35]	Web Ontology Language (OWL) Protégé 5.2	Personality and profile related information of students is gathered Ontology related to students and elective courses is constructed Semantic similarity between neighbours is calculated Sequence prediction is applied over the elective courses taken by the neighbourhood	Designed system recommends the elective courses at the university level students need to take to satisfy the university credits criteria
Domain ontology [36]	Web Ontology Language (OWL) Protégé 5.2	Data related to students, courses, university, and jobs are gathered Ontologies related to course, student, job are constructed Ontology mapping to link ontologies Semantic similarity calculation	Ontology Personalized Course Recommendation (OPCR) recommends courses at university level
Domain ontology [26]	Web Ontology Language (OWL) Protégé	Explicitly and implicitly collects data Develops ontology to model higher education institution, students, and employment Machine learning technique is applied Similarity between students and interests calculated	Proposed Majors and university based on identified student's requirements, preferences, interests, and capabilities

(Continued)

TABLE 5.2 Continued
E-learning Recommendation System Applying Ontology

Type	Means by Which Ontology Is Applied	Steps Involved	Recommendation Result
Formal ontology [31]	Web Ontology Language (OWL) Protégé	Learners and learning resources are modelled using ontology Similarity between neighbours is calculated Predicts ratings Collaborative filtering is applied to get N recommendations Sequence Pattern Mining is used to further filter to get k recommendations	Learning resources
Domain ontology [37]	Web Ontology Language (OWL) Resource Description Framework	Features like identifier, title, description, keywords, and learning resource type are extracted Pattern between instances of different ontologies found and classified by associative classifier Ontology matching is thereby applied	Learning material

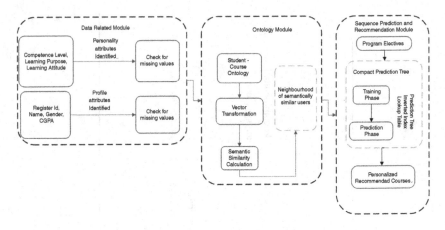

FIGURE 5.8   Ontology.

neighborhood similarity. Instead, it looks at the personality and profile aspects of a student. The architecture aims to overcome some of the traditional issues faced by recommender systems.

There are three primary modules in the proposed framework as shown in Figure 5.8. The first module is the data collection module; second, the ontology module, followed by the filtering step, i.e., the sequence prediction module. The proposed methodology overcomes the issue of lack of ratings by not depending on ratings. Instead, the work finds similarities between students based on their personality and profile information. Personality-related information like student's learning attitude, media preferences, etc. are collected. Profile information like student's name, register id, the cumulative grade point scored are some of the details collected. Once the data is collected, ontology is constructed. The proposed ontology has two classes: one class focuses on the student aspects and the other class looks into the course-related aspects. Object property assertions and data property assertions are defined for the instances which are the students in this case. After constructing the ontology in the ontology module, the next module finds the semantic similarity between the students. Related work in this area shows neighbourhood of similar users is found using cosine similarity, Euclidean distance, etc. Semantic similarity between concepts is a method to measure the semantic similarity, or the semantic distance between two concepts according to a given ontology. Once the semantically similar users are found, the students are sorted based on the similarity score. The similarity score leads to forming a similar semantic neighborhood.

The sequence of elective courses taken by this semantic set of similar users is then fed in as input to the next Sequence Prediction and recommendation module. Sequence prediction mining helps to predict the next possible item(s) of a given sequence. It finds relevance in many of today's real-life applications like language translation, forecasting, and medical imaging. The sequence prediction algorithm used in the given architecture is Compact Prediction Tree. There are three main data structure modules that the compact prediction tree uses: Prediction tree, lookup table, and inverted index.

For instance, a student had taken electives like Natural Language Processing, Web Mining, and Java Programming. With the number of neighbours set to five and on calculating the neighbourhood similarity, the student was found similar to a set of students who have taken the following electives: Java Programming, Human-Computer Interaction, Web Mining, Virtualization, Internet of Things, Cybersecurity, Natural Language Processing, Data Visualization, and Artificial Intelligence. On applying the compact prediction tree, PERKC recommended the following electives: Human-Computer Interaction, Internet of Things, Artificial Intelligence Cybersecurity, and Virtualization.

5.4 PERFORMANCE ANALYSIS

This chapter discussed the different ways recommender systems are designed in the e-learning domain. There are non-ontology-based and ontology-based recommender systems. Precision, recall, F1 measure, accuracy, mean average error, root mean square error are some of the evaluation metrics research in this area uses. The precision metric gives an idea of the items that got recommended correctly to all the items that were recommended. Figure 5.9 shows the average precision of some of the methodologies particularly employing the ontology discussed in this chapter. Onto+SPM [31] used the ontology with the sequence pattern mining to recommend learning material. The authors use ontology to model the learner and learning resource and thereby calculate ontological similarity to find suggestions.

Thereafter, the authors applied generalized sequence pattern mining to further filter and recommend more relevant learning materials. OPCR is the Ontology-based personalized course recommender system that recommends courses and universities [36]. OPCR uses a hybridization of methods. The author applies content-based

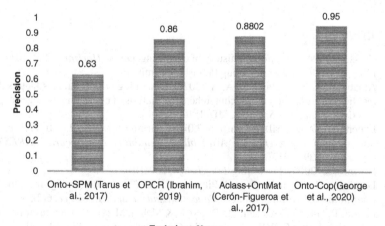

FIGURE 5.9 Performance.

recommendations along with collaborative-based recommendations. The collaborative recommendation methodology uses ontology. OPCR has an average precision of 0.86. The authors [37] applied an associative classifier to find patterns and perform ontology matching (Aclass+OntMat) among heterogonous ontologies to recommend learning material. The authors perform the initial pre-processing followed by extracting the needed features. Subsequently, the authors use an instance-based strategy to find similarities between entities. The system shows an average precision of 0.8802. Another work discussed in this chapter was shown in Figure 5.8. That method, Onto-Cop, utilizes semantic web ontology by coalescing ontology with sequence prediction mining to recommend personalized elective courses to students at the university level. Onto-cop gives an average precision of 0.95.

5.5 CONCLUSION

Most of the content available on the World Wide Web is designed in a way to make humans read and understand. There is no specific way for computers to comprehend the semantics behind web pages. It is the Semantic Web that aided in making the Internet smarter and allowing the computers to understand semantics as software agents browse through different web pages. This helps in carrying out more tasks for users in a decentralized manner. Among the different components of the semantic web is ontology. This chapter discusses ontology, its several applications, and its versatility if applied in e-learning-based recommender systems. The several advantages of ontology like improving the accuracy of searches on the Web, relating information on web pages to the related knowledge structure, and the reasoning enables to employ ontology as a stand-alone e-learning-based recommender system. As seen in several related works, ontology works well with several other technologies like Restricted Boltzmann Machine, neural network among others to bring out even more effective recommendations in the e-learning domain.

REFERENCES

1. Cross, J. (2004). An informal history of e-learning. *On the Horizon*, 12(3), 103–110. ISSN 1074-8121, Emerald Group Publishing Limited.
2. Alqahtani, A. Y., & Rajkhan, A. A. (2020). E-learning critical success factors during the covid-19 pandemic: A comprehensive analysis of e-learning managerial perspectives. *Education Sciences*, MDPI, 10(9), 1–16.
3. Ebner, M. (2007, April). E-Learning 2.0= e-Learning 1.0+ Web 2.0? In *The Second International Conference on Availability, Reliability and Security (ARES'07)* (pp. 1235–1239). IEEE.
4. Downes, S. (2005). E-learning 2.0. *ELearn*, 2005(10), 1.
5. Dominic, M., Francis, S. and Pilomenraj, A. (2014). E-learning in web 3.0. *International Journal of Modern Education and Computer Science*, 6(2), 8.
6. Favale, T., Soro, F., Trevisan, M., Drago, I., & Mellia, M. (2020). Campus traffic and e-learning during COVID-19 pandemic. *Computer Networks*, 176, 107290.

7. Almaiah, M. A., Al-Khasawneh, A., & Althunibat, A. (2020). Exploring the critical challenges and factors influencing the E-learning system usage during COVID-19 pandemic. *Education and Information Technologies*, 25, 5261–5280.
8. Radha, R., Mahalakshmi, K., Kumar, V. S., & Saravanakumar, A. R. (2020). E-Learning during lockdown of Covid-19 pandemic: A global perspective. *International Journal of Control and Automation*, 13(4), 1088–1099.
9. Aggarwal, C. (2016). *Recommender System the Textbook*. Switzerland: Springer International Publishing.
10. George, G. & Lal, A. M. (2019). Review of ontology-based recommender systems in e-learning. *Computers & Education*, 142(1), 103642.
11. Wu, E. H. K., Lin, C. H., Ou, Y. Y., Liu, C. Z., Wang, W. K., & Chao, C. Y. (2020). Advantages and constraints of a hybrid model K-12 e-learning assistant chatbot. *IEEE Access*, 8, 77788–77801.
12. Rubin, R., Redmond, A., Weber, G., & Guirez, K. (2017). Habits of highly successful professional learners and the corresponding online curriculum.
13. Pardos, Z. A., & Jiang, W. (2020, March). Designing for serendipity in a university course recommendation system. *In Proceedings of the Tenth International Conference on Learning Analytics & Knowledge* (pp. 350–359).
14. Polyzou, A., Nikolakopoulos, A. N., & Karypis, G. (2019). Scholars walk: A Markov chain framework for course recommendation. *International Educational Data Mining Society*, 396–401. https://eric.ed.gov/?id=ED599254
15. Jiang, W., Pardos, Z. A., & Wei, Q. (2019, March). Goal-based course recommendation. In *Proceedings of the 9th International Conference on Learning Analytics & Knowledge* (pp. 36–45).
16. Parameswaran, A., Venetis, P., & Garcia-Molina, H. (2011). Recommendation systems with complex constraints: A course recommendation perspective. *ACM Transactions on Information Systems (TOIS)*, 29(4), 1–33.
17. Shishehchi, S., Zin, N. A. M. and Seman, E. A. A. (2021). Ontology-based recommender system for a learning sequence in programming languages. *International Journal of Emerging Technologies in Learning*, 16(12), 123–141.
18. Mishra, D., Agarwal, A., & Kolekar, S. (2021). Dynamic identification of learning styles in MOOC environment using ontology based browser extension. *International Journal of Emerging Technologies in Learning (iJET)*, 16(12), 65–93.
19. Tarus, J. K., Niu, Z., & Kalui, D. (2018). A hybrid recommender system for e-learning based on context awareness and sequential pattern mining. *Soft Computing*, 22(8), 2449–2461.
20. El Aissaoui, O., El Madani, Y. E. A., Oughdir, L., & El Allioui, Y. (2019). A fuzzy classification approach for learning style prediction based on web mining technique in e-learning environments. *Education and Information Technologies*, 24(3), 1943–1959.
21. Bourkoukou, O., El Bachari, E., & El Adnani, M. (2017). A recommender model in e-learning environment. *Arabian Journal for Science and Engineering*, 42(2), 607–617.
22. Klašnja-Milićević, A., Ivanović, M., Vesin, B., & Budimac, Z. (2018). Enhancing e-learning systems with personalized recommendation based on collaborative tagging techniques. *Applied Intelligence*, 48(6), 1519–1535.
23. Salehi, M., Kamalabadi, I. N., & Ghoushchi, M. B. G. (2014). Personalized recommendation of learning material using sequential pattern mining and attribute based collaborative filtering. *Education and Information Technologies*, 19(4), 713–735.

24. Dean, A., & Jim, H. (2011). *Semantic Web for the Working Ontologist: Effective Modeling in RDFS and OWL*, United States of America: Morgan Kaufmann Publisher, Elsevier.
25. Berners-Lee, T., Hendler, J., & Lassila, O. (2001). The semantic web. *Scientific American*, 284(5), 34–43.
26. Obeid, C., Lahoud, I., El Khoury, H., & Champin, P. A. (2018, April). Ontology-based recommender system in higher education. In *Companion Proceedings of the Web Conference 2018* (pp. 1031–1034).
27. Al-Yahya, M., George, R., & Alfaries, A. (2015). Ontologies in E-learning: review of the literature. *International Journal of Software Engineering and Its Applications*, 9(2), 67–84.
28. Domingue, J., Fensel, D. and Hendler, J. A. eds. (2011). *Handbook of Semantic Web Technologies*. Berlin, Heidelberg: Springer Science & Business Media, Springer.
29. Gruber, T. R. (1993). A translation approach to portable ontology specifications. *Knowledge Acquisition*, 5(2), 199–220.
30. Mizoguchi, R., & Ikeda, M. (1998). Towards ontology engineering. *Journal-Japanese Society for Artificial Intelligence*, 13, 9–10.
31. Tarus, J. K., Niu, Z., & Yousif, A. (2017). A hybrid knowledge-based recommender system for e-learning based on ontology and sequential pattern mining. *Future Generation Computer Systems*, 72, 37–48.
32. Chimalakonda, S., & Nori, K. V. (2020). An ontology based modeling framework for design of educational technologies. *Smart Learning Environments*, 7(1), 1–24.
33. Makwana, K., Patel, J., & Shah, P. (2017 March). An ontology based recommender system to mitigate the cold start problem in personalized web search. In *International Conference on Information and Communication Technology for Intelligent Systems* (pp. 120–127). Cham: Springer. doi:10.1007/978-3-319-63673-3_15.
34. Nilashi, M., Ibrahim, O., & Bagherifard, K. (2018). A recommender system based on collaborative filtering using ontology and dimensionality reduction techniques. *Expert Systems with Applications*, 92, 507–520. doi:10.1016/j.eswa.2017.09.058.
35. George, G., & Lal, A. M. (2021). A personalized approach to course recommendation in higher education. *International Journal on Semantic Web and Information Systems (IJSWIS)*, 17(2), 100–114.
36. Ibrahim, M. E. (2019). An ontology-based hybrid approach to course recommendation in higher education (Doctoral dissertation, University of Portsmouth).
37. Cerón-Figueroa, S., López-Yáñez, I., Alhalabi, W., Camacho-Nieto, O., Villuendas-Rey, Y., Aldape-Pérez, M., & Yáñez-Márquez, C. (2017). Instance-based ontology matching for e-learning material using an associative pattern classifier. *Computers in Human Behavior*, 69, 218–225.

6 Exploring Reasoning for Utilizing the Full Potential of Semantic Web

Ayesha Ameen, Khaleel Ur Rahman Khan, and B. Padmaja Rani

CONTENTS

6.1 INTRODUCTION

The main objective of the semantic web is to give machines much-improved access to information so that they can act as information mediators in support of humans. The semantic web extends the world wide web to make it a web of data;

DOI: 10.1201/9781003309420-6

it combines knowledge representation formalism and logic with Web technologies in such a way that the contents currently represented on the web become machine-understandable and easily processable by machines. For making this possible, ontologies are recommended by W3C for representing formal semantics in Semantic Web. The semantic Web depends on ontologies for representing the data. Ontologies represent the data in terms of concepts and support reasoning mechanisms.

6.1.1 SEMANTICS IN SEMANTIC WEB

Understanding of Semantic Web starts with the word semantics which simply means meaning. If what the data means is known, then the data could be used more effectively. Semantic web establishes semantics through standardized connection among data by making data addressable and labeling data uniquely. The fundamental building block in the semantic web is a statement. A statement is a single piece of metadata that consists of a subject, predicate, and object for example a statement specifying Jack as a student of university is given "jack student university". The set of related statements constitutes metadata in the semantic web. These statements are linked together using constructs to form the semantics, which forms the meaning of the link, to provide a meaningful path instead of a user-interpreted one. Further, the statements can also incorporate logic that will facilitate further interpretation and drawing inferences. Many types of statements exist in the semantic web, concepts, individuals, restrictions, and logic are described by statements. Ontology comprises a set of statements defining concepts and relationships among the concepts. Statements describing individuals form the instance data. Statements can be inferred or asserted. A statement is said to be asserted if the application creates the statements directly and it is said to be inferred if a reasoner is used to infer additional statements logically [1]. Semantic Web vocabulary and language are used for identifying various types of relationships and statements.

6.1.2 SEMANTIC WEB ARCHITECTURE

The semantic web consists of layer cake architecture, where one layer is built on the other as depicted below in Figure 6.1. These layers specify the level of abstraction applied to the web.

The first layer consists of URI and Unicode. The standard for language character representation is Unicode, and URI is used for uniquely identifying resources on the web, it consists of a string of characters.

The second layer is XML which stands for Extensible Markup Language. XML is used for writing a web document in a structured way by using user-defined vocabulary. To identify data, organize and store XML tags are used. No information regarding semantics is provided by XML.

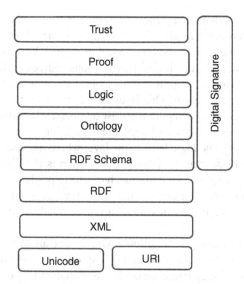

FIGURE 6.1 Layers of semantic web.

Consequently, the next layer does the semantic annotation of data and is called RDF (Resource Development Framework).

RDF is a World Wide Web Consortium (W3C) standard designed as a basis for metadata processing. RDF enables applications that exchange machine-understandable information on the web to be interoperable. Information becomes machine-understandable at this layer. RDF is used for writing statements about web resources.

RDFS is the next layer which is the vocabulary language for RDF. RDFS describes the semantics associated with the resources described by RDF. RDFS describes classes and properties. The classes and properties are arranged in the form of generalization and specialization hierarchies. For properties, domain and range can be defined using RDFS. The basic feature to define an ontology is supported by RDFS, but it is not able to represent the complex relationship among the classes. Thus, for representing the complex relationship among classes, the next layer named ontology vocabulary is added.

Ontology vocabulary represents ontologies in Web Ontology Language (OWL) which is further expressive than RDFS. OWL is a W3C standard for ontology representation and lets the description of complex relations like transitivity, symmetry, and cardinality constrain.

The inference mechanism is carried out by the next layer which is called a logic layer; it encompasses the inference provided by OWL by integrating reasoning potentiality necessary for inferencing knowledge and decision-support based on rules. Rules are used to represent the business policies, processes, and contracts. The explanation of the inference carried out in the logic layer is given by the proof layer. A deductive process to prove the answer obtained are right, validation of answers using a validation procedure is carried out at this layer. This layer also supports the representation of proof in various Web languages.

As information can be contributed from various sources in the semantic web, the next layer is used to access the acceptance of information as true when multiple information sources are present. This layer uses digital signatures and knowledge shared by trusted agents to compute the trust of a particular information source.

Among all the above-discussed layers, the logic layer is the prime focus of our research contribution. The logic layer is the layer where the inference mechanism comes into action to deduce the information, which is not directly available in the data. This layer relies on two main mechanisms to produce inference, the first one is the information content of the semantic web which is represented as ontology and the second one is the rules which are used to add additional logic to the semantic web.

This chapter is organized as follows. Ontologies are revisited, the importance of ontologies is discussed followed by the reasoning support offered by the ontologies. A web ontology language is discussed in detail followed by OWL-DL reasoners. The next part of the chapter focuses on the rules, and the relationship between ontologies and rules is specified. The significance of rules is highlighted. Rules types are discussed next followed by a few examples of rule languages supported in Semantic Web. Rule engine and what role it plays in adding additional reasoning capabilities to the semantic web application is discussed next, followed by a few freely available rule engines, at last, the measures used for calculating the performance of the rule engine are stated.

6.2 ONTOLOGIES

Ontology is a part of philosophy that is concerned with the nature of existence [2]. Ontology identifies and studies categories of things that exist in nature. In the last few decades with the growth in the volume and complexity of knowledge bases and the necessity to facilitate communication among heterogeneous devices, made the ontologies to be developed across many disciplines such as database theory and artificial intelligence. An ontology consists of a set of concepts and relations organized into generalization and specialization hierarchy which represents the appropriate concepts and relations extracted from the analysis of the system. Definitions of ontologies proposed for computer science are discussed as follows.

According to Guber, an ontology is defined as an "explicit specification of a conceptualization" [3]. Later Borst described an ontology as a "formal specification of a shared conceptualization" [4]. This definition emphasized that the conceptualization must convey a shared opinion among some parties and it must also be expressed as formal i.e., machine-readable format. Studer later combined earlier definitions as "An ontology is a formal, explicit specification of a shared conceptualization". With this definition, there is a need to understand the terms conceptualization, formal explicit specification, and shared.

An abstract basic view of the world that we want to present is represented by conceptualization. All knowledge base is dedicated to some conceptualization explicitly or implicitly. According to Genesereth and Nilsson [5] formally characterized knowledge is based on a conceptualization i.e., consists of concepts, objects, other entities, and relationships among them in some area of interest. A language must be used to represent the elements of conceptualization, and we must make sure that the symbols used for representing a language are interpreted under conceptualization committed to. The conceptualization must be made explicit. The formal language which is machine-readable must be used to represent the conceptualization. This conceptualization can be shared across various domains.

Knowledge formalization in ontologies is based on five components such as concepts, relations, instances, axioms, and functions. Concepts or classes comprises a group of individuals who share some common characteristics. Anything about which something is said can be a concept, it can describe a strategy, action, task, and reasoning process. Individuals who model concrete objects are called an instance. Relations: Relations describe the relationships between the concepts. Various relationships exist in ontologies such as specialization, partitive, and associative which relate concepts across tree structures. Relationships represent forms of interactions between concepts in the same domain. Classes or instances values are constraints by using axioms. Axioms represent general rules that must be followed always in an ontology.

6.2.1 ROLE OF ONTOLOGIES IN SEMANTIC WEB

In Semantic Web vision, ontologies have an imperative role; ontologies offer semantic vocabulary which is used to annotate a website in a meaningful manner that can be used for machine interpretation. This representation allows machines to process information and deliver semantically correct answers to the given queries. Ontologies not only present a sharable and reusable knowledge representation but also enable adding up of new knowledge based on the existing knowledge. Ontologies must be expressed using an ontology language.

Ontology language permits users to put down unambiguous, formal conceptualizations of the domain being modeled. There are a few requirements that an ontology language must abide by; It must have well-defined syntax and semantics; it must have sufficient expressive power and must support efficient reasoning. It must have ease of expression [6]. Ontology languages must hold well-defined syntax which stands as the basic requirement for the processing of information by machines. Formal semantics is needed to define the meaning of the language exactly. Formal semantic also permits to perform reasoning on ontologies as discussed below:

Equivalences of classes: If A is equivalent to X and X is equivalent to Y, then A is equivalent to Y. Where A, X, Y are classes.
Class membership: If 'a' is an instance of Class A and A is a subclass of B then, it can be inferred as 'a' is an instance of B.

Consistency: It checks the errors in the ontology that occur due to improper creation of class subclass hierarchies, violation of constraints imposed in ontologies. The instance created in an ontology must be assigned properly to class, if it is not done appropriately then consistency of the ontology might be in threat.

Classification*:* If a certain property value pair is sufficient for an individual/instance to be a member of class C, then any individual/instance, which satisfies this condition must be an individual/instance of C.

Hence, it can be concluded that formal semantics is essential for reasoning. In the semantic web, the reasoning mechanism is vital from the very definition of the semantics and the ability to add additional knowledge through automated reasoning procedures. There are various advantages of having a reasoning mechanism in the semantic web to state few; consistency of the ontologies can be checked by performing reasoning on ontologies; these checks ensure that no unintended relationships exist among classes. These consistency checks are very important when multiple people are involved in the creation of ontologies, reusing and sharing of ontologies across various domains becomes possible only with the consistent ontologies are created. Ontologies are represented in OWL as it is W3C standard language for representing machine-processable information on the web. OWL allows processing the information content rather than only the presentation of information to users. It provides improved machine interpretability of web content when compared to others. Existing reasoners like RACER and FACT are used to validate the consistency of OWL as it is partially mapped to description logic. Apart from the build-in reasoners Rule engines can also be deployed to infer additional knowledge by using the rule formalism. The rule can be given as input to the reasoner, it carries out inference based on the given rules. Rules can be written to add extra information about the domain that is being modeled or any business constraint.

6.3   WEB ONTOLOGY LANGUAGE (OWL)

Many noteworthy features of OWL make it a good choice for representing ontologies. The first feature is it extends the current web standards for example XML, RDF, and RDFS. Second, it is based on familiar knowledge representation idioms, and it is formally specified which means it describes the meaning of knowledge precisely. Third, OWL has the adequate expressive power to support automated reasoning mechanisms. Finally, it is easy to understand and use.

A domain is defined in terms of classes, individuals, properties represent the relationship and can be of two types object properties and datatype properties in OWL. Web resources properties can be described by OWL. Some features of OWL are it supports expressive operators including Boolean operators like union, intersection, and complement for concept description. Explicit quantifiers for properties and relationships exist in OWL. Two disjoint classes can be specified explicitly in OWL. Domain and range of properties can be specified using OWL and it also supports the local scope of properties which applies only to some classes. Cardinality restriction specifying the number of diverse values a property must or may have can be enumerated in OWL. Properties' special characteristics such as transitivity and inverse properties are supported by OWL which was not supported in RDFS.

For further understanding, an automobile ontology is constructed in OWL which specifies the general classification of automobiles [7]. An automobile is a self-propelled vehicle, used for the transportation of goods or passengers [8]. It is run with the help of an internal combustion engine which is usually powered by Diesel, Petrol, CNG, Electricity, etc. An Automobile ontology is constructed by making automobile as the root class then two subclasses are created Passenger_Vehicles and Goods_Vehicles and made disjoint. Although some goods vehicles can carry a limited number of people, for this ontology they are described as disjoint classes. Instances are created for Goods_Vehicles and Passenger_Vehicles. Properties that represent the relationship among classes are also added. Object and Datatype properties are supported in OWL were the first described the relationship between the instances and the second described the relationships between instances and literal data.

The header of OWL ontologies contains information about the namespace. rdf:RDF is the root element given as follows.

```
Header
<rdf:RDF
  xmlns="AutomobileOntology#"
  xml:base="AutomobileOntology"
  xmlns:rdfs="http://www.w3.org/2000/01/rdf-schema#"
  xmlns:owl="http://www.w3.org/2002/07/owl#"
  xmlns:xsd="http://www.w3.org/2001/XMLSchema#"
  xmlns:rdf="http://www.w3.org/1999/02/22-rdf-syntax-ns#">
```

After header collection of assertions for housekeeping is present in an OWL ontology. These assertions are organized under owl:Ontology elements and can contain information regarding the inclusion of other ontologies, comments, and version control illustrated below.

```
<owl:Ontology rdf:about="AutomobileOntology">
  <rdfs:label>AutoOnto</rdfs:label>
  <owl:priorVersion>None</owl:priorVersion>
  <owl:versionInfo>First Version</owl:versionInfo>
  <rdfs:comment>The Automobile ontology is constructed to
give an understanding of the classification of
automobiles</rdfs:comment>
  <owl:backwardCompatibleWith>None</owl:backwardCompatible
With>
  <owl:versionIRI rdf:resource="AutomobileOntology/1.0"/>
  </owl:Ontology>
```

owl: Class element defines class. The automobile class and its Subclasses are defined below.

```
<!-- AutomobileOntology#Automobiles -->
  <owl:Class rdf:about="AutomobileOntology#Automobiles"/>
    <!-- AutomobileOntology#Goods_Vehicles -->
  <owl:Class rdf:about="AutomobileOntology#Goods_Vehicles">
    <rdfs:subClassOf rdf:resource="AutomobileOntology#Automo
biles"/>
    <owl:disjointWith
rdf:resource="AutomobileOntology#Passenger_Vehicles"/>
  </owl:Class>
    <!-- AutomobileOntology#Passenger_Vehicles -->
  <owl:Class
rdf:about="AutomobileOntology#Passenger_Vehicles">
    <rdfs:subClassOf rdf:resource="AutomobileOntology#Automo
biles"/>
  </owl:Class>
```

Object property canTravelBy relates the instances of Passengers and Passenger_
Vehicles and is inverse of isUsedby. Similarly drives object property relates the
instances of driver and automobile and is inverse of isDrivenBy.

```
<!-- AutomobileOntology#canTravelBy -->
  <owl:ObjectProperty rdf:about="AutomobileOntology#canTrave
lBy">
    <rdfs:range
rdf:resource="AutomobileOntology#Passenger_Vehicles"/>
    <rdfs:domain rdf:resource="AutomobileOntology#Passeng
ers"/>
  </owl:ObjectProperty>
  <!-- AutomobileOntology#drives -->
  <owl:ObjectProperty rdf:about="AutomobileOntology#drives">
    <rdfs:range rdf:resource="AutomobileOntology#Automobiles"/>
    <rdfs:domain rdf:resource="AutomobileOntology#Driver"/>
    </owl:ObjectProperty>
  <!-- AutomobileOntology#isDrivenBy -->
  <owl:ObjectProperty
rdf:about="AutomobileOntology#isDrivenBy">
    <rdfs:domain rdf:resource="AutomobileOntology#Automobi
les"/>
      <rdfs:range rdf:resource="AutomobileOntology#Driver"/>
    <owl:inverseOf
rdf:resource="AutomobileOntology#drives"/>
  </owl:ObjectProperty>
  <!-- AutomobileOntology#isUsedBy -->
  <owl:ObjectProperty
rdf:about="AutomobileOntology#isUsedBy">
    <rdfs:domain
rdf:resource="AutomobileOntology#Passenger_Vehicles"/>
    <rdfs:range rdf:resource="AutomobileOntology#Passengers"/>
  <owl:inverseOf rdf:resource="AutomobileOntology#canTravelBy"/>
  </owl:ObjectProperty>
```

Datatype properties specified in automobile ontology are hasCapacity and has-Color which the domain as Automobile class and range as integer and integer, string respectively as shown as follows.

```
<!-- AutomobileOntology#hasCapacity -->
  <owl:DatatypeProperty rdf:about="AutomobileOntology#hasCap
acity">
    <rdfs:domain rdf:resource="AutomobileOntology#Automobi
les"/>
    <rdfs:range rdf:resource="&xsd;integer"/>
    <rdfs:range rdf:resource="&xsd;string"/>
    </owl:DatatypeProperty>

  <!-- AutomobileOntology#hasColor -->
  <owl:DatatypeProperty rdf:about="AutomobileOntology#hasCo
lor">
    <rdfs:domain rdf:resource="AutomobileOntology#Automobi
les"/>
    <owl:propertyDisjointWith rdf:resource="AutomobileOntolo
gy#isColorOf"/>
    <rdfs:range rdf:resource="&xsd;string"/>
    </owl:DatatypeProperty>
```

After creating an ontology, the correctness of the ontologies must be checked. Ontology constructed must represent knowledge properly without any ambiguities and inconsistencies. If there are inconsistencies in ontology then it means that errors or conflicts exist in an ontology resulting in misinterpretation of some concepts in the ontology. Inconsistencies can also create false semantic understanding and knowledge representation [9]. Hence, the consistency check must be performed before using ontologies to make sure that no discrepancies arise in knowledge represented by ontologies. Reasoners are used to checking the correctness of ontology. Reasoner is a program used to derive unknown facts from ontologies. The correctness of the created ontology is examined by any DL reasoner supported by OWL. Pellet, HermiT, FaCT++, RACER is widely used OWL-DL reasoners.

Description Logic is a formal knowledge representation language [10]. DL is further expressive than propositional logic. The core reasoning problems for DLs are generally decidable and efficient decision techniques are designed for them. DL is tailored to depict knowledge in the domain in a well-understood and organized manner. Artificial intelligence uses DL for describing and performing reasoning about the concepts in the domain. DL is particularly important because they offer logical formalism for ontologies. OWL and its profiles are based on DLs. DL describes the important things in the domain of interest as concepts also called classes, roles also called relations or properties, and individuals and their relationships. Axioms which are logical statements relating to roles and concepts are fundamental modeling concepts in DL. The knowledge base of DL represents terminology (Tbox) and data (Abox) separately. DL architecture is depicted as follows in Figure 6.2.

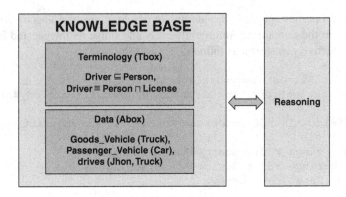

FIGURE 6.2   DL architecture.

Knowledge is represented as knowledge base KB= $\langle$ A, T $\rangle$ Where A is for A box which comprises of a set of assertions about named individuals e.g., Goods_Vehicle (Truck), Passenger_Vehicle (Car), drives (Jhon, Truck) specifies that truck is an individual of a goods vehicle, a car is an individual of passenger vehicle and drives object property relates the individuals of a driver to individuals of automobiles as shown in Figure 6.2 T is for terminology box which contains terminology definitions i.e., descriptions of concepts or roles e.g. Driver $\sqsubseteq$ Person, Driver $\equiv$ Person $\sqcap$ License, means a driver is a person and a driver concept is equivalent to a person having a license. Automobile ontology individuals for Goods_Vehicles, Passenger_Vehicles, Driver concepts are listed as follows.

```
    <rdf:type rdf:resource="&AutomobileOntology;Go
ods_Vehicles"/>
  </owl:NamedIndividual>
<owl:NamedIndividual rdf: about="&AutomobileOntology;Car">
    <rdf:type rdf:resource="&AutomobileOntology;Passen
ger_Vehicles"/>
  </owl:NamedIndividual>
  <owl:NamedIndividual rdf:about="&AutomobileOntology;Jhon">
    <rdf:type rdf:resource="&AutomobileOntology;Driver"/>
    <AutomobileOntology:drives rdf:resource="&AutomobileOnto
logy;Truck"/>
  </owl:NamedIndividual>
```

Reasoning Tasks of OWL resemble the standard description logic reasoning tasks that lead to infer novel facts from the knowledge base or validate the correctness of ontology [11]. The reasoning tasks can be separated as TBox reasoning and ABox reasoning tasks. TBox reasoning tasks are satisfiability, subsumption checking.

- Satisfiability assures that a class creates instances in line with the current ontology.

- Subsumption checks that class C subsumes a class D according to the current ontology, property subsumption is defined similarly, classification of the ontology to compute complete subsumption hierarchy of ontology is done.

ABox reasoning tasks usually occur during the run time of the ontology. Reasoning tasks for ABox are listed as follows:

- **Consistency check**: Checking whether ABox is consistent with respect to TBox.
- **Instance check**: Checks whether an assertion is required by the ABox.
- **Retrieval problem**: It retrieves all individuals who instantiate a class.
- **Property fillers**: It retrieve all individuals related to property. It can also retrieve all properties that exist between two individuals.
- **Conjunctive queries**: These are used as popular query formalism which is efficient in representing projection/selection/renaming/join relational queries.

6.3.1 DL Reasoners

A reasoner or reasoner engine is software capable to deduce logical consequences from a set of axioms or facts. Many DL reasoners are available among the most widely used are HermiT, FaCT++, RacerPro, Pellet.

HermiT
    HermiT is the first publicly accessible reasoner built on "hypertableau" calculus that offers effectual reasoning support compared to earlier reasoners [12]. It was developed by Oxford University in 2008. Ontologies that required hours and minutes to classify earlier can easily be classified by HermiT in seconds, it is successful in classifying certain ontologies which are too complex to be classified by any system. OWL is used in HermiT. HermiT checks OWL files to find whether it is consistent or not very quickly when compared to other reasoners, it finds the hierarchical relationship among classes and it has a fast process for classifying ontologies. HermiT is open-source, released under LGPL.
RacerPro
    RACER (Renamed ABox and Concept Expression Reasoner) RacerPro is the software commercial version [13]. RacerPro is very useful in implementing industrial projects built on RDF and OWL. Description logic reasoner is the backbone of RacerPro, it supports inference based on terminological knowledge along with a representation of knowledge concerning individuals. It is specifically tailored for supporting applications that are built on the exploitation of assertional reasoning (Aboxes) which are not static and generated on the fly. RacerPro architecture has large Aboxes kept in an AllegroGraph database. RacerPro

uses new optimization techniques and selects optimization based on the study of the queries and language of the input knowledge bases. It also has a built-in query and rule language support which makes it distinctive from other OWL reasoners.

FaCT ++

FaCT ++ is an updated version of FaCT. FaCT (Fast Classification of Terminologies) is developed by Horrocks, Ian [14]. It is a description logic classifier that is designed for testing modal logic satisfiability, it has two built-in reasoners, the first reasoner for SHF and other reasoners for SHIQ, reasoners use optimized implementations of complete and sound tableaux algorithms. The same algorithm is used in FaCT++ as that of FACT, but with a changed internal structure, it is implemented in C++ to maximize portability [15]. It can be used as a back-end reasoner or as a standalone DIG reasoner, now it is used as a built-in reasoner in Protégé 4 OWL editor. It is open-source and distributed with an LGPL license.

Pellet

Pellet is created by the Mind Swap group; it is an open-source OWL-DL reasoner [16]. It is developed in Java. It is built on the tableau algorithm and is the first reasoner which can perform reasoning on all OWL DL SHOIN(D) and OWL2(SHOIQ(D)). It provides an explanation of bugs. Pellet is the first sound and complete reasoner that supports reasoning with user-defined data types and individuals. It includes optimization for conjunctive query answering, nominals, and incremental reasoning. Pellet implements numerous extensions to OWL-DL like OWL/Rule hybrid reasoning and a non-monotonic operator. Pellet can used be along with OWL API libraries and Jena, and it can also be downloaded and included in other applications.

6.4   RULES IN SEMANTIC WEB

Recent advances in Semantic Web technologies have given rise to the need for sound and efficient methods for supporting reasoning on the knowledge scattered on the Web [17]. Ontologies support a basic form of reasoning as they are represented using OWL which is based on description logic, further reasoning support can be provided in Semantic Web by incorporating logical formalism and corresponding inference or rule engine.

6.4.1   NEED FOR RULES

The main intent of the logic layer is to enable the writing of rules [18]. The reasons for the need for rules in the Semantic Web are listed below [19].

- Rules are obvious in numerous applications such as service descriptions, specification of policies, business, database queries, etc. Indeed, it is necessary to design a rule language to express rules for web applications.

- Rules hide the details of the implementation by providing high-level descriptions. Rules are very brief and easier to write when compared to application code.
- Ontologies are used for describing the domain. Rules can be used for describing application-specific parts of the domain.
- Ontology languages fail to express certain characteristics of applications that are best expressible in the rule language.
- Rules are used to extract the logical inferences which cannot be extracted by OWL DL reasoners.
- Rules can be used to specify integrity constraints on the application domain.
- Statements that specify dynamic aspects of the domain can be expressed as rules.

Hence, it can be concluded that rules can be used to specify constraints, data transformation, perform updates on data, construct new data, event-driven actions. Rule are statements that can specify dynamic or static relationships between various data items, applications, and business logic in the enterprise.

6.4.2 LIMITATIONS OF ONTOLOGY FORMALISMS

Several limitations of OWL and Description Logic are revealed during the building of practical applications. Specifically, the expressiveness of Description Logic does not permit for the following characteristics [20]:

- Predicate definition of arbitrary arity.
- Use of variable quantifiers beyond tree-like structures of DL concepts.
- Expressive queries over DL knowledge base.
- Formalizing closed world reasoning in several forms over DL knowledge base.
- Expressing nonmonotonic knowledge.

The above-listed issues are acquiring much consideration in the Semantic Web community. In this regard, it is observed that many of the several representational capabilities that are missed in Description Logic require the support of nonmonotonicity of the primary logical formalism, which is in contrast with DL which is based on monotonicity. Knowledge-based systems based on various specializations of predicate logic such as OWL and RDF are committed to monotonic assumptions. This means that a description logic reasoner can never make inferences that will become invalid by the assertions of additional information ex. If we know that 'a' is an instance of A, then any assertions of more information of 'a' can never cause it to be NOT an instance of A. The property of being monotonic refers to the fact that on adding new premises a valid argument cannot be made invalid nor an invalid argument can be made valid. The next topic discusses how ontologies and rules collaborate to fulfill the required inference support needed in the Semantic Web.

6.4.3 Ontologies and Rules

Rules and ontologies have a substantial role, as they assign meaning and perform reasoning on data in Semantic Web. Both work together to realize the Semantic Web vision of extracting novel inferences from existing data. New inferences are derived from existing inferences based on vocabulary which consists of some information. In the Semantic Web, vocabulary is the ontological knowledge base. In the Semantic Web, the rule layer is not much researched and developed, but as of now much research is going on the rule layer, to identify its hidden potential which is not yet explored fully.

Even though ontologies and rules collaborate to support inference mechanisms there are few underlying differences between these two representation formalisms. The significant differences are highlighted as follows (Table 6.1):

- Ontologies are based on description logic, which is fragments of FOL, based on the open-world assumption (OWA).
- Rules are built on logic programming where closed world assumption (CWA) holds.
- In OWA, the knowledge denoted is considered incomplete and inferences that cannot be inferred from ontology are considered as unknown.
- In CWA, the knowledge which cannot be inferred from the ontology is considered false.
- DL is not based on Unique World Assumption (UNA) i.e., which enforces distinct terms to denote distinct objects.
- In contrast, rules are built on Unique World Assumption (UNA).
- Monotonic reasoning is supported by DL whereas rules support non-monotonic reasoning.

Hence, OWL is inadequate to offer the complete inference needed for Semantic Web application since it is founded on OWA and FOL principles that are insufficient to provide inference on applications that require comprehensive knowledge about the world. These shortcomings of OWL are conquered by integrating OWL with rules. Reasoning required in Semantic Web is depicted in Figure 6.3.

The integration performed must ensure the decidability preservation, i.e., the system derived from the integration of OWL and rules must preserve the

TABLE 6.1
Differences between Rules and OWL

Features	Rule	OWL
Reasoning	Non-monotonic	Monotonic
Assumption	CWA	OWA
Names	UNA	Non-UNA
Logic	Logic programming	First order logic

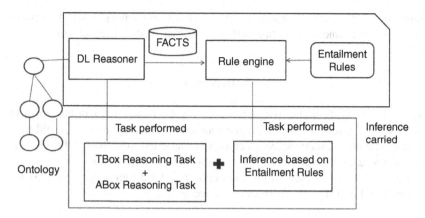

FIGURE 6.3   Reasoning in semantic web.

decidability. It is desirable to separate the reasoning performed by DL reasoner and rule-based reasoner in a modular way, which will create reasoning methods and engines over the top of deductive methods devised distinctly for DLs and rules.

6.4.4   IMPORTANCE OF RULES

Rules are the basis for reasoning that provides intelligent utilization and management of information [21]. Proper design of rules for the working of the Semantic Web was recognized as a significant design problem by Sir Tim Berners Lee et al. in [22]. Noteworthy issues in Semantic Web design are listed below:

> The challenge of the Semantic Web, therefore, is to provide a language that expresses both data and rules for reasoning about the data and that allows rules from any existing knowledge-representation system to be exported onto the Web.
>
> Adding logic to the Web {the means to use rules to make inferences, choose courses of action and answer questions} is the task before the Semantic Web community at the moment.

Hence, rules have a crucial part in comprehending the complete capabilities of the reasoning mechanism of the Semantic Web. Therefore, rule design is the prime focus of the current chapter. Ontologies are based on description logic, which is not capable of expressing chains of joins across different predicates which limits the types of inferences that can be drawn, and this is the major reason for the inclusion of the rule layer in Semantic Web architecture. Although ontologies support a limited form of reasoning, they are unable to provide the wide range of knowledge-based services for which the Semantic Web has been conceptualized.

6.4.5   RULES-BASED REASONING

Reasoning system based on rules comprises rules, rule language, and rule engine. The rule is the basis of inference execution in a knowledge-based system; in

Semantic Web, a rule represents a logical entailment among a set of formulas called premises and an assertion called a conclusion. Rule general formula is as follows:

$$A_1, A_2, A_n \rightarrow B$$

where A_i and B are atomic formulas. If any valid conclusion can be drawn by using a set of rules, then the inference solution provided by the rule is said to be complete; if no invalid conclusions can ever be asserted from the set of rules, then the inference solution provided by the ruleset is said to be sound.

Rules can be of different forms and can be used for different purposes [23]. For Example, R1 and R2 are two rules explained as follows.

Rule 1(R1): IF Automobile? A was manufactured between 1919 and 1930
THEN?A is Vintage Automobile
Rule 2(R2): ON request from Customer?C to buy Automobile
IF Customer?C is having valid documents
DO grant?C's request

R1 has two parts. The first part is IF part finds the automobiles manufactured between 1919 and 1930 and binds the variable A to automobiles. THEN part constructs new data based on the retrieved information. R2 has a distinct structure, ON part waits for an event i.e., the request to purchase an automobile from a customer to come in. The IF part R1 and R2 are similar i.e., perform checks for whether the customer willing to purchase an automobile has the required valid document or not. If the customer has the needed documents, then it grants the customer request.

Various types of rules specify diverse requirements to the implementors. For instance, Rule R1 is simpler than Rule R2. Rich language and complex execution semantics are needed by R2, because its ON part needs the support for the identification of events, and the DO part requires the provision for executing actions. Both R1and R2 have conditions in the IF part; this common feature is present in various kinds of rules suggesting a means to develop a rule interchange format comprising common parts (IF part) followed by different features of various types of rules.

6.4.5.1   Types of Rules

To gain more insights into various types of rules, they are categorized into three types; deductive rules, reactive rules, and normative rules as presented by Boley et al. [24].

Deductive rules
    Deductive rules are statements that specify how to use logical inference to derive knowledge from another knowledge. They are also termed as derivation rules in a business rule group, views in databases,

and constructive rules by logicians. They specify static dependencies between things that can be utilized to deduce extra implicit knowledge from explicit facts in the knowledge base. They are specified as head ← body, head specifies the data that has to be created or inferred and body performs a query on knowledge base data. Head and body parts of the rule usually have common variables. The body part holds the variable bound to data and the head part uses this binding for carrying out inference. In the earlier example, R1 is a deductive rule, whose body is presented by IF and head by THEN. Head and body have a common variable A. 'A' is a placeholder for an Automobile object. The rule syntax depends on the rule language.

Reactive rules

Reactive rules are used to specify reactive behavior and are used to implement reactive systems which automatically perform specified actions on the occurrence of events of interest happening or specific conditions become true. Other names for reactive rules are dynamic or active rules. Contrasting to deductive rules, reactive rules specify the changes in state. Events corresponding to variations in the state of the rules and world require additional changes that should occur in response to the occurring events. They generally have the forms of production rules or Event-Condition-Action (ECA) rules. ON Event IF Condition Do Action is the general form of ECA rules, it means that the action must be performed on the occurrence of an event, only if the condition hold. The event part is used for detecting the occurrence of the desired event and selecting the data item from their representation by binding the variables. Many types of events are represented from low-level to high-level and data-driven. A rule language can support a single occurrence of an event or composite events consisting of a temporal combination of events. Different types of actions can be carried by reactive systems such as performing updates on knowledge base data, modifying the rules in the ruleset, triggering new events, making procedure calls, different actions can also be combined to implement complex applications. Rule R2 is a reactive rule which specifies the event of a request of purchase of automobile from customer and uses C as a binding variable which is a placeholder for Customer object IF part specifies the condition to check whether the customer has the needed documents to purchase the automobile, DO part performs the action of granting the customer request provided that IF part condition is true.

Normative rules

Normative rules are used to put conditions on data or the logic of an application. They are used to make sure that the modifications made to the knowledge base do not create any inconsistencies and must abide by the business rules of a company. In business rule groups these rules are called structural rules and in databases, they are called integrity constraints. They define forbidden inconsistencies instead of inferencing novel knowledge. For example, every customer should have an exclusive identification number, two identification numbers for a customer indicate errors in the knowledge base. Either deductive rules or reactive rules are used for the implementation of normative rules. The choice of which rules to use for implementing normative rules depends on the existing support for diverse types of rules and applications.

6.4.5.2 Rule Languages

There are many rule languages available, each has its own syntactic and semantic peculiarities. In Semantic Web, the rule language allows to define relations that are not described by description logic, it even permits for sharing and reuse of prevailing rules across various organizations [25].

6.4.5.2.1 Features of Rule Languages

Different features supported by rule languages are discussed below.

- Logical quantifiers are used to specify the scope of variables in rule formulation. There aser two kinds of quantifiers: existential and universal quantifiers. Existential quantifiers require that predicates may be satisfied by at least one member, universal quantifiers specify that the predicates must be satisfied by every member in the domain.
- Logical operators are used to link the rule terms. Negation(not) and implication(if-then) conjunction(and), disjunction(or).
- Built-in functions compute the value of the variable in rules. Boolean functions, mathematical functions, and string functions are examples of Built-in functions.
- Data types can be supported in rule language as RDF data may hold statements with precise values which has an associated data type, these data types specify constraints on the allowed values. Numerical values such as float, integer, etc., and strings, data, and time data types are some examples of data types.
- Rules can be grouped into rule-set. A rule-set contains a set of related rules which can be used for extending the scope of variables in a rule that can be reused by another rule belonging to the same rule-set.
- Rules can also be named. A rule name specifies a name to the rule that can be used in proof explanation or retrieving rule by mentioning the name.
- Production operations specify the actions that must be performed by the inference engines when the conditions occur, actions are specified in the conclusion of a rule. These actions can change the knowledge base by asserting a fact, modification of fact, and withdrawing a fact and instance displaying. Few rule languages forbid removing of facts from the knowledge base.
- Rule languages must support interchange formats for exchanging rules in diverse rule-based systems. Rule languages must aid in exporting rules to JSON and XML.

Not all of the above-mentioned features are supported by all rule languages. Most of the rule languages support a set of features which are listed in Table 6.2.

6.4.5.2.2 Few Rules Languages

Semantic Web rule languages include those languages devised by different inference engines and those introduced by W3C. Below are the most commonly used rule languages in Semantic Web [26].

Rule ML

Rule ML stands for Rule Markup Language created by W3C. RuleML Markup Initiative created Rule ML to cover diverse ways of writing rules [27]. Rule ML is a markup language created to convey a lineage of Web rules in XML that can be used for deduction, reaction rewriting in addition to transformational, inferential, and behavioral

TABLE 6.2

Features of Rule Languages

	Rule Languages					
Features	**RuleML**	**FOL-RuleML**	**Notation3**	**RIF**	**Jena Rule**	**SWRL**
Rule name	SUP	SUP	N SUP	SUP	SUP	SUP
Rule-set	N SUP	N SUP	N SUP	SUP	SUP	N SUP
Production operations	Assertion, retraction	Assertion, retraction	Assertion	Assertion, retraction, modification	Assertion, retraction	Assertion, retraction
Built-in functions Supported (logic, math, string, URI, time list)	All	All	All	All	Math, string, time, list	All
Explicit logical operators	C, N	C, D, N	–	C, D, N	–	C, D, N
Explicit logical quantifiers	Universal	Universal, existential	Universal, existential	Universal, existential	–	Universal, existential
Data-types	Supported	Supported	Supported	Supported	Supported	Supported
Interchange format	XML	XML	–	XML	–	XML

SUP, Supported; N SUP, Not Supported; C, Conjunction; D, Disjunction; N, Negation.

tasks. Rule ML develops a hierarchical specification of diverse kinds of rules encompassing facts, queries, reaction rules, production rules, derivation rules, integrity constraints, in addition to transformation and tools to and fro other rule systems. Rule ML expresses the rules in XML. A group of statements in the rule must be written inside an atom in Rule ML. The topmost element is Imp (Implies) which characterizes the rule implication in IF-Then form. Datalog Rule ML is defined on individual constants and data constants with a non-compulsory attribute for URI.

FOL-RuleML

First-order Logic Rule Markup Language is a sublanguage of RuleML and is used for articulating First-order logic in Web [28]. Each rule in FOL-RuleML contains a set of statements termed an atom. Atoms consist of objects which can be variables, individuals, and relations among objects. FOL-RuleML has disjunction, conjunction, negation, universal and existential operators. For representing two-way implication rule equivalence using If and Only If is provided, along with rule implication (If-Then). Top-level elements in FOL-RuleML are 'Imp', 'Query' and 'Fact'. Imp element signifies rule implication. Fact element specifies that its inner entities are facts or axioms. Query element specifies that the inner entity is the query. The XML format is used for the Serialization of FOL-RuleML.

SWRL

Semantic Web Rule Language is founded on the union of OWL-Lite and OWL-DL sublanguages of OWL with Unary/Binary Datalog RuleML [29]. It was developed by W3C in 2004. SWRL proposal allows Horn-like rules to be joint by OWL axioms by extending the OWL axioms. Syntax of SWRL is quite like RuleML, both can interoperable. RuleML content can be a part of SWRL. Quantification and logical operators' support of SWRL is the same as RuleML. Rules in SWRL are of a form of an implication linking an antecedent(body) and consequent(head). The meaning of the rule can be presented as every time the conditions stated in the body hold, then the conditions stated in the head must also hold. The head and body can contain zero or more atoms. Multiple atoms in antecedent or consequent are treated as a conjunction of rule atoms. OWL property, IRI, a data range, or a built-in function can be a relation. Variable, a literal value, an individual, or a blank node can be objects. Built-in functions like mathematical, string functions are supported by SWRL. Like RuleML, XML presentation syntax is used for SWRL which combines elements from SWRL and RuleML namespaces. SWRL is syntactically simpler and easy to learn and it also supports the users in the creation and edition of SWRL rules by available rule editors. SWRL rules can also be used for querying OWL ontologies using SQL extended library of SQL influenced built-in functions. Basic counting, ordering, aggregation, and duplicate elimination operators are provided in the SWRL query library. It does not impose any restriction on how a reasoning process is performed using the rules, thereby any reasoning engines can be used to implement the reasoning process.

Notation3 or N3

N3 is an assertion, logic language and is a superset of RDF [30]. N3 encompasses the RDF data model with an addition of rule formulae, logical implications, variables, and functional predicates, in addition to textual syntax alternatives which are compact and readable compared to RDF/XML syntax. Statements in N3 are of the subject-predicate-object triple form. A node can also be a formula. N3 aim is to optimize the expression for logic and data in the same language, to make rules integrated easily with RDF, to permit quoting: statements about statements, to be symmetrical, readable, and as natural as possible. These aims are accomplished by URI abbreviation using prefixes that are bound to a namespace like XML. Quantification and variables are allowed in rules and by using simple and consistent context-free grammar. N3 supports both existential and universal quantifiers. Built-in functions can also be represented as predicates.

Jena rule

Jena inference engine uses its rule format in the form of Jena Rules [31]. Jena Rule language syntax is built on RDF(S), it makes use of a triple representation of RDF descriptions, similar to N3. A rule can be named in Jena. No formula notation is supported in the Jena Rule language and built-in functions are formulated in function terms. A rule for inference engine is described by a Java Rule object which consists of premises (list of body terms), conclusions (a list of head terms). Each term is either a call to a built-in function, triple pattern, an extended triple pattern. Rule name acts as an identifier of rule and rules can have optional reasoning mode specification. The forward rule has an arrow from the premises to conclusions, the backward rule has an arrow from the conclusions to premises and the hybrid rule has arrows in each direction from premise to conclusions and conclusions to premises.

RIF

RIF is a standard developed by the Rule Interchange Format (RIF) Working Group. RIF standard is designed for rule exchange between web rule engines [32]. RIF focuses on exchange instead of focusing on developing a general-purpose rule language. Rule exchange was documented as an unnerving task before the development of the RIF Working Group. The RIF Working Group proposed a family of languages named dialects which has precisely specified syntax and semantics. These dialects are planned to be uniform and extensible, where uniformity means that the dialects are expected to share existing syntax and semantics as much as possible, extensibility means it must be possible to design new dialects as an extension to the existing dialects with new elements representing the required additional functionality. The main concept behind RIF is that various systems will support syntactic mapping from their native language to RIF dialects and back. Mapping must be semantics-preserving and hence the rule sets can be communicated from one system to another on the condition that both systems support appropriate dialect which can be used for communication. Different dialects for RIF are Basic Logic Dialect (RIF-BLD), the RIF Core Dialect, and a Production Rule Dialect (RIF-PRD) which is only rules-with action dialects defined by Rule Interchange Format (RIF) Working Group. RIF is built on first-order logic. RIF rules can comprise both facts and rules. RIF defines extra rule elements like 'Group' that is used to group set of rules; 'Document' consists of one or more groups; 'import' is used for bringing other documents to create a multi-document object; 'external' is used to call externally defined document content. For writing common predicates shorthand symbols are provided in RIF.

An example rule is taken, and it is expressed different rule languages as follows: Example rule state that if 'a' is an automobile manufactured by 'c' company which manufactures specific automobiles such as Passenger vehicles only or Goods vehicles only then the type of vehicle information can be inferred from manufacturing company type. Here it is assumed that a manufacturing company either manufactures passenger vehicles or goods vehicles but not both.

Rule:
$$\forall a \forall c \forall v \; \text{manufacturedBy}(a, c) \wedge \text{manufactoringCompanyType}$$
$$(c, v) \rightarrow \text{isTypeOfVehicle}(a, v)$$

RuleML Datalog (implication) rule:

```
<Imp>
<head>
<And>
  <Atom>
    <Rel> manufacturedBy
</Rel>
    <Var>a</Var>
    <Var>c</Var>
  </Atom>
  <Atom>
  <Rel> manufactoring
CompanyType</Rel>
    <Var>c</Var>
    <Var>v</Var>
  </Atom>
</And>

</head>
<body>
  <Atom>
   <Rel> isTypeOfVehicle
</Rel>
```

In **FOL-RuleML** will be rewritten with an 'Implies' connective whose 'closure' attribute will be generated from the inner close attribute of the parent connective, usually an 'And':

```
<Implies closure="universal">
  <head>
  <And>
    <Atom>
      <Rel> manufacturedBy </Rel>
      <Var>a</Var>
      <Var>c</Var>
    </Atom>
    <Atom>
      <Rel> manufactoring
CompanyType </Rel>
      <Var>c</Var>
      <Var>v</Var>
    </Atom>
  </And>
  </head>
<body>
    <Atom>
```

```
        <Var>a</Var>
        <Var>v</Var>
      </Atom>
    </body>
  </Imp>
```

SWRL Rules concrete
syntax
```
<ruleml:imp xml:base="#">
  <ruleml:_body>
<swrlx:individual
PropertyAtom
swrlx:property="
manufacturedBy ">
  <ruleml:var>a</
ruleml:var>
    <ruleml:var>c</
ruleml:var>
  </swrlx:individual
PropertyAtom>
  <swrlx:individual
PropertyAtom
swrlx:property="
manufactoringCompanyType
">
    <ruleml:var>c</
ruleml:var>
    <ruleml:var>v</
ruleml:var>
  </swrlx:individual
PropertyAtom>
  </ruleml:_body>
  <ruleml:_head>
  <swrlx:individual
PropertyAtom
swrlx:property="
isTypeOfVehicle ">
    <ruleml:var>a</
ruleml:var>
    <ruleml:var>v</
ruleml:var>
  </swrlx:individual
PropertyAtom>
  </ruleml:_head>
</ruleml:imp>
```

```
    <Rel> isTypeOfVehicle
  </Rel>
    <Var>a</Var>
    <Var>v</Var>
  </Atom>
  </body>
  </Implies>
```

Notation3
```
@prefix : <#>.
@forAll?a?c?v
{?a : manufacturedBy?c.?c :
manufactoringCompanyType?v. } =>
{?a : isTypeOfVehicle?v. }.
```

Jena Rule
```
@prefix : <#>.
[ RuleVehicleType: (?a :
manufacturedBy?c) (?c :
manufactoringCompanyType?v) ->
(?a : isTypeOfVehicle?v) ]
```

RIF
```
Document (Prefix(<#>)
Group (ForAll?a?c?v
And(: manufacturedBy (?a?c) :
manufactoringCompanyType
(?c?v))
:- : isTypeOfVehicle (?a?v)))
```

6.4.5.3   Rule Engine

Rule engines or inference engines are application software used for extracting new facts from knowledge bases [33]. Rule engine for Semantic Web comprises of high-performance reasoning algorithms. Rule engine supports expressive rule languages having built-in functions. They provide interchangeable syntax and are compatible with the prevailing Semantic Web standards. A rule engine performs reasoning with the rules on data to derive new facts from existing. It checks for a data match in the rule condition, if a match occurs then the rule engine can make changes to the knowledge base such as fact retraction, assertion, or execution of functions to display the derived facts. Various factors affecting the performance of rule engines are production operation, join complexity, built-in functions, negation, and dependencies among rules.

6.4.5.3.1   Comparison of Rule Engines

Semantic Web rule engines are compared based on the following criteria as follows:

Reasoning strategies
> The procedures used by the rule engine to execute reasoning tasks are called reasoning strategies. There are two major reasoning strategies forward and backward chaining. Some of the rule engines also support hybrid that is a mixture of backward and forward chaining.

Reasoning algorithms
> Optimized algorithms and procedures can be supported for enhancing the performance of the rule engine.

Reasoning feature
> Other than performing inference a rule engine can support other functions like proof tracing and explanation, it can also provide various configuration options for reasoning.

Expressivity for reasoning
> Rule engines may provide various subsets of logic in reasoning. The common level of reasoning for Semantic Web data is RDFS, OWL subsets, and user-defined rules.

User-defined and built-in functions
> Some rule engines permit the creation of additional custom functions by users. Whereas built-in functions like mathematical functions, boolean, strings are supported by most of the rule engines.

Rule language support
> The rule engine may support standard Semantic Web Rule Languages e.g., RIF, N3 or define specific rule language like Jena rule language designed especially for Jena rule engine.

6.4.5.3.2   Rule Engine Examples

There are many rule engines available in the market but in this section, only the reasoners, which are available under freeware license and have user community are selected and discussed.

BaseVISor

BaseVISor 2.0.2 was designed by VIStology. BaseVISor is a forward reasoning engine that processes the RDF triples [34]. It has a highly effective forward reasoning engine which is enhanced for ontological and rule-based reasoning. It can be deployed as a web service, as a plugin to TopBraid Composer 3.4. It can be implanted into existing applications and can also be programmed to extend the functionalities as per the application-specific requirements. Application rules can be changed without changing the applications underlying compiled code base. BaseVISor allows data sources and business processes to map to each other through a common meta-model which results in a reduction of data redundancy. The inference supported are retraction, assertion, and procedural control functions. It also provides some optimizations like caching, indexing, and join optimization. It provides the users with many programmable built-in functions.

Jena inference engine

Jena inference engine is a tool of the Jena Semantic Web framework [35]. Jena Semantic Web framework is built on Java. Jena is an open-source application framework created for Semantic Web applications. Jena has some predefined reasoners such as an OWL-lite reasoner, an RDFS reasoner, and a general-purpose rule engine that permits users to write rules specific to their applications in the Jena rule syntax. General-purpose rule engine provides forward, backward chaining, and hybrid execution strategies. The general-purpose rule engine comprises two inner rule engines, a forward engine built on the RETE algorithm and a tabled Datalog backward engine. Many built-in functions are supported by the Jena rule engine and can also be extended as per the requirements of applications. Additional features supported are pre-processing, proof explanation, and proof tracing.

Euler YAP engine

Euler YAP Engine (EYE) is a backward reasoning engine, it works on backward-forward-backward chaining [36]. It is built on the Prolog YAP engine. EYE can support clients in several languages e.g., C#, Java, JavaScript, Prolog, and Python. It is configurable with various options for reasoning and supports user-defined plugins. It also provides useful information on reasoning such as debugging, warning logs, and proof explanation. Features supported by various rule engines are listed in Table 6.3.

The performance of a rule engine plays a vital role in choosing among several rule engines. One of the requirements of an inference engine is to perform reasoning on a huge volume of data, i.e., it must be highly scalable and efficient. The time required to perform reasoning must not be more. Apart from these general requirements, an application can also have other application-specific requirements because every application has its own required expressive power, data size, and dependencies among data. Rule engine performance can be measured in terms of the following [37] in Table 6.4.

TABLE 6.3
Features of Rule Engines

| | | Rule Engines | |
| | | Jena Inference | |
Features	**BaseVISor**	**Engine**	**Euler YAP Engine**
Rule language	RuleML, Its rule language	Its rule language	N3
Reasoning RDFS/OWL	OWL2-RL, R-entailment	RDFS, OWL-lite	Selected predicates of RDFS/ OWL
Programming languages/ application programming interface	Web Service, Java	Java	C#, Java, JavaScript, Prolog, Python, Web Service
Production operations	Assertion, retraction	Assertion, retraction	Assertion
Reasoning strategies	Forward chaining	Forward chaining Backward chaining	Forward chaining Backward chaining
Reasoning Algorithms	RETE	RETE and datalog backward engine	Euler path detection, prolog demand-driven Indexing

TABLE 6.4
Reasoning-Performance measure

S. No.	Reasoning-Performance Measure	Description
1.	Loading time	The time needed for loading the ontology in the rule engine.
2.	Configuration time	The time required for the configuration of the reasoning engine.
3.	Reasoning time	The time needed for performing reasoning on ontology.
4.	Memory consumption	The total memory consumed to perform inference. Memory consumption is important because if an ontology is large then it cannot be processed by the rule engine.

6.5 CONCLUSION

In this chapter, reasoning capabilities of the Semantic Web are highlighted with a special emphasis on reasoning based on rule engines. Web Ontology Language is revisited and supported DL logic reasoners are discussed. The importance of rules in the Semantic Web is emphasized, types are rules are demonstrated using Automobile ontology as an example. Similarities and differences between OWL DL reasoning and reasoning using rules are presented. In addition, rule languages and rule engines examples are reviewed and compared about the features supported. Followed by rule engine reasoning performance measures.

Future work includes devising an integrated framework to promote interoperability between different rule languages and rule engines. Rule priorities that help in selecting the best rule to fire when there is more than one applicable rule will be focused on in the future, to develop a conflict resolution approach that will resolve the conflicts based on rule priorities. Most of the researchers have exploited forward-chaining reasoning capabilities of a rule engine, much work is not done on backward-chaining reasoning capabilities and hybrid reasoning capabilities yet. In the future, backward-chaining and hybrid reasoning capabilities will be explored to enhance the reasoning potential of Semantic Web applications.

REFERENCES

1. Hebeler, J., M. Fisher, R. Blace, and A. Perez-Lopez. "Semantic web programming." *Notes* 3 (2009): 4.
2. Nguyen, Van. "Ontologies and information systems: A literature survey." *Defence Science and Technology Organisation* (2011). Available at: https://apps.dtic.mil/sti/pdfs/ADA546186.pdf.
3. Gruber, T. R. A translation approach to portable ontologies. *Knowledge Acquisition* 5, no. 2 (1993): 199–220.
4. Borst, W. Construction of engineering ontologies. PhD thesis, Institute for Telematica and Information Technology, University of Twente, Enschede, The Netherlands, 1997.
5. Genesereth, M. R. and Nilsson, N. J. *Logical Foundations of Artificial Intelligence.* Morgan Kaufmann, Los Altos, CA, 1987.
6. Antoniou, G., and F. Van Harmelen. "Web ontology language: Owl." In *Handbook on Ontologies*, pp. 67–92. Springer, Berlin, Heidelberg, 2004.
7. https://what-when-how.com/automobile/general-classification-of-automobiles/.
8. https://learnmechanical.com/classifications-or-types-of-automobile/.
9 Abburu, S. "A Survey on Ontology Reasoners and Comparison." *International Journal of Computer Applications* 57 (2012): 33–39. ISSN: 0975-8887.
10. https://en.wikipedia.org/wiki/Description_logic.
11. Bock, J., P. Haase, Q. Ji, and R. Volz. "Benchmarking OWL reasoners." In *ARea2008-Workshop on Advancing Reasoning on the Web: Scalability and Commonsense.* Tenerife, 2008.
12. Shearer, Rob, Boris Motik, and Ian Horrocks. "HermiT: A highly-efficient OWL reasoner." *In Owled* 432 (2008): 91.
13. Haarslev, V., K. Hidde, R. Möller, and M. Wessel. "The RacerPro knowledge representation and reasoning system." *Semantic Web* 3, no. 3 (2012): 267–277.
14. Horrocks, I. "Benchmark analysis with FaCT." In *International Conference on Automated Reasoning with Analytic Tableaux and Related Methods*, pp. 62–66. Springer, Berlin, Heidelberg, 2000.
15. Tsarkov, D., and I. Horrocks. "FaCT++ description logic reasoner: System description." In *International joint conference on automated reasoning*, pp. 292–297. Springer, Berlin, Heidelberg, 2006.
16. Sirin, E., B. Parsia, B. C. Grau, A. Kalyanpur, and Y. Katz. "Pellet: A practical owl-dl reasoner." *Journal of Web Semantics* 5, no. 2 (2007): 51–53.
17. Horrocks, I. "What are ontologies good for?" In *Evolution of Semantic Systems*, pp. 175–188. Springer Berlin Heidelberg, 2013.

18. Lisi, F. A. "Building rules on top of ontologies for the semantic web with inductive logic programming." *Theory and Practice of Logic Programming* 8, no. 3 (2008): 271–300.

19. Maluszyński, J. "On integrating rules into the semantic web." *Electronic Notes in Theoretical Computer Science* 86, no. 3 (2003): 1–11.

20. Rosati, R. "Integrating ontologies and rules: Semantic and computational issues." In *Reasoning Web International Summer School*, pp. 128–151. Springer, Berlin, Heidelberg, 2006.

21. Alferes, J. J., C. V. Damásio, and L. M. Pereira. "Semantic web logic programming tools." In *Principles and practice of Semantic Web reasoning*, pp. 16–32. Springer, Berlin, Heidelberg, 2003.

22. Berners-Lee, T., J. Hendler, and O. Lassila. "The semantic web." *Scientific American* 284, no. 5 (2001): 28–37.

23. Report: Rule-Based Intelligence on the Semantic Web, edited by Paul Smart, Electronics & Computer Science, University of Southampton, 2007.

24. Boley, H., M. Kifer, P.-L. Pătrânjan, and A. Polleres. "Rule interchange on the web." In *Reasoning Web International Summer School*, pp. 269–309. Springer, Berlin, Heidelberg, 2007.

25. Rattanasawad, T., K. R. Saikaew, M. Buranarach, and T. Supnithi. "A review and comparison of rule languages and rule-based inference engines for the Semantic Web." In *2013 International Computer Science and Engineering Conference (ICSEC)*, pp. 1–6. IEEE, 2013.

26. Mehla, Sonia, and Sarika Jain. "Rule languages for the semantic web." In *Emerging Technologies in Data Mining and Information Security*, pp. 825–834. Springer, Singapore, 2019.

27. Boley, H., B. Grosof, and S. Tabet. "RuleML tutorial." *The RuleML Initiative* (2005). Available at: http://ruleml.org/papers/tutorial-ruleml-20050513.html.

28. Boley, H., M. Dean, B. Grosof, M. Sintek, B. Spencer, S. Tabet, and G. Wagner. FOL RuleML: The first-order logic web language (2005). Available at: http://www.w3.org/Submission/2005/SUBM-FOL-RuleML-20050411/ and http://www.ruleml.org/fol/.

29. Horrocks, I., P. F. Patel-Schneider, H. Boley, S. Tabet, B. Grosof, and M. Dean. "SWRL: A semantic web rule language combining OWL and RuleML." *W3C Member submission* 21, no. 79 (2004): 1–31.

30. Berners-Lee, T., and Connolly, D. Notation3 (N3): A readable RDF syntax. W3C (2008). Available at: http://www.w3.org/TeamSubmission/n3/.

31. Available at: https://jena.apache.org/documentation/inference/

32. Hawke, S. and Polleres, A. RIF In RDF (Second Edition). W3C Working Group Note, Feb. 2013. Available at: http: //www.w3.org/TR/rif-in-rdf/.

33. Rattanasawad, T., M. Buranarach, K. R. Saikaew, and T. Supnithi. "A comparative study of rule-based inference engines for the semantic web." *IEICE TRANSACTIONS on Information and Systems* 101, no. 1 (2018): 82–89.

34. VIStology, "BaseVISor by VIStology, Inc.," 2013. [Online]. Available at: http://www.vistology.com/basevisor/basevisor.html

35. Apache Software Foundation, "Apache Jena - Reasoners and rule engines: Jena inference support," [Online]. Available at: https://jena.apache.org/documentation/inference/.

36. J. De Roo, Euler Yet another proof Engine, 2015. Available at: http://eulersharp.sourceforge.net/.

37. Van Woensel, W., N. Al Haider, P. C. Roy, A. M. Ahmad, and S. S. R. Abidi. "A comparison of mobile rule engines for reasoning on semantic web-based health data." In *2014 IEEE/WIC/ACM International Joint Conferences on Web Intelligence (WI) and Intelligent Agent Technologies (IAT)*, Vol. 1, pp. 126–133. IEEE, 2014.

7 Ontology Modeling
An Overview of Semantic Web Ontology Formalisms and Engineering Approaches with Editorial Tools

Olaide N. Oyelade

CONTENTS

DOI: 10.1201/9781003309420-7

7.1   INTRODUCTION

In previous chapters, it is understood that ontology supports the process of conceptualization based on domain knowledge. This chapter is aimed at describing the how of ontology modeling. The emphasis here is the need to understand the correct procedure for the formalization of knowledge in ontological structure. This formalism is often built on logical representations based on some forms of logic. The suitability and expressivity of such logical constructs underpinning the ontological structure provide for automated reasoning of ontologies. Studies have applied the ontological structure to knowledge-intensive systems in breast cancer diagnosis [1,2], coronavirus diagnosis [3], e-commerce [4], self-driving cars [5], and other domains. We however seek to dwell much on the fundamental approach for ontology modeling rather than its application.

In the parlance of knowledge engineering, ontology modeling, also referred to as ontology engineering, describes the elicitation and semantically association of all possible concepts and taxonomy which adequately describes a domain. We note that drawing up these concepts and taxonomy in a domain often requires domain experts who support the knowledge engineer with sufficient and correct information with a wider acceptance among other domain experts. Ontology modeling, therefore. refers to a representational approach supporting the formalization of the engineering of relevant domain knowledge based on the specification of concepts. The outcome of such modeling is to provision for the storage of the formalized knowledge and allow for interoperability among applications. The supported applications in turn are often applied to process managing tasks and those in decision-making domain. For concise and precise representation, ontology engineers often adopt a layered approach in the modeling process. The layered approach sometimes involves using separated ontological models to describe a domain in a manner that general domain and vocabulary are modeled in an ontology while some specific domain knowledge is model using another.

In whatever case, a good engineering procedure should result in an ontology that demonstrates the existence of semantic relationships among concepts and relations in the ontology structure. These ontological structures are often applied to knowledge representation using other human-understandable formats such as visualization approach. They could also be used in query answering systems and in interactive query generation, inferring new knowledge which is not explicitly declared in the ontology. Ontological models are also known to support reasoning through some reasoners such as Pellet or Hermit to check consistency and other related ontological operations. Information retrieval and data mining techniques have now been supported by well-modeled ontological structures. Some systems agglomerate domain ontologies to provide web service to other applications. These ontological structures are considered resourceful for different uses because of the nature of their application independence. In most cases, the complete conceptualization of a domain may appear daunting in the first round of ontology modeling. This often then requires that some maintenance strategy be planned into the process of ontology modeling to ensure that continuous efforts allow to sufficiently represent the domain.

The following sections are structured to provide the reader with an understanding of some ontology modeling methods. A good ontology engineer (OE) must adopt or develop a standard modeling method for the development of ontologies. This will ensure that an ontological structure serves its purpose and represents the domain in an accurate manner. Meanwhile, we described different standardized formalisms, constructs, and ontology languages available for OE with emphasis on the most relevant formalism. The editorial tools which support the use of these formal languages are detailed in the following sections. A comparative description of these modeling tools is presented to the reader to allow for an informed decision in the choice of editor.

7.2 ONTOLOGY MODELING METHODOLOGY

Approaching the task of ontology modeling without a concrete proven procedural approach will be unappealing. Considering that ontology modeling is an engineering task, it is required that well-crafted engineering paradigms are followed in provisioning users with useful ontological structures. We found several ontology modeling methods which have been presented in the literature. Therefore, an attempt is made to enumerate these methods and draw a conclusion through an argumentative approach on the selection of the appropriate approach. These modeling methods are often categorized into two as noted by [6]. These are:

a. A collaborative approach in comparison to a one-team approach. The thought around the collaborative approach is that it allows for developing a richer, complex, and well-thought-through ontological structure. On the other hand, the single-team ontology is believed to allow an OE to draw up an ontological structure with the hope that time is conserved using a stand-alone development method.
b. The second category compares the micro-level with the macro-level design of an ontology. The micro-level approach is focused on exploiting all details required to achieve formalization of ontology, while the macro-level approach is concerned with following a systematic process for the information system.

In the following sub-sections, we present some notable ontology modeling methodologies.

7.2.1 SAMOD

The simplified agile methodology for ontology development (SAMOD) was proposed in ref. [7]. This method adopts the agile approach of software engineering techniques. It is conceived from the understanding that ontologies can be developed using the small steps with a continuous iteration on those small steps until a complete and qualitative ontological structure is achieved. The SAMOD method describes three steps in the process of ontological design:

• Using some informal rules, the ontology engineer models domain ontology in collaboration with a domain expert who is expected to supply domain knowledge. The resulting ontology is then tested in some cases on the domain to test the correctness of its ontological structure. Once tested, gaps in the current ontology are addressed by iteratively following the same process to achieve a new and better version of the ontology.
• The former ontology is then combined with a new ontology that currently satisfies the agreed formalism between the domain expert and ontology engineer.
• The combined ontology is then further reviewed by the ontology engineer to ensure that the newly added part of the ontology does not render it inconsistent.

The interesting thing about this approach is the provision for revisiting the created ontology in such an iterative manner. This allows for adjusting wrongly modeled concepts and relations in the taxonomy-box (T-BOX).

7.2.2 TOVE

The TOVE ontology development method was derived from the principles learned while developing ontologies that were engineered for commercial and public enterprises in the project referred to as the Toronto Virtual Enterprise (TOVE) [8]. The authors proposed the following procedure in modeling ontology considering their experience in TOVE.

- Ontology modeling must first draw a clear understanding of the scenarios or domain for which it is anticipated for application or use.
- Second, thoughtful consideration of the availability of all required resources or expertise needed for the engineering process needs to be done.
- Specification of domain terminology is carried out based on information gathered through domain expert.
- An examination of domain terminology is carried out to ensure consistency.
- The specification of axioms or logical statements which describes the domain is specified and formalized by the ontology engineer.
- Completeness theorems are applied to the resulting ontological structure.

7.2.3 PATTERN-BASED

The pattern-based ontology development (PBOD) method is unique in its approach considering the use of an Excel spreadsheet for pattern encoding which in turn is used for ontology development [9]. Once the initial ontology is developed, the same encoded pattern is applied for the refinement task on the ontology. Meanwhile, the method allows ontology engineers to work with domain experts in a collaborative manner to substantiate the authenticity of the resulting ontology. The basic procedures required are:

- Pattern encoding in Excel spreadsheet document
- Encoded data is passed into a PBOD tool which then parses the Excel file for the ontology modeling process using Expert2OWL.
- The ontology output by the process is made available to the tooling process which allows the ontology to be queried using SPARQL and DL.

7.2.4 METHONTOLOGY

This is another ontology modeling method that uses a user-defined description for authoring ontology [10]. The user is expected to detail all activities needed in building the ontology. Following the IEEE specification for software development, the automated process can build the domain-specific ontology. The steps below are followed in the Methontology method:

- The planning phase provides users to list out all activities for ontology building.

- The specification of the aim of the ontology with a defined scope.
- The conceptualization of all concepts, taxonomy, and relations in the domain of consideration.
- The formalization of the conceptualization in the previous step.
- The integration of the newly created ontology with other ontologies for improved ontology.
- Ontology implementation based on a selected ontology formalism
- The evaluation of the implemented ontology for consistency checking
- The documentation of the correct and consistent ontology.
- The maintenance of ontology in the future to keep it usable.

7.2.5 Lexicon-Based

Another study [11] proposed the use of a lexicon-based approach in the modeling of ontology. The idea of lexicon arose from the need to consider a language-centered process. The language extended lexicon (LEL) mechanism is proposed for the ontology development method. The following are the procedures applied:

- Identification of relevant information to support the ontology development process.
- The concepts and relations resulting from the information sourcing are then classified in lexicon form.
- The resulting lexicon formed from the last step is verified and validated for correctness.
- Reordering of the verified lexicon is carried out in alphabetical order. This list is then further divided into concepts, relations, and axioms.
- The ontology is created from the lists in the last step and concepts are then rearranged according to their is-a relationship.
- The ontology is verified for consistency.

7.2.6 KACTUS

The KACTUS project birthed an ontology modeling method which has proven to be useful to ontology engineers [12]. This method approached the ontology development process using modular design, redesign strategy, and provision for reuse. Hence, it was proposed that these three perspectives be incorporated into ontology design to capture such meta-level viewpoint as it relates to a domain. The KACTUS team proposed the use of specification languages to support the process of ontology modeling. These are the STEP, EXPRESS, and Ontolingua languages to achieve increased expressivity while modeling the ontology. The following describes the formal procedure for the engineering task:

- Clear definition of the domain of application of the proposed ontology.
- Basic design of the ontology to enhance the engineering process.

- The resulting ontology is refined and structured by revalidating with domain specification.
- Continuous modular design of the current ontology and its reorganization is achieved through ontology refactor operation.

7.2.7 HORROCKS METHOD

Ian Horrocks proposed a simple and unambiguous step to ontology design. We found this method interesting and capable to guide even amateur ontology engineers. The development process follows:

- Knowledge gathering in the domain of ontology application
- Elicitation of domain concepts, taxonomy, and relations (also determine the domain and ranges of those relations).
- Concepts derived from the last steps should be categorized according to their type so that instances of each concept or class are further derived.
- Associate relations to the respective classes or concepts.
- Iterate over the last three steps to refine the ontology obtained.
- Determine if the ontology is consistent.
- Populate the ontology with instances for each class and associate their corresponding relations.
- Determine the consistency of the ontology.

7.2.8 DEVELOPING ONTOLOGY-GROUNDED METHODS AND APPLICATIONS (DOGMA)

Another method for ontology development as proposed by [13] is the DOGMA approach. The DOGMA approach provides integrated techniques in model theory, natural language processing, and database semantics to achieve ontology modeling. Their approach was presented as a framework that supports knowledge engineers in modeling formal ontology. The framework has two major components, one which enables the preparation of relevant domain knowledge and the other which supports the engineering. The following describes the steps for ontology engineering using the DOGMA approach:

- Define the aim of the ontology creation
- Investigate the feasibility of engineering such ontology and consider if there are resources worth deploying for the ontology engineering task
- Create a plan to achieve the project of the ontology creation so that the scope of the task is factored into the whole process
- Gather domain-related information to support the process
- Conceptualize the domain to collect and define concepts, taxonomy, and relations with the support of domain expert.
- Refine the last step so that the outcome of the conceptualization agrees with the domain of application of the ontology.

7.2.9 FUZZY ONTOLOGY DEVELOPMENT METHOD

Recently studies have identified the need to eliminate vagueness in ontology to support complete automation of reasoning. Several methods have been proposed to address this challenge, but fuzzy logic appears to have gained much attention. As a result, different fuzzy-ontology engineering methods have been proposed. For instance [14], described a new method for formalizing domain knowledge using fuzzy ontologies to reduce vagueness. They applied the method to developing ontology in medicine, specifically breast cancer ontology. The ontology modeling method consists of the following steps:

- Representation of domain concepts using fuzzy concepts
- Representation of domain properties or relations (object and data properties) using fuzzy relations and data types
- Domain-based formalism of fuzzy modifier, concepts, data types, and relations
- Selection of type of fuzzy logic to be used for the fuzzification of the ontology
- Check for consistency of the resulting ontology

Similar to this method are those of Three-Layer Fuzzy Ontology Model, s Fuzzy Ontology Map (FOM) [15], and Fuzzy Inference Mechanism (FIM) approach.

We have so far presented nine different ontology modeling methods though some other methods exist which share some similarities in terms of approach with those presented here. An example of such similarities is the need for a good comprehension of domain knowledge. Another is the need for conceptualization of the knowledge in the domain with the support of domain expert. Next is the process of formalizing or representing the ontology follows while the need to check for consistency of the ontology often ends the process. These steps are considered foundational in the ontology engineering process. However, one critical step which we felt is mostly not stated in the methods is the selection of the formalization language. To demonstrate that this step is important, we present a summary of different ontology formalisms and languages in the next section and show that they differ in representational approach.

7.3 ONTOLOGY MODELING FORMALISMS

The formalization of ontology design allows for the use of ontology in knowledge-intensive and intelligent-based applications. This formalization allows for interoperability and support machine readability of the ontology across different applications. Ontologies have been deployed to knowledge-intensive applications such as expert systems, semantic web, question and answering systems, eLearning applications, and several others. The formalization process of ontology helps the ontology engineer to select an ontology language that optimally supports some operations such as expressivity and automated reasoning.

Ontology models are therefore ontologies, specified using a language with rich and symbolic representational constructs to describe the design of the ontology. A variety of formalisms for representing ontology are considered in the following subsections with emphasis on features relevant to guide the selection of a language. Specifically, we highlight ontology representation languages, rule and query languages as well.

7.3.1 ONTOLOGY LANGUAGES

Ontology languages with their corresponding rule and interchange formats are considered and discussed in this section.

7.3.1.1 XML

The extensible markup language (XML) is widely used and supports so many applications and information interchange interfaces. The language was designed to promote the interoperability of machines through readable text files structured in a markup style. The language is made of a set of structured tags that gives interpretation to the text in a digital document. XML is considered an enhancement to the classical web page formatting language, hypertext markup language (HTML). The XML, combined with its affiliated XML schema is capable of describing the semantics and structure of a digital document in a domain. It shares similarities with mainstream ontology languages given its support for semantics and knowledge interchange. In addition, XML has been successfully used for auto-generating domain ontologies in other more expressive and high-level ontology languages such as OWL. Serialization and representational format in OWL often assume the structure of XML. Hence, the usefulness of XML in both knowledge interchange and ontology modeling operations. The popular semantic web cake often presents the XML as being foundational to the design of the semantic web. However, the language suffers from the inability from achieving inferencing operation. As a result, the XSL Transformation language (XSLT) is used in conjunction with XML to achieve inferential operation through the transformation of the XML structure. In addition, the XML is limited because it may require different syntax to represent meta-data. It also has a limitation in its capability to represent relationships in schemas and ontologies with respect to objects. This has further motivated the proposal for better ontology languages.

7.3.1.2 DAML and OIL

The DARPA agent markup language (DAML) and ontology inference layer (OIL) are two different languages, separately developed which later combine to form the DAML+OIL ontology language. The DAML was motivated by the success of the resource description framework (RDF) and extensible markup language, as a result, it became an extension of both languages. The language was aimed to address the lack of expressivity in those early ontology languages. It was a build-up from the DAML-ONT language which was also earlier proposed to support RDF to model complex class definitions. Meanwhile, OIL development was inspired by constructs in XML, RDF, and Open Knowledge Base Connectivity (OKBC). The OIL language was motivated by the need to add expressivity to ontology languages in its building block through advancing their syntax and semantics. It supports the use of constructs that are seen in the objected-oriented paradigm and frame-based ontologies and assumes its semantics from the description logic with the aim of improving automated reasoning on its ontologies. Combining DAML and OIL into DAML+OIL provided ontology engineers with a plethora of language constructs with high ontology expressivity supporting machine understandability. The pooled efforts of DAML and OIL soon became the foundation for the W3C's proposal for OWL language. Its wide acceptance of expressivity is associated with the use of class, properties, axioms, and definition of complex rules as constructs in the language. The DAML+OIL is limited by the reduction in decidability when the concept of cardinality is added to properties. Some of its shortcomings are now being remedied by OWL.

7.3.1.3 RDF+RDFS

Resources description framework (RDF) was proposed to argue XML by allowing for the representation of semi-structured information. The language allows for making triple-format statements with respect to resources in a directed graph-like format. Meta-data representation is being corroborated by the use of universal resource identifier (URI) to locate resources such as objects or digital documents on webpages. In addition, RDF is considered to provide a rectification mechanism in modeling domain knowledge. To support the semantic power of RDF, a lightweight schema language named RDF schema (RDFS) was proposed. The aim is to enable the combination of RDF/RDFS to provide a structured document with classes and properties. The use of RDF/RDFS is aimed to achieve is-a relationship or subsumption of classes and properties. This abstract mechanism also gives support to further specifying the domain and range of classes defined in an RDF/RDFS document. Considering its competitiveness with XML, the RDF/RDFS was able to allow a better knowledge interchange and interoperability among machines using the XML format and related formats. One of the limitations of RDF/RDFS is its inability to specify the type of object and relation its classes are allowed to have. Likewise, its properties are simply represented as individuals which may only have a subsumption feature describing it further. Also, RDF alone is insufficient to describe the semantics of a domain except it is being supported by RDFS which provides a definition of vocabularies to specify class and property hierarchies with statements modeled as triple format (subject-predicate-object). Data typing expression for enumerations is some of the drawbacks seen in RDF/RDFS. These limits the expressivity of the language thereby not allowing for a detailed and semantic-rich description of data. However, this weakness in RDF/RDFS makes it short of being considered a standard knowledge modeling tool for artificial intelligence (AI) systems, hence the motivation for better ontology languages.

7.3.1.4 KIF

Knowledge Interchange Format (KIF), as the name suggests, provides facilities for knowledge sharing and interchange among machines. Apart from its support of knowledge interchange, the KIF language promotes modeling domain knowledge in a semantically rich manner. Using logical (semantically predicate, and syntactically LISP) sentences, the language is able to allow ontology engineer to make knowledge representation decision explicit so that new constructs with semantic meaning are added to it. Although not an ontology language in itself, it provides ontology language with the mechanism for representing them in different formalisms. For instance, the combined use of KIF and frame ontology forms the language of ontolingua – a representation of frame ontology defining classes, slots, facets, and lists.

7.3.1.5 RIF

Whereas KIF is concerned with knowledge interchange, the rule interchange format (RIF) provides standard formats for exchanging rules among disparate rule-based systems. Its adoption and use in the semantic web by the W3C allow for information processing using the rule-based approach with representation in two dialects namely basic logic dialect (BLD) and production rule dialect (PRD). The use of rules in ontology modeling has proven to be supportive and enrich the knowledge representation formalism using ontological structures. The RIF was drafted to therefore achieve standardization of knowledge representation and information processing. It has been used particularly to support OWL increase semantics and enhance the richness of ontologies modeled using RIF. It can be represented using XML syntax with support for defining and merging rule sets. In addition to its support for ontology languages, RIF supports interchange among rules represented using different rule languages and inference.

7.3.1.6 SADL

The Semantic Application Design Language (SADL) is a well-rounded language with constructs for modeling domain-based semantics, rules, and queries. It is an English-like language that makes ontology representation in OWL and rule formalism simple and usable by non-experts in the knowledge engineering field. The Language is shipped with an editing tool that provides an all-in-one platform for controlling the knowledge represented using SADL so as to allow for validating and maintaining knowledge captured by SADL.

7.3.1.7 OWL

The acronym OWL was coined from web ontology language and represents ontology language widely used, derived from description logic (DL). The language was an improvement on the RDFS capability with the aim of enhancing expressivity and automated reasoning. OWL ontologies can be serialized into different ontology formats such as the XML, functional, triple, and Manchester formats. It provides a rich set of constructs for modeling classes, object properties, data properties, enumeration, union and disjoint classes,

---

TABLE 7.1

Basic Constructs of Ontology Languages

Construct	Notation
Classes	Sets, concepts, topics, types, kinds
Instances	Individuals, entities, members, records, things
Relation Properties	Relations, predicates
Descriptive Properties	Attributes, descriptors
Values	Data

---

subsumption of classes and properties, named and derived classes, and a lot of constructs. The language supports mainstream reasoners such as the Pellet, Hermit, Fact++, and ELK. The degree of expressivity of the OWL language is often decided by its variants which include the OWL Lite, OWL Full, and OWL DL. The OWL Lite variant is applied to building taxonomy or lexicon of a domain while the OWL DL is positioned for increased expressivity and support for DL-based reasoners. The OWL Full variants are all-encompassing in the sense that most constructs abridged in the OWL Lite and OWL DL variants are all combined in OWL Full. Meanwhile, there are two versions of the OWL language owing to the increased need to allow for better expressivity of the language. These are the OWL and OWL2 versions with the latter supporting more constructs that allow for expressing difficult expressions such as negative property assertions.

Of course, there are other ontology languages such as LOOM, OCML, FLogic, OKBC, Onto OKBC, XOL, and SHOE. The common constructs and components associated with the syntax of most ontology languages are listed in Table 7.1. More recent languages are however adding up new constructs to support complex axiom formalization in ontologies.

The motivation for the design of new ontology language has always been the need for complete automation of the reasoning process and increased expressivity. Expressivity is seen as key in enabling ontologies to be deployed for AI-driven systems. Expressivity allows for describing axioms and cases which are considered complex and highly logical for humans but support machines understandability and inferring new knowledge. Whereas the need for automation of reasoning is to allow reasoners to parse ontologies and check for consistency, ontology engineers must complement their ontology design task by selecting the best ontology language which will allow for accurate formalization of the domain knowledge considered in their projects. This is the aim of knowledge representation and reasoning.

7.3.2 Semantic Query and Rule Languages

We have previously mentioned the RIF protocol which supports the interchange of rules across platforms and the interoperability of rules coded using different languages. Although the RIF support ontology, its expressivity and constructs are suitable for the constructs of programming languages. To overcome this, semantic web-based rule and query languages have been proposed to allow them to demonstrate high-level suitability in representation and construct usage as seen in OWL. Examples are the SPARQL, SWRL, and SQWRL. There are quite a number of query and rule languages, we however chose those considering their suitability with OWL.

7.3.2.1   Query Languages

The language SPARQL Protocol and RDF Query Language (SPARQL) is designed to retrieve information contained in an RDF data model. Though there are several other XML query languages used in querying XML, considering that XML is lower in the semantic web technology stack compared to RDF, it becomes pertinent to have different query languages for RDF. The SPARQL leverages the RDF-based serialized representation of its knowledge to hide the complexity of language constructs, and execute queries. Moreover, there is no canonical RDF serialization for some OWL constructs. SQWRL was built upon SWRL and it replaces the consequent part of SWRL rule with a query. Using the built-in functions of SWRL, the SQWRL is able to extend the rule language as an effective query mechanism. Its advantage over SPARQL when used to query OWL implies that the OWL-based ontology need not be serialized to RDF since the constructs of OWL applies to those of SQWRL.

7.3.2.2   SWRL

The rule language, semantic web rule language (SWRL) works in handy with query languages such as SQWRL. This rule language is an appropriate rule language for Semantic web. It adds simple Horn-style rules to OWL Defeasible Logic DL in a syntactically and semantically coherent manner and supports the representation of rules using the constructs of OWL. Constructs used for classes, properties, and individuals in OWL are all supported by SWRL. Rules implemented with SWRL are saved in the same ontology in which they are being used and they can work with reasoners.

7.3.2.3   SQWRL

The query language squirrel or semantic query web rule language (SQWRL), is an extension of OWL/SWRL. This query language natively understands OWL and, like RDQL, allows the querying of individuals only. Its operators assume the forms of the SQL syntax that assumes the closed world (CWA) and has an extensible collection of built-ins, and that of arithmetic operations. There are two categories of query operators allowed in SQWRL: core and collection operators. The query language also allows for the comparison of values of different properties on the same individual but does not permit querying from different ontologies across the repositories. SQWRL Query Tab, a subsystem of SWRL API, provides a user interface approach to the execution of queries on ontologies through Protégé and a separate standalone application. Users of this tab deploy the use of SQWRL Query API which makes it possible for the queries to be executed, although, queries executed on the underlying ontology do not in any way modify it. SQWRL and DL Query are available for use in Protégé – an ontology authoring and editing tool.

7.3.2.4 Semantic Web Reasoners and Rule Engines

Rules are written to specify arguments that are valid based on the asserted statements in an ontology. Rules are necessary because they allow for the use of explanation facility and promotes the change of logic in such an easy way. One may then ask if the knowledge representation using ontology (like OWL) is not sufficient such that we will need to use rules? Let us note that the expressivity of ontology languages like OWL was controlled to allow for computation (decidability) on the ontology. So, rule languages were provided to argue the language expressivity of ontology since rules support decidability and good reasoning. As a result, when rules are added to ontology, the ontology can express statements that OWL may not provide but enriches the ontology through entailment provided by the added rule. For instance, the OWL language which is an application of description logic (DL) is based on classical logic while rule language like SWRL operates based on logic programming. The classical logic supports monotonicity (open-world assumption OWA, and negation as failure), while logic programming supports non-monotonicity (closed-world assumption CWA, and classical negation). Using both OWA and CWA features in a single-point knowledge representation is an advantage. Moreover, there is the benefit of additional knowledge generated by the inference rule. Thus, inferencing with rules connotes another form of knowledge representation in support of ontology.

While we note that rule inference support ontology enrichment means to represent some information which ontology language may not be adequate due to concern for decidability. Rule inference presents its own little challenges to the process. The minor challenge with rule inference is that except the rule is carefully specified, assertions derivable from applied rules might contradict existing restrictions in the ontology. In addition, often time, rule inference is separated from OWL classifier (OWL reasoner like Pellet and Hermit) and the derived knowledge arising from the entailment from the classifier might benefit the rule inference process. So, we see that while the result of the classifier might benefit the rule inference, the result of the rule inference might contradict the ontology thereby affecting the classification process. Hence both rule inference and reasoning in OWL must be done with this consideration. A summary of some rule engines and reasoners is presented in Table 7.2 to help to guide the selection of the best options to use with ontology, query, and rule languages.

The use of the rule engines and reasoners is beyond the scope of this chapter, as intended to provide readers with a plethora of options available for supporting the ontology engineering process when rules and consistency checking are required.

TABLE 7.2

Semantic Web Reasoners and Rule Engines

Rule Engine/Reasoner	Description
Drools	The Drools is a JBoss Rules engine which uses a modified form of the Rete algorithm.
Algernon	The Algernon engine provides access to ontology classes and instances with support for both forward and backward rule chaining.
InfoSapient	The use of fuzzy logic for rule expression is provided with this engine
RDFExpert	RDF-driven expert system shell. The RDFExpert software uses Brian McBride's JENA API and parser. A simple expert system shell that uses RDF for all of its input: knowledge base, inference rules, and elements of the resolution strategy employed. It supports forward and backward chaining.
Jena 2	Jena2 has a reasoner and inference engine components for modeling and executing rules on RDFS and OWL Lite ontologies.
Pellet OWL Reasoner	To support OWL DL ontologies, a Java-based and tableaux algorithm-based reasoner, Pellet is often used. It is available in Jena and OWL API libraries. Using this reasoned, the validation, entailments, and consistency of ontologies are checked.
SweetRules	To promote interoperability among rules and ontologies modeled with different languages, SweetRules is developed for use among ontology engineers. Built around the RuleML, SweetRules supports backward and forward inferencing.
IRIS Reasoner	Support for rule-based languages to carry out reasoning tasks is provided in Integrated Rule Inference System (IRIS) reasoning engine.
Jess	Jess is a RETE-based rule engine for rule encoding and execution using the Java platform. Rules are encoded using either the Jess rule language or XML-based representation.

7.4 ONTOLOGY MODELING TOOLS

Ontology modeling is an engineering process that follows the rigorous procedure in building knowledge representing the correct interpretation of domain representation. The engineering methodology process and the formalism language are critical in this procedure. However, implementation of the ontology design often presents a tedious task when handcrafted. As a result, several ontology authoring and editing tools have been proposed and applied to the task. An explorative discussion of the most relevant ontology editors and development environment from a plethora of such tools is presented. Each editor and development environment will be discussed around their performance, available features and plugging, popularity, and ontology model that they support. A comparative analysis of the modeling tools and editors is presented. This is expected to guide ontology engineers to make an informed decision in choosing ontology modeling tools. The following are some major features that characterize ontology engineering editors: authoring, editing, visualization, browsing of ontologies, transformation of ontologies into different representational formats, use of plugins to provide additional functionalities, merging of ontologies, importation, and exportation of ontologies into and out of an existing ontology, rule and query encoding, and also ontology annotation.

7.4.1 SWOOP

SWOOP follows the Model-View-Controller (MVC) approach to provide both a web OWL ontology browser and editor. The browser effect of SWOOP presents users with the means to interact with data and documents on the web. SWOOP was implemented using Java and is maintained in subversion data storage with support for other related plugins and APIs like WonderWeb OWL API. This ontology editor is credited to render its content with great hype in the use of multimedia and visible in almost all activities supporting easy navigation of OWL entities and comparing and editing related entities. The editor also provides referencing other ontologies and search within an ontology, and mechanism for annotating with multimedia. It supports multiple ontology browsing and editing and makes provision for six different plugins which aid its rendering capability. SWOOP enables these plugins to present both the ontology and entities in different presentations: ontology renderer for information and entity renderers for the concise format, OWL abstract syntax, turtle, and RDF/XML. SWOOP allows multiple ontologies to be opened at the same time with the possibility to save each ontology in different formats (XML, RDF, and Stanford) presentations of OWL. There is provision for semantic search for other ontologies which might be related with the current ontology that is been viewed. In addition, it enables data markup, ontology refactoring and debugging, collaborative annotation by using the Annotea framework so that another Annotea Server may be employed to publish and distribute the annotations created in SWOOP. Usually fixing errors or bugs in ontology could be a nightmare for the ontology engineer. Developers of SWOOP however help deal with these issues by building different rendering styles, formats, and icons to highlight key entities and relationships that are likely to be helpful to debugging process. Reasoning with SWOOP is possible using OWL reasoner which can be integrated to achieve consistency checking.

7.4.2 APOLLO

Apollo is a Java-based multi-lingua ontology engineering editor with support for English and Czech. It has a growing community of users, and of course a user-friendly interface with full consistency check during editing, open design based on views, special dialog for quick creation of anonymous instances, I/O plugin, export plugins to CLOS and OCML, and Java-based user-interface. Apollo is based on the Apollo knowledge-modeling tool, which was developed at KMI. In addition to being an ontology modeling tool, Apollo is also a genome annotation-editing tool. The open-architecture nature of Apollo's user interface allows for the implementation of additional views. Apollo internally model ontologies as frames in accordance with the internal model of the OKBC protocol and does a full consistency check while editing. Furthermore, Apollo provides means for setting user preferences and full undo of all edit operations. Apollo has been reported to contain relatively few bugs and does not follow the conventional ontology editors with graph view for ontologies. Similarly, Apollo does not support information extraction and multi-user capabilities or collaborative processing. Unlike Protégé and SWOOP which are bound to ontology languages like OWL, Apollo can be adapted to support different storage formats. Plugins characterizes most of the ontology editors, and this is also peculiar with Apollo. Some plugins that can be used in Apollo are OCML plugin, XML plugin, Bidirectional relation plugin, and Diff repository plugin. Web Apollo, a variation of Apollo, is a web-based genomic annotation editing tool. Sequencing technology in genomics necessitated this to analyze and describe the features of a genome in real time. Apollo has the support for users to work with large and robust ontologies with little or no inhibition; it also allows the ontologies to be imported and exported in different file formats. Apollo allows ontology consolidation and editing – enabling other users to participate in the development of ontology even if the active collaboration of ontology development is not supported. Recall the concept of inheritance from object-oriented programming; the frame-based representation of Apollo allows it to support ontologies to be inherited from other ontologies and can be used as if they were their own ontologies. A class represents a concept in the ontological meaning and an instance also represents an instance in the ontological terminology. However, a slot could represent a slot of a class or instance, and an instance of a facet implies a property of a slot, with the capability to create multiple facets for each slot. Other components that can be modeled in Apollo are relation, rule, procedure, function, and axiom.

7.4.3 OntoEdit

OntoEdit is a collaborative ontology development environment and was developed by Ontoprise. This editor supports both the developmental and maintenance stages of ontology and it is a collaborative tool allowing team members to participate in the phases of ontology development. The editor provides for support for this role of ontology creation and maintenance through both graphical means and as textual means. Like OntoStudio, OntEdit uses OntoBroker as a means for carrying out inference making. The peculiarity of OntoEdit lies in the layout of its components namely user interface, OntoEdit core, and Parser. These ontology development phases include requirements gathering and formalization, validating ontology with domain knowledge, and evaluation of ontology. The editing tool has a feature that allows all participating team members who contribute to the ontology developmental phases in the OntoEdit version to their contributions. This feature coordinates the requirements gathering for ontology development, continuous refinement of the ontology, and ontology evaluation capability. The requirement gathering phase brings together domain experts and ontology engineers to deliberate on fundamental issues in the ontology. Furthermore, the ontology refinement phase enables the team to create competency questions (CQ) which provides partial development of the ontology and then uses OntoEdit to complete the process. Finally, the ontology evaluation phase is aimed at completing the process of ontology development by modeling all information gathered during the requirement phase. OntoEdit enables all these three phases to be automated and parceled into the ontology development process. This explains the other view of the collaborative capability of OntoEdit. Meanwhile, whereas the former discussion of the collaboration prowess of OntoEdit promotes the concept of ontology sharing, this later description of collaboration with respect to OntoEdit enforces the development of ontology that adheres to standard. Moreover, when we consider the issues of ontology sharing – by ontology engineers – client/server architecture could as well have made this collaborative nature possible. This suggests the need for extending OntoEdit to client/server architecture. OntoEdit has a flexible plugin framework that extends its functionality in a number of ways as in the cases of import/export, inference making, and ontology brainstorming during collaboration. Some of these plugins are Mind2Onto, Instance editor plugin, Disjoined concepts plugin, OntoEdit Base Textual Rule Editor plugin, and Graphical Rule Editor plugin.

7.4.4 KAON

The Karlsruhe Ontology and Semantic Web framework (KAON) is a Java-based and open-source ontology authoring and management tool. In KOAN, ontology definition is closely described in an easy pattern which makes the ontology understood. This makes the ontology modeled to be a high-level representation. The editor, which is also a framework, provides services for ontology and metadata management and the interfaces needed to create and access web-based semantics-driven e-services. Meanwhile, the framework also plays the role of ontology engineering, visualization while at the same time allowing for interfacing services that apply semantic web technologies to e-commerce and B2B applications. KAON has an API which is at the core of its Model-View-Controller (MVC) architecture. The editor has a feature for persistent RDF storage optimized for concurrent engineering and direct access. In addition, KAON has a persistent mechanism similar to relational databases, and demonstrates similarity with the IBM SNOBASE system, and provides a single user and server-based solutions. KAON is a propriety extension of RDFS and may not be improved for KAON2 since KAON2 is typed for OWL-DL and F-Logic. KAON2 architecture which is an infrastructure for OWL-DL, F-logic, and SWRL has provision for file persistency and connectivity to relational database management systems. It also allows for using programming approach to manage OWL-DL, SWRL, and F-Logic ontologies. APIs that allow to read OWL XML and OWL RDF ontology formats, DIG, and reasoning provides means for KOAN2 to interface with semantic web applications and other ontology consumption points. A stand-alone server providing access to ontologies in a distributed manner using an inference engine and SPARQL query for answering conjunctive queries is provided. It has the capability to extract ontology instances from relational databases and manipulate F-Logic ontologies. For reasoning, KAON2 supports the description logic (DL) fragment of SHIQ(D) and has tolerance for a subset SWRL which is DL-safe.

7.4.5 WebOnto

WebOnto is both an ontology server and ontology editor that is mostly controlled through a browser-based interface coupled as Java with access to richer functionalities through web server. In WebOnto, editing instances and associated attribute values, concept taxonomies, and relations in the knowledge models are done through the browser interface. This interface can be rendered in the form of Tadzebao client and WebOnto for broadcasting and receiving clients. In this client/server-based architecture, ontologies are represented as open configuration and management layer (OCML), with support for collaborative ontology management. This collaboration is achieved through locking ontology to ensure that conflicting edit operations, such as two users changing the same OCML structure, is avoided. The architecture of the web server consists of LispWebHTTP server, Tadzebao server, OCML libraries, WebOnto server, Notepad library, and broadcaster. The client sides are the Tadzebao, WebOnto, WebOnto broadcaster and receiver clients. The LispWeb server passes on HTTP requests from the WebOnto and Tadzebao clients to the appropriate servers. The WebOnto server interacts with the ontology libraries represented in OCML and is responsible for loading requested libraries, running applications, converting OCML structures into ASCII strings which are then sent through HTTP. The Tadzebao server is responsible for maintaining the notepad library, which contains the content of all the notepads and delivering notepads to requesting Tadzebao clients.

7.4.6 OntoStudio

OntoStudio is another relevant ontology engineering environment/editor deplorable as a web application and standalone application. OntoStudio is an improvement on OntoEdit and presents as the front-end to OntoBroker inference-making engine which supports F-logic. This inference engine allows OntoStudio to carry out reasoning on the ontologies been engineered. The inference-making feature clearly indicates that OntoStudio will provide means for rule creation, execution/ invocation and application, and dynamic integration of data sources. The rules in OntoStudio can be modeled using graphical objects and textual mode. OntoStudio provides functionalities to aid intuitive ontology development and has support for matching data from different sources into a new ontology file. This ability to create ontology from different data formats makes OntoStudio to be very powerful knowledge modeling tool. Legacy knowledge warehouses will benefit from this ontology editor with support for moving the systems from the classical systems into the modern information systems. OntoStudio provides for interfacing with plugins and support for setting preferences to increase personalization. Rule Editors, Debugging, and Query are some of the plugins that have been used on

OntoStudio. As peculiar with knowledge modeling tools, OntoStudio has support for some knowledge modeling formats such as RIF and ObjectLogic which can be exported to DAML+OIL, F-Logic, OWL, RDF(S), and OXML. It supports collaborative development of ontologies, and for easy connection of databases or knowledge bases using a graphical mapping tool.

A variant of the editor is the lean version, Web OntoStudio, accessible through a browser. Web OntoStudio is based on client/server architecture, enabling large, distributed team members of ontology engineers to collaboratively build a knowledge base using the client end of the browser interface. This collaborative feature is made possible because of OntoBroker enhancement collaborative server. Like Apollo, OntoStudio has support for multilingual development, and the knowledge model is related to frame-based languages. There is a provision for API for accessing ontologies in an object-oriented fashion. Functionalities available for ontology engineers in OntoStudio include editing of the hierarchy of concepts or classes, creation of classes and properties, graphical and especially the textual creation of rules, and as well as the creation of advanced mappings.

7.4.7 ONTOLINGUA

Ontolingua, often considered an ontology language, is an ontology authoring and management tool large variety of user preferences for controlling the behavior of the user to facilitate the development and sharing of ontologies. It is a web-based ontology editor that aids with basic development tasks, facilitate collaboration, and ease of use. It provides a mechanism for writing ontologies in a traditional format so that they can be easily translated into a variety of representation and reasoning systems. Ontolingua system consists of a suite of tools for authoring ontologies from existing fragments of ontologies. The tool supports ontology inclusion, circular dependencies, and polymorphic refinement. It enables the creation and separation between the presentation and representation of ontology. One mechanism employed by Ontolingua in eliminating conflict arising from ontology merging is by making the symbol vocabulary of every ontology disjoint from the symbol vocabulary of all other ontologies. As earlier reported in some ontology tools, Ontolingua also provides users with features for collaboration. The collaborative work is facilitated by user and group access control through multi-user sessions. The collaborative nature of Ontolingua is made obvious by its use of the world-wide web to enable wide access and through the provisioning for users with the ability to publish, browse, create, and edit ontologies stored on an ontology server. Unlike other ontology editors, Ontolingua Server currently does not provide many inferential capabilities nor does it have some plugins to aid ontology reasoning. However, it does provide some support for using ontologies. One way to use ontologies developed with the Ontolingua Server is to translate the ontology into the representation language of another system such as CLIPS, LOOM, or Prolog. Ontolingua editor enables users to be able to browse, edit, publish, and collaborate on ontologies. The browsing environment is being able to quickly jump from one term in the ontology to another through the aid of hyperlinks. Each term or concept represented in the interface is displayed in an object-oriented or frame-based form. Also, Ontolingua Server provides features to assist with ontology maintenance. Another feature enabled in Ontolingua is support for splitting a large ontology into several smaller ontologies that may include each other. Ontology sharing is another feature in Ontolingua with the primary mechanism for supporting ontology reuse.

7.4.8 RDFEDIT

RDFEdit is a browser-based RDF tuple management tool that uses the Django application and other support packages and libraries. RDFEdit was aimed at managing legal resources metadata in the Publications Office of the European Union Cellar database through the enablement of their search, display, and edit. Using RDFEdit, creation and manipulation of RDF instances are made possible through Django, rdflib and jQuery DataTables. Other supportive libraries used by RDFEdit are Maven, Spring MVC, Jena for RDF processing, Jackson for JSON processing, RDForms, Bootstrap, and Jquery. RDFEdit provides its users to manipulate ontologies through the following: uploads and editing of RDF/XML files, use SPARQL-Endpoint-URI to parse a given number of triples, and to allow for exporting the edited data as RDF/XML. Through the RDForms and RDF template, the display and edit of RDF triple are made possible. In addition, RDFEdit uses the Semantic web search engine Sindice to transform literals into RDF URIs. RDFedit can import literals from other RDF storage and add the triples to the local graph being worked on by the user.

7.4.9 SWIDE

SWIDE is an ontology editor which integrates other applications to enable the creation and reuse of ontologies with the aim of problem-solving tasks. These problem-solving tasks are majorly knowledge-intensive systems. Building Semantic web application in SWIDE is made possible owning to an integration of tools with editing capability, logic reasoner, and semantic search engine. Furthermore, SWIDE is designed to support iterative development, where there are cycles of revision to the ontologies and other components of the knowledge-based system. Unlike protégé which has dedicated tabs for classes and instances, SWIDE makes it easier to work simultaneously with both classes and instances. Thus, a singular instance can be used on the level of a class definition, and a class can be stored as an instance. As common with other ontology editing software, SWIDE possesses features interesting which include semantic search engine, and consistency checker. In addition, it has the capability for ontology export in the following formats: RDF/XML, RDF/XML Abbreviated, N3, N-Triple, OWL/XML, OWL Functional and Manchester. Other features of SWIDE are ontology visualization, querying using SPARQL. The logic reasoner in SWIDE enables the ontology engineer to be able to check the truth of an instance against the underlying knowledge base. In addition to the features of SWIDE, ontology mapping capability is provided in SWIDE to support the process of harmonizing ontologies from different sources. SWIDE provides openings for some of these features through their APIs. These APIs includes Watson package APIs, Jena API, OWL API, OWL2Prefuse API and a query engine named ARQ.

7.4.10 DODDLE-OWL

Domain Ontology rapiD DeveLopment Environment – OWL extension (DODDLE-OWL) is a Java-based domain ontology construction tool specifically for OWL with support for the construction of both taxonomic relationships and non-taxonomic relationships in ontologies. It is an improvement on DODDLE-II which was built for ontology construction for typical knowledge systems. The ontology editing tool consists of five (5) modules namely: input, construction, refinement, visualization, and translation modules. The refinement module, using an interactive indicator, has the capability to keep the ontology engineer informed about the section of the ontology that needs to be refined. In addition to this feature, DODDLE-OWL provides semi-automatic generation of the initial ontology using domain-specific documents and general ontologies. DODDLE-OWL allows the reuse of existing ontologies that may be represented in Wordnet and other formats. Supportive libraries such as WordNet, Japanese WordNet, Japanese Wikipedia Ontology, Apache Jena, Lucene-Gosen, extJWNL, and GNU getopt – javaport, InfoNode Docking Windows, SQLite JDBC Driver, Apache POI, Apache PDFBox, Lombok, Stanford Parser, Silk Icons, and exewrap are used as plugins in the tool.

7.4.11 TopBraid Composer

TopBraid Composer is an eclipse-based ontology editor and manager with support for RDFS and OWL. It is sometimes referred to as TopBraid Suite with an accompanying application namely supporting Flex programming language so that users may use it as either a programming environment and/or an ontology editor. TopBraid Enterprise Vocabulary Net (EVN) provides a browser interface for a web solution to support developing and maintaining vocabularies. Within TopBraid Composer is the TopBraid Live which serves as a semantic web server in providing a quick platform for the development and deployment of ontology-driven applications. TopBraid Composer itself has good support for the visual modeling environment of ontology languages like RDF and OWL, with interoperability with UML databases and XML schema. RDFS, RDFa, GRDDL, and query language SPARQL, SWRL are supported in TopBraid Composer. In addition to supporting ontology languages mentioned earlier, file type such as **.sdb** and **.tdb** which support connecting data for Jena SDB or TDB database; **.oracle** which support connecting data for Oracle RDF database; **.s2r** which support connecting data for Sesame 2 databases; **.ag4** which support connecting data for AllegroGraph; **.rdfa** which support connecting data for RDFa sources; **.smg** which support connecting data for SPARQLMotion Graphs; **.mds** and **.rds** which support connecting data for Microdata and RDFa Sites).

7.4.12 IODT

Integrated ontology development toolkit (IODT) is an ontology development environment with a focus on OWL. It is used for semantic data management and allows for the integration of server with provision for ontology reasoning engine. IODT is based on Ontology Definition Meta model (ODM) and with features for OWL ontology storage, inference, and query using SPARQL and relational database management systems (RDBMS). Ontology visualization mechanism is allowed in IODT to enable ontology engineers follow the development process.

7.4.13 LINKFACTORY WORKBENCH

LinkFactory workbench is a terminology and ontology management tool that makes the creation and management of complex, multilingual, and formal ontologies possible.

7.4.14 PROTÉGÉ

Protégé is the most widely used open-source ontology editor with an intuitive nature. It was developed at Stanford University majorly for medical planning purposes. It is based and developed with Java and will run efficiently on computers having Java installed. In addition, it includes a programming development kit (PDK) which enables the design and deployment of knowledge base applications that rely on the knowledge base modeled in the Protégé. It has support for different plugins which aids the development and maintenance of ontology engineering process. Although Protégé was originally designed for frames-based ontologies in accordance with the Open Knowledge Base Connectivity protocol (OKBC), further development and versioning of Protégé have made it possible for ontology languages like RDF and OWL to be conveniently used. Another interesting thing about Protégé is its ability to handle ontology versioning and imports. The need for versioning became necessary given the evolving nature of ontology engineering, and the need for ontology import was necessitated by the promotion of ontology reuse. Also, Protégé has good support for ontology visualization – a feature that quickly gives the knowledge engineer a pictorial view and relationship in the ontology. More so, it provides support for project management, software engineering and other models which the ontology modeled may serve as underlying knowledge base. Protégé ontologies can be exported into a variety of formats including RDF(S), OWL, and Extended Mark-up Language (XML) Schema. Reasoning with ontology developed by the ontology engineer is one of the major tasks to be carried out on ontologies. There are number of reasoners that seamlessly interface with the Protégé. Some of them include Pellet, HermiT, Racer, and FaCT++. These are description logic reasoners and provide support for checking to classify and check the consistency of ontologies

The architecture of protégé allows the use of several auxiliary plugins. Some of these plugins includes OWL plugin which extends the Protégé itself and can be used to edit ontologies in the Web Ontology Language (OWL), to access description logic reasoners, and to acquire instances for semantic markup. Other plugins that are available are JSave which is an application-based plugin to generate Java class definition stubs for Protégé classes. Protégé Web Browser is a Java-based collaborative tool web application that allows users to share Protégé ontologies over the Internet. The WordNet plugin provides Protégé users an interface to WordNet knowledge base. The UML plugin is also a backend plugin that provides an import and export mechanism between the Protégé knowledge model and the object-oriented modeling language UML. DataGenie is an import/export plugin that allows reading and creating a knowledge model from relational databases using JDBC. Furthermore, the need for ontology editor like Protégé to be programmable and open-up outlets for other languages and shells paved way for rule engines like Jess and Algernon. JessTab is a rule engine that supports forward chaining and the use of production rules. Algernon also supports forward and backward chaining rules, much like CLIPS or JESS. However, it is easier to use this feature with Protege because it operates directly on Protege ontology knowledge bases rather than requiring a mapping operation to and from a separate memory space. Other Protégé plugins are the PROMPT plugin OWL-S Editor plugin. Often times, Protégé supports the use of tabs in other to expose access to the plugins. Some of these tabs in Protégé are the DL Query Tab, ManageHierarchyTab, OtherTerminologiesTab, NotesTab, UserDefinedTab, MetadataTab, IndividualsTab, ClassesTab and OWLViz tabs. The OWLViz tab shows a graphical representation of the class subsumption hierarchy. DL Query requires an arbitrary class description can be entered and the reasoner is queried for the sub/super classes, and inferred members.

Other ontology editing tools are: Anzo for Excel which includes an RDFS and OWL ontology editor within Excel for generating ontology; Be Informed Suite which is a tool for building large ontology-based applications which includes visual editors, inference engines, and export to standard formats; CmapTools is an ontology editor that supports numerous formats; Dot15926 is a Python-based ontology editor with provision for Python scripting and pattern-based data analysis for data compliant to engineering ontology standard; EMFText is an OWL2 Manchester syntax editor which is Eclipse-based with support for Pellet integration; Fluent ontology editor is built for OWL and SWRL supporting OWL, RDF, DL with functional rendering; Model Futures OWL editor known to be capable of working with very large OWL files; OBO-Edit built for editing biological ontologies; TODE.Net for ontology development and editing; others are KMgen Ontology editor, Knoodl, Menthor Editor, OWLGrEd, Thesaurus Master, Transinsight, and TwoUse Toolkit.

In Table 7.3, we present a comparative analysis of some of the ontology editors with highlight on features available with each tool.

Each of the ontology editor considered in this study was built with specific interests. These interests usually influence the features that are made available with the tool. While some editors were collaborative in nature, others were single-user driven. Several ontology editors were not designed strictly for an

TABLE 7.3

Comparison of Ontology Editors/Development Environment

Patient No.	Collaborative	Ontology Libraries	User Acceptance	Visualization	Ontology Storage	Web-based	Plugin Extensibility
Protégé	Yes	Yes	Very wide	Very Rich	Files & DBMS	Yes	Yes
SWOOP	Yes	No	Wide	No	HTML models	Yes	Yes
Apollo	No	Yes	Low	No	Files	No	Yes
OntoEdit	Yes	No	Medium	Yes	Files & DBMS	Yes	Yes
KAON	Yes	Yes	Low	Yes	Files & DBMS	Yes	Yes
WebOnto	Yes	Yes	Medium	Yes	Servers	Yes	Yes
OntoStudio	Yes	Yes	Wide	Yes	DBMS	Yes	Yes
Ontolingua	Yes	Yes	Medium	Yes	Server	Yes	Yes
RDFedit	No	No	Low	No	Server	No	Yes
SWIDE			Medium				Yes
DODDLE-OWL	No	Yes	Medium	Yes	DBMS	No	Yes
IODT	No		Low		Files, RDBMS	No	-
LinkFactory	Yes	Yes	Low	No	DBMS	Yes	Yes
TopBraid	No	Yes	Low	No	DBMS	No	Yes
Composer							

ontology language, allowing ontology engineers access to tools for use across different languages. A few of these tools were developed especially for some ontology languages like OWL, RDF, and F-Logic. Another feature to note when comparing these ontology authoring tools is the provision for visualization of ontologies. Inference generating and reasoning mechanisms are interesting features that ontology engineers will want to be shipped with ontology editing tools. These allow for building rich and AI-supportive ontologies. The continuous need for collaborative research cutting across multidiscipline is a major motivation for this feature to be embedded in ontology editing tools. Also, provision for the creation and execution of queries and rules as supportive tools in an ontology creation and management tool is an added advantage capable of influencing user acceptability. We found the Protégé tool to provide most of these features in an elegant manner. These features and some others present the ontology engineer with information for deciding on the editor suitable for authoring an ontology. Meanwhile, the domain of knowledge for an ontology, and prospective applications that will consume the ontology may play a major role in the selection of ontology editor. Another factor that may influence the choice of an ontology editing tool is its open-source nature and support for plugins. These plugins will help in extending the features of the editor.

7.5 CONCLUSION

In this chapter, we have focused on ontology modeling with emphasis on the methodology applied to the development process, the choice of ontology language and ontology editing tool. Discussion has also been presented to guide the ontology engineer in the process of selecting a formalism for representing an ontology and the appropriate use of editing tools. This demonstrates the research paradigm has been applied to the challenge of ontology engineering in several fields.

REFERENCES

1. O. N. Oyelade, E. A. Irunokhai, O. Najeem and A. Sambo, "A Semantic Web Rule and Ontologies Architecture for Diagnosing Breast Cancer Using Select and Test (ST) Algorithm", *Computer Methods and Programs in Biomedicine Update*, vol. 1, p. 100034, 2021.
2. O. Oyelade and A. F. D. Kana, "OWL Formalization of Cases: An Improved Case-Based Reasoning in Diagnosing and Treatment of Breast Cancer", *International Journal of Information Security, Privacy and Digital Forensics (IJIS)*, vol. 3, no. 2, pp. 92–105, 2019.
3. O. N. Oyelade and A. E. Ezugwu, "A Case-based Reasoning Framework for Early Detection and Diagnosis of Novel Coronavirus", *Informatics in Medicine Unlocked*, vol. 20, p. 100395, 2020.
4. O. N. Oyelade, S. Junaidu and A. A. Obiniyi, "Semantic Web Framework for E-Commerce Based on OWL", *International Journal of Computer Science Issues (IJCSI)*, vol. 11, no. 2, pp. 145–153, 2014.

5. S. K. Aina and O.N. Oyelade, "An Intelligence Aided VANET System: A Development of Its Ontology Knowledgebase and Rule set", *International Journal of Scientific & Engineering Research*, vol. 10, no. 3, pp. 749–758, 2019.

6. M. Keet, *An Introduction to Ontology Engineering*, Cape Town: University of Cape Town, 2021.

7. S. Peroni, "A Simplified Agile Methodology for Ontology Development", *OWLED 2016, ORE 2016: OWL: Experiences and Directions – Reasoner Evaluation*, pp. 55–69, 2017.

8. M. Gruninger and M. Fox, "The Design and Evaluation of Ontologies for Enterprise Engineering", *In Proceedings: DA94*, 1994.

9. K. Tahar, J. Xu and H. Herre, "Expert2OWL: A Methodology for Pattern-Based Ontology Development", *Stud Health Technol Inform*, pp. 165–169, 2017.

10. M. Fernández-López, A. Gomez-Perez and N. Juristo, "METHONTOLOGY: From Ontological Art Towards Ontological Engineering", *AAAI-97 Spring Symposium Series, Stanford University, EEUU*, pp. 24–26, 1997.

11. K. K. BreitmanJulio and C. S. d. P. Leite, "Lexicon Based Ontology Construction", *SELMAS 2003: Software Engineering for Multi-Agent Systems II*, pp. 19–34, 2003.

12. G. Schreiber, B. Wielinga and W. Jansweijer, "The KACTUS View on the 'O' Word", *IJCAI Workshop on Basic Ontological Issues in Knowledge Sharing*, p. 21, 1995.

13. P. Spyns, Y. Tang and R. Meersman, "An Ontology Engineering Methodology for DOGMA", *Applied Ontology*, vol. 3, pp. 1–27, 2008.

14. O. N. Oyelade and A. E.-S. Ezugwu, "Enhancing Reasoning through Reduction of Vagueness Using Fuzzy OWL-2 for Representation of Breast Cancer Ontologies", *Neural Computing and Applications*, vol. 34, pp. 3053–3078, 2021.

15. T. H. Lam, "Fuzzy Ontology Map – A Fuzzy Extension of the Hard-Constraint Ontology", *2006 IEEE/WIC/ACM International Conference on Web Intelligence (WI 2006 Main Conference Proceedings) (WI'06)*, pp. 506–509, 2006. doi:10.1109/WI.2006.85.

8 Semantic Annotation of Objects of Interest in Digitized Herbarium Specimens for Fine-Grained Object Classification

Zaenal Akbar, Wita Wardani, Taufik Mahendra, Yulia A. Kartika, Ariani Indrawati, Tutie Djarwaningsih, Lindung P. Manik, and Aris Yaman

CONTENTS

DOI: 10.1201/9781003309420-8

8.1   INTRODUCTION

Machine learning techniques, especially supervised ones, have been widely used in numerous applications, including natural language processing, biological image classification, stock market analysis, self-driving cars, and precision agriculture [1]. Furthermore, with the increased availability of data and computational resources, techniques in many applications have been widely accepted [2]. In biodiversity research, machine learning techniques have been used for multiple purposes, such as plant sciences, mainly on images of herbarium specimens [3]. Plant morphology, growth and development, the ecological interactions of plants with herbivores, and other related cases may now be analyzed quickly, precisely, and easily using machine learning. These techniques could also be used to assess the global change biology by processing herbarium data known to have biases over space, time, and phylogeny [4]. Machine learning can also help the basis of biological research, such as improving the accuracy of species identification [5]. The discovery of phenological patterns on unprecedented scales has also been made possible by machine learning and combined data from herbarium specimens and spatiotemporal data retrieved from specimen labels [6].

When it comes to herbarium specimens, it is widely known that this is the most valuable data source for biodiversity research. From about 3,100 herbaria around the World, there are a total of 390 million botanical specimens.[1] Extinct, uncommon, endemic, and common botanical specimens are all preserved in herbarium collections to serve as a reference for future study. Many efforts have been performed to digitize the specimens and share them, so they can be used by researchers all over the world. An example of the portals that provide digitized specimen data is the Integrated Digitized Biocollections (iDigBio) Portal, which has a collection of approximately 131 million digitized specimens.[2]

Without a doubt, these data sources offer a great potential to help us learn more about biodiversity, and machine learning techniques are one approach to achieve that. Multiple works have been done in this area. For example, to identify leaves and other components from digitized herbarium specimens [7], and to discover patterns and trends of plant-herbivores interactions [8–11]. Further biological analysis can also be performed, for example, to determine the driver of shifting interactions such as phenological change, distributional shifts, and urbanization [10]. Another example is using machine learning to segment plant tissues in herbarium specimen images and removes the background pixels [12].

For supervised training of machine learning models, large training datasets are necessary [13]. Data variability is an essential aspect of machine learning, especially for object identification or object classification tasks. Therefore real-world image variation is essential to ensure the results [14]. But on the other hand, using uncontrolled natural images might be seriously misleading, dangerously leading in the incorrect direction. Labeled data is made up of a large set of representative photos that have been labeled or highlighted with the relevant features. An extensive and accurate labeled dataset, the ground truth, is required for training the algorithm [15]. Recent advancements in machine learning approaches, such as deep learning techniques, can generate features automatically, which saves feature engineering costs, but in return, may require larger volumes of labeled data [16]. The size and quality of training datasets will affect the quality of the trained models [17]. Due to those requirements, it is now essential to share the raw data and, most importantly, the annotated/labeled data.

8.1.1 ANNOTATION OF DIGITIZED HERBARIUM SPECIMEN

Figure 8.1 shows two images of annotated digitized herbarium specimens. Each image consists of at least two types of information:

1. The sheet on the right bottom side of each image depicts the specimen's label, which includes information about the spatiotemporal dimension of the specimen, as well as when and where the specimen was collected. It also contains information about the person who has collected it and taxonomic data on the specimen.
2. The images themselves depict parts of a plant such as leaves, branches, flowers, and fruits. The yellow rectangles alongside the annotations indicate where the images have been annotated. The first image has six annotations, while the second image has 21 annotations scattered around their leaves. As can be seen, the annotated areas' size is not uniform, and two annotations can overlap the others.

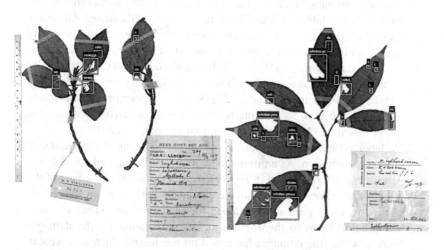

FIGURE 8.1 Two examples of annotated digitized herbarium specimens.

In this work, we focus on the second type of information, namely parts of the specimen that has been annotated. The annotations in this example will be used to identify objects of interest within images of herbarium specimens. Furthermore, using machine learning techniques, the annotations will be utilized to build a model for automatic object classification.

8.1.2 MOTIVATION

Dealing with such a vast amount of data requires a systematic approach. Machine learning techniques are sometimes unfamiliar to scientists who work with data. Having tools that assist scientists in processing data and revealing its potential is critical. For example, plant scientists may use an object detection application programming interface (API) to assist an object detection pipeline in detecting morphological features of a plant specimen [18]. A workflow for generating high-quality image masks for segmentation tasks can also be made available to scientists outside of the domain to help them [12]. Another example is providing a tool to annotate an image with specific pre-defined labels [11].

Despite the efforts to make digitized herbarium specimens more accessible, there are still several issues and challenges that remain. One challenge is finding efficient pre-processing techniques to produce a learning system that can deal with data collected from various sources. This problem arises as a result of data being scattered over multiple areas, systems, and applications. The "meaning" of the data may differ from one source to another, which may significantly impact the quality of the machine learning outcomes [2]. The discrepancies label in the annotated images is one example. In order to perform correctly, machine learning requires a sufficient amount of data training with the proper label. There are several challenges introduced by labeling data that have motivated our work, as follows:

1. Various annotation data formats. There are a number of picture annotation tools available, such as LabelMe[3] [19] and VGG Image Annotator (VIA)[4] [20], each with its own data format. There are a couple of widely used data formats, such as the JSON-based Common Object in Context (COCO) data format[5] and the XML-based Visual Object Classes (VOC).[6]
2. Labels tend to be noisy. Label noise can significantly impact the performance of deep learning models [21]. Since erroneous predictions might influence decision, so labeling requires domain expertise with a wide range of knowledge and the capacity to make precise label.
3. A label is applied individually without considering the relations to another annotation. As an illustration, in a digital image of a herbarium specimen, as can be seen in Figure 8.1, we can annotate the part of the plants like leaves and the damage caused by herbivores on the leaves. Both annotations are self-contained for each case and do not consider the relation between them. However, the annotation for the damages indicates that the damages happened on the leaves. Such that we can infer the relations that the damages are parts of the leaves.

The rest of the paper will be organized as follows: Section 8.2 lists and discusses a few related works and outlines our contribution. After that, our research methodology will be explained in Section 8.3. Finally, Section 8.4 presents our results and discusses our findings before concluding our work and explaining a few future works in Section 8.5.

8.2   RELATED WORK

In this section, we describe a few related works from various topics. Then, we align our work to two broad research topics: ontology and semantic annotation for biodiversity research. Finally, at the end of this section, we outline our contributions.

8.2.1   ONTOLOGY FOR BIODIVERSITY RESEARCH

Biodiversity science, like most other fields, has been flooded with a huge amount of data. Biological specimens that have been collected in various herbaria across the world have become a valuable data source for biodiversity research. Furthermore, technology has also introduced a new data source, so-called born-digital [22], in which data is collected digitally without collecting a specimen first. Combining these digital data with traditional data sources like data from in situ and remote sensors, community data resources, biodiversity databases, and data from citizen science have pushed this field into Big Data Era [23]. Besides the Volume of the available data, this research area also deals with the Variety and Veracity of the data. The two latter mentioned dimensions are related to and determine the quality of the data. Data collected by multiple organizations and stored in various formats are two examples of how semantic technologies such as ontology, could play a key role in big data analytics.

Ontologies play a crucial role in improving data aggregation and integration across the biodiversity domain in this area. They can be used to describe physical samples, sampling processes, and biodiversity observations that involve no physical sampling [24]. It has been predicted that the data will become less centralized, but the need for cross-species queries will become more common [25]. That is why ontologies would help scientists to achieve that. For example, Plant Ontology (PO) has been widely used to describe plant anatomy and morphology, as well as stages of plant development [26]. A simplified version of PO also can be used to drive a question answering dialog between non-expert users and a knowledge-base about Capsicum [27]. Another widely used ontology is the Darwin Core (DC), a standard for sharing data about the occurrence of life on earth and its associations with the environment [28]. It provides terminology for describing multiple types of information from an organism, such as taxonomic, location, and sampling protocol. It can be used to not only record the occurrence of a species at a specific time and location but also to manage alien species [29] and as a hub to connect data across multiple biodiversity information systems [30].

8.2.2 SEMANTIC ANNOTATION FOR BIODIVERSITY RESEARCH

Data annotation provides multiple advantages. First, it allows for data enrichment by embedding more information. Resulting in more efficient data discovery, so the data can be found, accessed, integrated, and re-used. Digital images of specimens from natural history collections must be bound to semantically rich data across museums to provide a unified user experience [31]. Semantic enrichment of herbarium specimen digital data would reduce the orthographic data variances, particularly for person and place names [32]. Further, data standardization and harmonization can be accomplished simply by using dictionary mapping to annotate data set columns [33]. It will improve discovery, interoperability, re-use, traceability, and reproducibility of the data.

Data annotation also can be used to encode knowledge into data by labeling objects of interest from texts, digital images, or videos as an example. The encoded labels can be used to enhance the data as well as serve as training material for automatic data extraction and classification. For example, multiple annotations of named entities in historic biological texts have been used to fine-tune a machine-learned classifier [34]. In that case, an annotation framework was developed based on a modified version of the Model-Annotate-Model-Annotate (MAMA) cycle. As a result, links between the exploration of biodiversity literature and document retrieval can be provided. The utilization of optical character recognition (OCR) would automate the acquisition process of herbarium specimen metadata [35]. Combined with other specimen image analysis services, the approach would provide a high degree of automation for information extraction from herbarium specimens. An approach for semi-automated extraction of named entities from natural history archival collections also can be developed by using the semantic annotation of the collections [36]. In that case, digital images of the field book will be annotated by drawing a bounding box over the image and attaching additional information. A tool for marine image annotation would increase efficiency and effectiveness in the manual annotation process [37]. In another case, images can be annotated by selecting features of interest (e.g., flowers, birds, or patterns) as tokens with bounding boxes from the images [38]. Finally, tokens can be associated with properties or traits (e.g., colors, behaviors) derived from pre-defined domain ontologies.

8.2.3 CONTRIBUTIONS

In line with the two research areas discussed above, we outline our contributions as follows:

1. In contrast with existing works that primarily introduced new annotation tools, our solution would utilize existing annotation tools. We focus more on the annotation format produced by each annotation tool and align them to a unified schema to achieve a common annotation. To the best of our knowledge, none of the existing works have looked into this situation.

2. Similar to other existing approaches, our solution would also allow annotating digitized images based on the entities and properties of pre-defined
ontology. But, instead of treating the annotations as independent entities,
our solution would also consider the relationship between them. These
inferred relations can be used to fine-tune the performed classification
tasks.

8.3 METHOD

This section describes our research methodology, followed by our method to
develop a uniform schema, and how to align annotations produced by various
tools to the schema through mapping rules.

8.3.1 METHODOLOGY

Figure 8.2 shows our research methodology. It consists of three main activities
as follows:

1. Schema development activity. It is a method for identifying and formalizing information taken from digitized herbarium specimens, as well as
its structure. Interviewing domain experts, in this case botanists, is how
the activity is carried out. As a result, a schema will be created that covers plant characteristics.
2. Labeling activity. It is a process of marking all regions of interest and
their relevant labels for each herbarium specimen image. Domain
experts can perform this task using any existing tools they usually use.
This activity produces two types of information, are regions of interest
(typically by using bounding boxes) and their relevant labels.
3. Mapping activity. It is a process to align marked objects of interest with
a pre-defined schema. As a result, at the end of the process, a unified
annotation will be obtained.

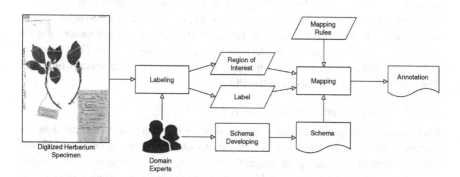

FIGURE 8.2   Research methodology.

It is important to note that the produced labels will be highly task-dependent. It will depend on the objective of the classification tasks at hand. For example, when the classification task considers only damages caused by herbivory, there is no need to label the other cause of damages. However, if we also consider the spatiality of the damages, it will be necessary to label parts of the plant, such as leaves where damages are found.

8.3.2 SCHEMA DEVELOPMENT

An annotation schema will be created to represent annotations uniformly across numerous annotation tools and to enrich the annotation by defining relationships between entities within an annotated object. The majority of the concepts are derived from Plant Ontology (PO), a widely used ontology for describing plant anatomy, morphology, and developmental stages [26].

8.3.2.1 Entities

As the representation of objects or things in the domain of interest, we identify several types of entities that can be extracted from digitized herbarium specimens.

8.3.2.1.1 Plant Morphological Entities

Plant morphology refers to the physical appearance of a plant. Physical characteristics of a plant that can be found in a herbarium specimen include leaves, fruits, flowers, and so on. Leaves, in particular, are an essential feature for species delimitation and recognition [18]. Further, trait information about area, perimeter, shapes, colors, textures of leaves can provide important insight into plant species' ecology and evolutionary history [7]. The morphology of plants, such as leaves, flowers, fruits, bark, and branches is ideal for image-based plant species identification [5]. The challenges lie in the diversity of similar features. For example, leaf morphologies of plants native to specific regions will be different from the plant from other regions [11]. Therefore, it is necessary to have a standardized way to share information about these features to ensure they can be consumed and appropriately re-used.

The PO adopt a Gene Ontology (GO) data model to cover flowering plants in general. The PO is ideal for sharing knowledge among scientists who know the issue but is not always understandable by non-expert users [27]. It has been widely used as a common reference ontology for plant structures and development stages. In the same vein, we created a modest but powerful ontology that met our requirements. We started with a simple ontology to improve Capsicum species literacy.[7] The ontology is a small subset of the PO.

8.3.2.1.2 Plant-Animal Interactions Entities

Plant-animal interactions can be seen as a process that has immediate and delayed effects on both entities that interact [39]. An interaction entails an enormous diversity of outcomes depending on interaction type (predation, symbiosis, parasitism, mutualism, commensalism) [40]. Herbarium specimens contain additional information, like nutrients, defense compounds, herbivore damage, disease lesions, and sign of physiological processes that capture ecological and evolutionary responses [4]. In particular, herbivory damage interactions data can be utilized to uncover various global change drivers across a diversity of insect herbivore-plant associations.

Examples of plant-animal interactions, in this case the damage caused by herbivory, are shown in Figure 8.1. As we can see, several damages on leaves can be identified visually. Based on these interactions, further analysis can be performed, for example, to identify different types of interactions and can be associated with different types of animals that contribute to the interactions [11].

8.3.2.2 Entity Relationships

Besides super-class and sub-class relationships, we would like to outline several critical other types of relationships.

Table 8.1 shows three essential relationships that can be used to represent how entities within digitized herbarium specimens are related to each other. While the first relation relates to morphological relationships, the last two relationships relate to spatial proximity between entities.

8.3.3 MAPPING RULES

Declarative mapping, and more precisely mapping rules, establish relationships between various schemas of multiple data sources and a common schema. The relationships align data elements from each source to a single common target, as well as the appropriate structural and data type transformations. Declarative mappings are available in a variety of formats, including the widely used and language-independent tables and spreadsheets [41] to highly structured formats such as R2RML[8] and RML [42]. RML[9] defines mapping rules from heterogeneous data structures and serializations other than relational databases, for example, CSV, XML JSON, to the RDF dataset.

8.4 RESULT

In this section, we list and discuss our results. First, we describe the characteristics of our dataset, followed by the description of our schema. After that, the mapping procedure will be explained before discussing our findings.

TABLE 8.1
Entity Relationship

No.	Relation	Meaning
1	part_of	This relation represents if an entity is part of another entity. It is a core relation to describe a part and its whole.
2	adjacent_to	This relation represents if an entity is in contact with or in spatial proximity to another entity.
3	located_in	This relation represents if the location of an entity is within the location of another entity.

8.4.1 DATASET

To construct our dataset, we digitized the herbarium collection from the Herbarium of Bogoriense.[10] The digitization began with the identification of a collection of specimens suspected to interact with insects, represented by various damages on the specimen. The dataset can be explained as follow:

1. We digitized herbarium specimens from *Excoecaria agallocha*, which belongs to the genus *Excoecaria of Euphorbiaceae*. In total, we obtained 244 specimen sheets.
2. From each digitized sheet, we asked experts to annotate them with three types of damages:
   a. pre-processing (damages that occur before the specimen were collected)
   b. during-process (damages that occur during specimen collection and or during drying and mounting on a sheet)
   c. insects (damages that occur due to insects during preservation)
3. The VGG Image Annotator was used to conduct the annotation. An annotator would draw a bounding box to indicate damage and choose one possible source of the damage.
4. It is important to note that the number of damages on each sheet can be multiple and of different types, overlapping damages in a sheet are also possible.

Table 8.2 shows the size of our multi-labeled dataset. Each sheet of herbarium specimen in our dataset can have several labels. In most cases, all labels occur. Our dataset has around three thousand labels in total, which are divided into three categories. The distribution of label count for each type of label is shown in Figure 8.3. The first three figures depict the distribution of labels for every type of damage, pre-processing, during processing, and insects, respectively. As we can see, most of the damages are up to six, mainly for every type of damage. Finally, the last image depicts the distribution of damages, and as we can see, the average damages are 6, 4, and 3 for every type of damage, respectively.

TABLE 8.2
Collection of Datasets

No.	Types of Labels	# Object of Interest
1	Damages during pre-processing	1,420
2	Damages during process	1,069
3	Damages caused by insects	882
	Total	3,371

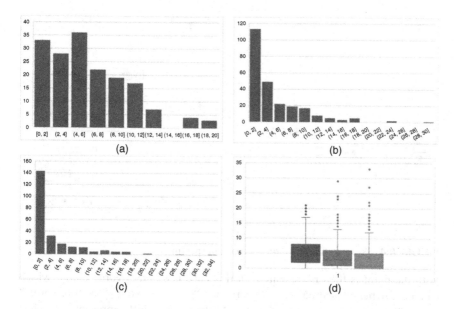

FIGURE 8.3 Label distribution based on damages (a) pre-processing (b) processing (c) insect (d) distribution of all three labels.

8.4.2 SCHEMA

We developed our schema by re-using terms from multiple existing ontologies as follows:

1. The plant structures, we adopted the schema from the OntoCapsicum,[11] an ontology to improve species literacy of Capsicum, which covers morphological characteristics of seeded plants in general [27].
2. The animals, since our research object is plants, we focused on herbivores, animals whose primary food source is plant-based. Herbivores can be classified further into frugivores (fruit eaters), granivores (seed eaters), nectivores (nectar feeders), and folivores (leaf eaters).
3. The interaction refers to the interaction between herbivores and plants which can be characterized by defense mechanism mark or damage on the specimens.
4. The interaction mark, the specimen's spatial dimension can be viewed as a mark of interaction (based on the image perspective).
5. The temporal dimension of the interaction represents the time when an interaction took place.

Figure 8.4 depicts our schema for annotating herbarium specimens, modeled using the Protégé Editor [43]. The core object in a herbarium specimen is the "Plant, " which will be described by the plant's main morphological traits, such as "Flower, " "Leaf, " "Fruit, " and "Stem". The animal that interacts

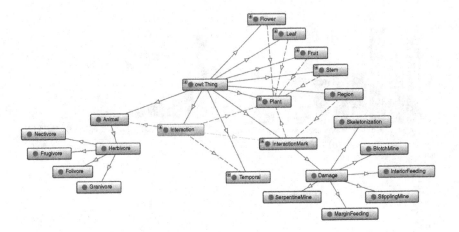

FIGURE 8.4   Schema for annotating herbarium specimens.

with the plant is then represented, in this case is a "Herbivore" entity, which can be further classified as a "Nectivore, " a "Frugivore, " a "Folivore, " and a "Granivore". Further, an entity "Interaction" will represent the interaction between a herbivore and a plant as reflected in a herbarium specimen. Next, "InteractionMark" can be used to identify the interaction on the specimen, which can be "Damages" or "DefenceMechanism". Multiple types of damages can be identified, such as "MarginFeeding", "InteriorFeeding", "Skeletonization", "BlotchMine" and so on. To reflect its spatial location inside the specimen image, the marks will be represented as "Region.". An interaction can also be annotated further with "Temporal" to represent when the interaction happened. The schema currently has 44 entities, 6 object attributes, and 14 data properties in its initial version.

8.4.3   DATA MAPPING

As mentioned earlier in this section, we would like to annotate the damages found on herbarium specimens and classify the damages based on the time when the damage occurred. Three classes were defined: prior-processing, during-processing, and after-processing (damages caused by insects).

```
<#InteractionMapping> a rr:TriplesMap;
  rml:logicalSource [
  rml:source "Batch-1-Updated.json";
  rml:referenceFormulation ql:JSONPath;
  rml:iterator "$._via_img_metadata.[*]"
];
  rr:subjectMap [
  rr:template "http://lipi.go.id/herbarium/{filename}";
  rr:class hso:Interaction;
];
```

```
rr:predicateObjectMap [
rr:predicate hso:hasRegion;
rr:objectMap [ rr:parentTriplesMap
<#InteractionMarkMapping>;
  rr:joinCondition [ rr:child "filename"; rr:parent
"filename"; ];
];
].
<#InteractionMarkMapping> a rr:TriplesMap;
  rml:logicalSource [
  rml:source "Batch-1-Updated.json";
  rml:referenceFormulation ql:JSONPath;
  rml:iterator "$._via_img_metadata.[*]"
];
  rr:subjectMap [
  rr:template "http://lipi.go.id/herbarium/
{filename}-{size}";
  rr:class hso:InteractionMark;
];
  rr:predicateObjectMap [
  rr:predicate hso:hasRegion;
  rr:objectMap [ rr:parentTriplesMap <#RegionMapping>;
  rr:joinCondition [ rr:child "filename"; rr:parent "shape_
attributes.id"; ];
];
].
```

Figure 8.5 shows a snapshot of our mapping rules using RML in combination with the Function Ontology (FnO).[12] We use JSON files generated by the VGG Image Annotator for the input files. We ran into a few issues when generating mapping for the files, mainly because the RML has limited support for nested data, such as nested objects in a JSON object [44]. We found it challenging to map objects in an array because no specific field can distinguish between members of the array and map them to their parent object. To solve this issue, we pre-processed the input files by inheriting the identification field from the parent object into members of the array in the child object. In this way, the parent object can be linked to each array member. As shown in Figure 8.5, several mappings are defined as a collection of "TriplesMap". A "logicalSource," a "subjectMap," and one or more "predicateObjectMap" are all defined in each definition. The relationships between entities were defined using "parentTriplesMap" from "objectMap" of the source to other related definitions.

The RMLMapper[13] is used to generate the annotation based on the updated input files. Figure 8.6 shows a snapshot of the annotation produced by the mapper, and Table 8.3 list the number of corresponding statements/triples. Out of 244 annotated digital herbarium specimens, we generated 21,058 triples using the current mapping rules. The majority of the triples are linked to the spatial information of damages detected on specimen images.

```
<#InteractionMapping> a rr:TriplesMap;
  rml:logicalSource [
    rml:source "Batch-1-Updated.json";
    rml:referenceFormulation ql:JSONPath;
    rml:iterator "$._via_img_metadata.[*]"
  ];
  rr:subjectMap [
    rr:template "http://lipi.go.id/herbarium/{filename}";
    rr:class hso:Interaction;
  ];
  rr:predicateObjectMap [
    rr:predicate hso:hasRegion;
    rr:objectMap [ rr:parentTriplesMap <#InteractionMarkMapping>;
      rr:joinCondition [ rr:child "filename"; rr:parent
"filename"; ];
    ];
  ].

<#InteractionMarkMapping> a rr:TriplesMap;
  rml:logicalSource [
    rml:source "Batch-1-Updated.json";
    rml:referenceFormulation ql:JSONPath;
    rml:iterator "$._via_img_metadata.[*]"
  ];

  rr:subjectMap [
    rr:template "http://lipi.go.id/herbarium/{filename}-{size}";
    rr:class hso:InteractionMark;
  ];

  rr:predicateObjectMap [
    rr:predicate hso:hasRegion;
    rr:objectMap [ rr:parentTriplesMap <#RegionMapping>;
      rr:joinCondition [ rr:child "filename"; rr:parent
"shape_attributes.id"; ];
    ];
  ].
```

FIGURE 8.5 A snapshot of our mapping rules.

```
@prefix hso: <http://lipi.go.id/herbarium/> .

hso:2021_03_17_11_52_560001.jpg a hso:Interaction;
hso:hasRegion hso:2021_03_17_11_52_560001.jpg-7739588 .

hso:2021_03_17_11_52_560001.jpg-7739588 a
hso:InteractionMark;
hso:hasRegion _:00106841-cc4a-4674-8851-1a72d4a1a828,
   _:0c688ab1-b924-4301-b0ed-264e73d314a5 .

_:00106841-cc4a-4674-8851-1a72d4a1a828 a hso:Region;
hso:height "362"^^xsd:int;
hso:width "410"^^xsd:int;
hso:x "3606"^^xsd:int;
hso:y "4373"^^xsd:int.
```

```
@prefix hso: <http://lipi.go.id/herbarium/> .

hso:2021_03_17_11_52_560001.jpg a hso:Interaction;
  hso:hasRegion hso:2021_03_17_11_52_560001.jpg-7739588 .

hso:2021_03_17_11_52_560001.jpg-7739588 a hso:InteractionMark;
  hso:hasRegion _:00106841-cc4a-4674-8851-1a72d4a1a828,
                _:0c688ab1-b924-4301-b0ed-264e73d314a5 .

_:00106841-cc4a-4674-8851-1a72d4a1a828 a hso:Region;
  hso:height "362"^^xsd:int;
  hso:width "410"^^xsd:int;
  hso:x "3606"^^xsd:int;
  hso:y "4373"^^xsd:int .

_:0c688ab1-b924-4301-b0ed-264e73d314a5 a hso:Region;
  hso:height "197"^^xsd:int;
  hso:width "330"^^xsd:int;
  hso:x "1151"^^xsd:int;
  hso:y "3260"^^xsd:int .
```

FIGURE 8.6   A snapshot of our annotation.

TABLE 8.3
Produced Annotation

No.	URL	# Triples
1	http://lipi.go.id/herbarium/Interaction	246
2	http://lipi.go.id/herbarium/InteractionMark	246
3	http://lipi.go.id/herbarium/Region	3,396
4	http://lipi.go.id/herbarium/hasRegion	3,642
5	http://lipi.go.id/herbarium/x	3,382
6	http://lipi.go.id/herbarium/y	3,382
7	http://lipi.go.id/herbarium/width	3,382
8	http://lipi.go.id/herbarium/height	3,382

```
_:0c688ab1-b924-4301-b0ed-264e73d314a5 a hso:Region;
hso:height "197"^^xsd:int;
hso:width "330"^^xsd:int;
hso:x "1151"^^xsd:int;
hso:y "3260"^^xsd:int.
```

8.4.4   DISCUSSION

We have introduced a solution to annotate images of herbarium specimens semantically. The annotation can be used as data training for herbivory classification tasks. Unfortunately, most of the existing tools perform the data labeling process individually and have its own format, which is different from the others.

As the consequence, finding a common technique for sharing annotations from one tool to another is difficult. In this paper, we demonstrated how our strategy may solve the annotation discrepancy and become the bridge for multiple tools. Furthermore, the relationship between objects inside annotated specimens is taken into account by our system. As a result, the generated annotation can recognize the objects in the specimens as well as their relationships.

Our schema represented the processes (i.e., the interaction between organisms) as entities. In this case, objects are integrated with the processes, where a process consumes inputs (i.e., parts of a plant) and produced output (i.e., interaction marks such as damages on leaves). This approach is similar to other modeling approaches in multiple domains. For example, the General Formal Ontology [45] in biological and biomedical areas, the OntoDM [46], and the Data Mining Optimization Ontology (DMOP) [47] for data mining processes. A biological interaction was viewed as a process, with actors (such as herbivory) performing actions (such as consuming the part of plant) and cause something (i.e., damages on parts of the plant). It is also important to mention that a set of processes is linked to spatial and temporal data. Each specimen contains the location where the specimen was collected for the spatial information. Objects within specimen images also include the region information where they are found. For the temporal information, each specimen also contains the time when it was collected. Furthermore, performed acts should be described in terms of when they occurred as points in time. Interactions should be distinguishable based on their places and time references. Moreover, as our annotation focused on multiple objects of interest on images, most of them are presented as nested objects. Therefore, we believe that preserving a unique identity for each item is essential for mapping definition. Instead of identifying objects by their position (such as index of an array), it will be better to have an attached identification scheme for consistency throughout the mapping process. This object of interest identification approach would make it easier to keep track of relationships between herbarium specimens, plant parts, and objects of interest contained inside the parts of plant.

8.5   CONCLUSION

Herbarium specimens have become the primary data source for biodiversity research. Multiple organizations collected specimens from various locations and kept them in herbaria all around the globe. The attempt to digitize specimens and share them publicly has piqued the interest of scientific communities, allowing scientists from all around the world to analyze them. As a result, millions of digitized herbarium specimens are available online. From this digital data collection, images are the primary data, accompanied by labels on the specimens (such as taxonomy, spatial information about where the specimens were collected, temporal information about when the specimens were collected, the person who has collected the specimen).

A herbarium specimen holds great valuable information, including spatial and temporal information about the specimen, and other additional information that can be found from it. For example, plant structures (such as the shape of leaves, and stems) are characteristics that can be extracted from images of herbarium specimens. This type of information can be used to develop intelligent applications, such as a computer vision-based application for automatic species identification. More than that, images of the specimen could also hold the interaction between plant and animal as indicated by the mark of damages or defense mechanisms found on the specimen. The latter type of information can be used further for advanced analytics, such as analyzing invasive species, and global warning indicators.

When the number of digitized herbarium specimens grows exponentially, scientists optimize the data analysis process by automating most of the steps. Artificial intelligence techniques such as machine learning are one option to make it happen. Machine learning techniques, especially supervised ones, require data training to discover the patterns from the data and use them to perform data classification tasks. In this case, machine learning algorithms would use the pattern to classify unknown data. It is widely known that a machine learning algorithm needs a sufficient amount of data training with high quality to produce the best model with highly accurate results. Unfortunately, this kind of data is not always publicly available. Multiple labeling technologies were used to create the majority of the shared data. Therefore, the challenge has shifted from data acquisition to labeling data in cases when there is a label discrepancy.

This work proposes a method to produce high-quality digitized herbarium specimens using semantic annotations. Annotations will be used to identify objects of interest in images and how they are related to one another. The annotation was achieved by employing an ontology to uniformly represent labels of images in a consistent way that is aligned with the goal of any classification tasks at hand. We started by identifying entities found in herbarium specimens before defining relations among them. As a result, the constructed ontology can uniformly represent objects of interest in digitized herbarium specimens. After that, we aligned the ontology with labels generated by multiple image labeling tools through declarative mapping rules. As a result, annotations from digitized herbarium specimens were obtained.

We evaluated our proposed method for an herbivory classification task, where images were labeled with three pre-defined classes. During the mapping process, we discovered that annotations were successfully created with only minimum pre-processing. The main goal of the evaluation was to investigate if we could extract data for machine learning tasks while maintaining the links between objects in the annotation that needed to be semantically represented. Furthermore, by using a shared ontology and declarative mapping rules, we can accommodate a variety of categorization tasks. This work is our first attempt to encode knowledge into machine learning workflows, which remains under-investigated to the best of our knowledge. Moreover, this work is another endeavor to contribute to big biodiversity data management and foster research in this area to move forward faster. In the future, we would like to extend our work by including numerous types of annotation in a variety of categorization tasks across diverse domains.

ACKNOWLEDGMENT

This research was supported by the Research Organization for Life Sciences, National Research and Innovation Agency, Indonesia under the national research priority program "Exploration and Utilization of National Biodiversity", the fiscal year 2021. We would like to thank all members of the Knowledge Engineering Research Group, Research Center for Informatics and the Plant Systematic Research Group, Research Center for Biology for their valuable suggestions and feedback.

REFERENCES

1. S. Dargan, M. Kumar, M. R. Ayyagari, and G. Kumar, "A survey of deep learning and its applications: A new paradigm to machine learning," *Arch. Comput. Methods Eng.*, vol. 27, no. 4, pp. 1071–1092, Sep. 2020. doi:10.1007/s11831-019-09344-w.

2. J. Qiu, Q. Wu, G. Ding, Y. Xu, and S. Feng, "A survey of machine learning for big data processing," *EURASIP J. Adv. Signal Process.*, vol. 2016, no. 1, p. 67, May 2016. doi:10.1186/s13634-016-0355-x.

3. P. S. Soltis, G. Nelson, A. Zare, and E. K. Meineke, "Plants meet machines: Prospects in machine learning for plant biology," *Appl. Plant Sci.*, vol. 8, no. 6, Jun. 2020. doi:10.1002/aps3.11371.

4. E. K. Meineke, C. C. Davis, and T. J. Davies, "The unrealized potential of herbaria for global change biology," *Ecol. Monogr.*, vol. 88, no. 4, pp. 505–525, Nov. 2018. doi:10.1002/ecm.1307.

5. J. Wäldchen and P. Mäder, "Machine learning for image based species identification," *Methods Ecol. Evol.*, vol. 9, no. 11, pp. 2216–2225, Nov. 2018. doi:10.1111/2041-210X.13075.

6. K. D. Pearson et al., "Machine learning using digitized herbarium specimens to advance phenological research," *Bioscience*, vol. 70, no. 7, pp. 610–620, Jul. 2020. doi:10.1093/biosci/biaa044.

7. W. N. Weaver, J. Ng, and R. G. Laport, "LeafMachine: Using machine learning to automate leaf trait extraction from digitized herbarium specimens," *Appl. Plant Sci.*, vol. 8, no. 6, Jun. 2020. doi:10.1002/aps3.11367.

8. E. K. Meineke, A. T. Classen, N. J. Sanders, and T. Jonathan Davies, "Herbarium specimens reveal increasing herbivory over the past century," *J. Ecol.*, vol. 107, no. 1, pp. 105–117, Jan. 2019. doi:10.1111/1365-2745.13057.

9. C. Beaulieu, C. Lavoie, and R. Proulx, "Bookkeeping of insect herbivory trends in herbarium specimens of purple loosestrife (Lythrum salicaria)," *Philos. Trans. R. Soc. B Biol. Sci.*, vol. 374, no. 1763, p. 20170398, Jan. 2019. doi:10.1098/rstb.2017.0398.

10. E. K. Meineke and T. J. Davies, "Museum specimens provide novel insights into changing plant–herbivore interactions," *Philos. Trans. R. Soc. B Biol. Sci.*, vol. 374, no. 1763, p. 20170393, Jan. 2019. doi:10.1098/rstb.2017.0393.

11. E. K. Meineke, C. Tomasi, S. Yuan, and K. M. Pryer, "Applying machine learning to investigate long-term insect–plant interactions preserved on digitized herbarium specimens," *Appl. Plant Sci.*, vol. 8, no. 6, Jun. 2020. doi:10.1002/aps3.11369.

12. A. E. White, R. B. Dikow, M. Baugh, A. Jenkins, and P. B. Frandsen, "Generating segmentation masks of herbarium specimens and a data set for training segmentation models using deep learning," *Appl. Plant Sci.*, vol. 8, no. 6, Jun. 2020. doi:10.1002/aps3.11352.

13. A. Krizhevsky, I. Sutskever, and G. E. Hinton, "ImageNet classification with deep convolutional neural networks," *Commun. ACM*, vol. 60, no. 6, pp. 84–90, May 2017. doi:10.1145/3065386.

14. N. Pinto, D. D. Cox, and J. J. DiCarlo, "Why is real-world visual object recognition hard?," *PLOS Comput. Biol.*, vol. 4, no. 1, pp. 1–6, 2008. doi:10.1371/journal.pcbi.0040027.

15. N. Zhou et al., "Crowdsourcing image analysis for plant phenomics to generate ground truth data for machine learning," *PLOS Comput. Biol.*, vol. 14, no. 7, pp. 1–16, 2018. doi:10.1371/journal.pcbi.1006337.

16. Y. Roh, G. Heo, and S. E. Whang, "A survey on data collection for machine learning: a big data – AI integration perspective," *IEEE Trans. Knowl. Data Eng.*, vol. 33, no. 4, pp. 1328–1347, Apr. 2021. doi:10.1109/TKDE.2019.2946162.

17. C. Shorten and T. M. Khoshgoftaar, "A survey on image data augmentation for deep learning," *J. Big Data*, vol. 6, no. 1, p. 60, Dec. 2019. doi:10.1186/s40537-019-0197-0.

18. T. Ott, C. Palm, R. Vogt, and C. Oberprieler, "GinJinn: An object-detection pipeline for automated feature extraction from herbarium specimens," *Appl. Plant Sci.*, vol. 8, no. 6, Jun. 2020. doi:10.1002/aps3.11351.

19. B. C. Russell, A. Torralba, K. P. Murphy, and W. T. Freeman, "LabelMe: A database and web-based tool for image annotation," *Int. J. Comput. Vis.*, vol. 77, no. 1, pp. 157–173, May 2008. doi:10.1007/s11263-007-0090-8.

20. A. Dutta and A. Zisserman, "The VIA annotation software for images, audio and video," in *Proceedings of the 27th ACM International Conference on Multimedia*, 2019, pp. 2276–2279. doi:10.1145/3343031.3350535.

21. D. Karimi, H. Dou, S. K. Warfield, and A. Gholipour, "Deep learning with noisy labels: Exploring techniques and remedies in medical image analysis," *Med. Image Anal.*, vol. 65, p. 101759, 2020. doi:10.1016/j.media.2020.101759.

22. R. Kays, W. J. McShea, and M. Wikelski, "Born-digital biodiversity data: Millions and billions," *Divers. Distrib.*, vol. 26, no. 5, pp. 644–648, May 2020. doi:10.1111/ddi.12993.

23. S. S. Farley, A. Dawson, S. J. Goring, and J. W. Williams, "Situating ecology as a big-data science: Current advances, challenges, and solutions," *Bioscience*, vol. 68, no. 8, pp. 563–576, Aug. 2018. doi:10.1093/biosci/biy068.

24. R. L. Walls et al., "Semantics in support of biodiversity knowledge discovery: An introduction to the biological collections ontology and related ontologies," *PLoS One*, vol. 9, no. 3, p. e89606, Mar. 2014. doi:10.1371/journal.pone.0089606.

25. R. L. Walls et al., "Ontologies as integrative tools for plant science," *Am. J. Bot.*, vol. 99, no. 8, pp. 1263–1275, Aug. 2012. doi:10.3732/ajb.1200222.

26. L. Cooper et al., "The plant ontology as a tool for comparative plant anatomy and genomic analyses," *Plant Cell Physiol.*, vol. 54, no. 2, p. e1–e1, Feb. 2013. doi:10.1093/pcp/pcs163.

27. Z. Akbar et al., "An ontology-driven personalized faceted search for exploring knowledge bases of capsicum," *Futur. Internet*, vol. 13, no. 7, pp. 1–17, 2021. doi:10.3390/fi13070172.

28. J. Wieczorek et al., "Darwin Core: An evolving community-developed biodiversity data standard," *PLoS One*, vol. 7, no. 1, p. e29715, Jan. 2012. doi:10.1371/journal.pone.0029715.

29. Q. Groom et al., "Improving Darwin Core for research and management of alien species," *Biodivers. Inf. Sci. Stand.*, vol. 3, p. e38084, Oct. 2019. doi:10.3897/biss.3.38084.

30. Z. Akbar, Y. A. Kartika, D. Ridwan Saleh, H. F. Mustika, and L. Parningotan Manik, "On using declarative generation rules to deliver linked biodiversity data," in *2020 International Conference on Radar, Antenna, Microwave, Electronics, and Telecommunications (ICRAMET)*, 2020, pp. 267–272. doi:10.1109/ICRAMET51080.2020.9298573.

31. R. Hyam, "Semantically linking specimens and images," *Biodivers. Inf. Sci. Stand.*, vol. 3, p. e35343, Jun. 2019. doi:10.3897/biss.3.35343.

32. D. Röpert, F. Reimeier, J. Holetschek, and A. Güntsch, "Semantic annotation of botanical collection data," *Biodivers. Inf. Sci. Stand.*, vol. 3, p. e36187, Jun. 2019. doi:10.3897/biss.3.36187.

33. S. M. Rashid et al., "The semantic data dictionary – An approach for describing and annotating data," *Data Intell.*, vol. 2, no. 4, pp. 443–486, Oct. 2020. doi:10.1162/dint_a_00058.

34. A. Lücking, C. Driller, M. Stoeckel, G. Abrami, A. Pachzelt, and A. Mehler, "Multiple annotation for biodiversity: Developing an annotation framework among biology, linguistics and text technology," *Lang. Resour. Eval.*, Aug. 2021. doi:10.1007/s10579-021-09553-5.

35. A. Kirchhoff et al., "Toward a service-based workflow for automated information extraction from herbarium specimens," *Database*, vol. 2018, 2018. doi:10.1093/database/bay103.

36. L. Stork et al., "Semantic annotation of natural history collections," *J. Web Semant.*, vol. 59, p. 100462, 2019. doi:10.1016/j.websem.2018.06.002.

37. D. Langenkämper, M. Zurowietz, T. Schoening, and T. W. Nattkemper, "BIIGLE 2.0: Browsing and annotating large marine image collections," *Front. Mar. Sci.*, vol. 4, p. 83, 2017. doi:10.3389/fmars.2017.00083.

38. G. S. Mai, F. C. Yang, and M.-N. Tuanmu, "Annotating out the way to the linked biodiversity data web," *Biodivers. Inf. Sci. Stand.*, vol. 1, p. e20270, 2017. doi:10.3897/tdwgproceedings.1.20270.

39. E. W. Schupp, P. Jordano, and J. M. Gómez, "A general framework for effectiveness concepts in mutualisms," *Ecol. Lett.*, vol. 20, no. 5, pp. 577–590, May 2017. doi:10.1111/ele.12764.

40. P. Jordano, "The biodiversity of ecological interactions: Challenges for recording and documenting the Web of Life," *Biodivers. Inf. Sci. Stand.*, vol. 5, p. e75564, Sep. 2021. doi:10.3897/biss.5.75564.

41. A. Iglesias-Molina and D. Chaves-Fraga, "Towards the definition of a language-independent mapping template for knowledge graph creation," Nov. 2019. doi:10.5281/ZENODO.3526141.

42. A. Dimou, M. Vander Sande, and P. Colpaert, "RML: A generic language for integrated RDF mappings of heterogeneous data," in *Workshop on Linked Data on the Web*, Apr. 2014, p. 5.

43. M. A. Musen, "The Protégé project: A look back and a look forward," *AI Matters*, vol. 1, no. 4, pp. 4–12, 2015. doi:10.1145/2757001.2757003.

44. T. Delva, D. Van Assche, P. Heyvaert, B. De Meester, and A. Dimou, "Integrating nested data into knowledge graphs with RML fields," in *Proceedings of the 2nd International Workshop on Knowledge Graph Construction (KGCW 2021)*, 2021, vol. 2873, p. 16, [Online]. Available: http://ceur-ws.org/Vol-2873/paper9.pdf.

45. H. Herre, "General formal ontology (GFO): A foundational ontology for conceptual modelling," in *Theory and Applications of Ontology: Computer Applications*, R. Poli, M. Healy, and A. Kameas, Eds. Dordrecht: Springer Netherlands, 2010, pp. 297–345.

46. P. Panov, L. Soldatova, and S. Džeroski, "Ontology of core data mining entities," *Data Min. Knowl. Discov.*, vol. 28, no. 5, pp. 1222–1265, Sep. 2014. doi:10.1007/s10618-014-0363-0.

47. C. M. Keet et al., "The data mining optimization ontology," *J. Web Semant.*, vol. 32, pp. 43–53, 2015. doi:10.1016/j.websem.2015.01.001.

NOTES

1 Index Herbariorum, http://sweetgum.nybg.org/science/ih/, Accessed: 05/11/2021.

2 https://www.idigbio.org/portal/, Accessed: 05/11/2021.

3 http://labelme2.csail.mit.edu/, Accessed: 05/11/2021.

4 https://www.robots.ox.ac.uk/~vgg/software/via/, Accessed: 05/11/2021.

5 https://cocodataset.org/\#format-data, Accessed: 05/11/2021.

6 http://host.robots.ox.ac.uk:8080/pascal/VOC/, Accessed: 05/11/2021.

7 https://ricover.hpc.lipi.go.id/ontocapsicum/, Accessed: 08/11/2021.

8 https://www.w3.org/TR/r2rml/, Accessed: 11/11/2021.

9 https://rml.io/, Accessed: 11/11/2021.

10 http://www.biologi.lipi.go.id/botani/index.php/about-bo, Accessed: 06/11/2021.

11 https://ricover.hpc.lipi.go.id/ontocapsicum/, Accessed:14/11/2021.

12 https://fno.io/, Accessed: 20/11/2021.

13 https://github.com/RMLio/rmlmapper-java, Accessed: 20/11/2021.

9 UpOnto
Strategic Conceptual Ontology Modeling for Unit Operations in Chemical Industries and Their Retrieval Using Firefly Algorithm

Ayush Kumar A and Gerard Deepak

CONTENTS

9.1 INTRODUCTION

Unit operations are the most important part of any manufacturing industry because they can be used to study even the most complex processes with simplicity. They are the general building blocks of a process and show what exactly is going on in a process plant. Unit operations and processes build up most of the chemical engineering data in the literature because of their sheer size and diversity because of which, there is no one reliable source for referencing them. To represent and retrieve data and the underlying relationships, an ontology encompassing domain knowledge of chemical engineering can be constructed to streamline the learning and archiving process. An ontology is a collection of conceptualized topics in a shared fashion, and it incorporates the domain knowledge in a very specialized way. One of the most significant advantages of an ontology over the conventional data structures is its ability to prevent redundancy and foster sharing of attributes between the classes and other components. For technically vast domains like chemical and process engineering, redundancy of the information stored is one of the biggest problems which does not help the readers to grasp the unique relationships between the widely different yet related subclasses and also increases the storage requirement for storing and the computational power required to perform any operation on a domain-related data. In this paper, an ontology for the chemical processes and equipment is modeled using the framework of Methontology. This ontology perhaps is one of its kind because of the scarcity of ontologies based on Chemical Engineering related topics. This ontology tries to conceptualize all the unit processes which are involved in various branches of chemical engineering. Building such an ontology requires an exceptional amount of domain knowledge especially in a highly theorized way because of the countless number of different industries and the processes used in them. Here, regarding this work, the domain is chosen to be the various unit operations. This work encompasses the gathering of domain knowledge, conceptualization, implementation of the knowledge, building the RDF structure of the ontology, and finally, the evaluation of the created ontology using the semiotics approach. Retrieval of the ontology is then discussed with biological-based machine learning algorithms. A hybrid model of the firefly algorithm is used in this paper which retrieves the concepts and relationships present in the ontology depending on the user test case. These algorithms are essential for the application purpose of the ontology into real-time usable applications in domains like e-learning, course recommendation, data archiving and storage, etc.

> **Motivation**: The basic idea of the work is to conceptualize the dispersed knowledge of the unit processes and the unit operations which are involved in any chemical industry. Chemical industries play a very important role in one's life as everything from a tiny pin to a gigantic car involves a chemical engineer as they are the ones who are involved in optimizing the processes and increasing their efficiency in order to maximize the profit and make the best use of the resources available. This ontology incorporates all the necessary unit operations, which are modeled into a hierarchy using XML, and then the method of transition from the nature of the ontology from being XML to RDF is proposed, and the ontology is modeled using Web Protégé.
>
> **Contribution**: The most fundamental objective of this work is to develop a model which can encompass all the unit processes used in various

industries and the strategic model for the same has been proposed. Out of the various available ontology development methodologies, Methontology has been used for this as it is appropriate to design the ontologies which contain a wide variety of knowledge and are humongous in terms of classes. But this model is used with certain negligible modifications to suit the nature of the ontology. The knowledge about the unit processes is obtained by the large literature available on chemical engineering and also by conducting several visits to the manufacturing industries and studying the processes operated there. The assembled knowledge is conceptualized, axiomatized, scrutinized, and modeled into an Ontology. Finally, the ontology is evaluated using the Semiotics approach and by choosing the suitable metrics to facilitate the quantitative and qualitative analysis of the ontology.

Organization: The rest of the chapter is organized as follows. Section 9.2 discusses the Related Works and the scope of further development. Section 9.3 deals with the Proposed Methodology for Ontology Development and the Conceptualization of the domain knowledge. The Implementation stage has been dealt with briefly in Section 9.4 accompanied by the RDF schema, development, and visualization phase of the Ontology. The assessment of the Ontology has been done in Section 9.5 and the paper is concluded in Section 9.6.

9.2 RELATED WORKS

Guo et al. [1] worked on the integration of big data for ontology retrieval systems. They also formulated a MapReduce-based system for efficient ontology retrieval on parallel distributed nodes. Highlighting the significance of the semantic web in daily life, this technology, if scaled to a large level can potentially change the way the Internet operates. Wang et al. [2] formulated an ontology that can evaluate the chemical toxicity of a substance based on its Chemical Species Phenotypic Profile. Vandana et al. [3] proposed a methodology to amalgamate the IoT with semantic technology with a resource description format using ontologies for better than state-of-the-art works. Since the semantic web has huge potential, it can be used to integrate the enormously automatic technologies of AI with that of the conventional web forming a true web of knowledge. Farazi et al. [4] formulated an ontology called OntoKin which incorporates the intricacies of Chemical Kinetics and the semantics of reaction mechanism into an Ontology that is validated by applying a reasonable tool. Syamili et al. [5] have built an ontology that depicts the Greek Mythology characters in a strategic conceptual way. The ontology thus framed can also answer the questions based on the domain. Kim et al. [6] have proposed a way for distilling any material synthesis ontology which can convert any large paragraph which was written in past tense and passive voice into sharp, concise steps which can enhance machine readability and could help in saving time. Dooley et al. [7] have developed a harmonized food ontology that can improve the ability to trace the global food production, preservation, and consumption and depict more data as it has derived its vocabulary from the world wide web consortium. Zhou et al. [8] have applied the chemical genomic to integrate the yeast cell to mammalian pathway analysis incorporating semantic web and ontologies. Cao et al. [9] have worked on a novel framework for the ontological description of the process industry to eliminate the immense amount of redundant knowledge in the Chemical engineering domain. There are several instances in the modern chemical industry where the same data or theory is used in different domains, and the knowledge becomes superfluous. To eliminate this, ontological models should be used where the multiple formats of storing domain knowledge like images, video, text, and table formats can be interlinked with each other based on only the inherent data and information present and not the formats. For this, the division of an industrial plant into raw materials, immediate materials, processes, and products has been suggested by them, and a knowledge model has been created. A semantic web-based knowledge graph can be useful in several instances, including validation research. When there are inconsistencies in the already existing models, one should go through them from scratch to find out the potential points of improvements and failures. For this, the interlinked data model of ontologies can be highly beneficial in order to query across the huge literature available and to determine a particular point of concern. Hence, for chemistry and chemical engineering purposes, knowledge graphs pertaining to several sections like kinetics and thermodynamics can be highly useful in research and validation purposes. Farazi et al. [10] have extended their work on OntoKin to capture the underlying information and the basic semantic characteristics of chemical reactions and processes.

Industrial equipment and machines are central to any process plant where any production process is going on. One of the major problems with the equipment is their maintenance and support requirements as they begin to operate and age. Based on chemical engineering principles like cause-effect correlation, measurement-identification options, and boundary conditions, the mechanism of fouling of industrial equipment can be studied. Ansaldi et al. [11] have worked on an ontology for modeling of the aging phenomenon of the process equipment and their deterioration mechanisms. These ontologies can be linked to already existing domain knowledge ontologies to provide a deeper amalgamation of industry-academia knowledge where the root cause of a problem can be determined with the help of semantic models. Mann et al. [12] have designed an ontology for predicting the outcomes of chemical reactions using ontological models. They have worked on a grammar-based demonstration of the molecules involved in the chemical reaction in a machine translation framework. With an accuracy of 80.1%, they have approached the problem with a source to target linking mechanisms relating to the reactants and the products of the concerned reaction. In [13–16] several Ontology-based applications and paradigms are discussed in support of the proposed approach.

9.3 METHODOLOGY

The ontology which is to be modeled is created using the methodology which is defined by Methontology. According to this approach, ontology development encompasses four steps viz: the gathering of domain knowledge, the conceptualization of the domain data collected from the literary sources, assimilation of the conceptualized ontology into machine-readable functional ontology by defining the knowledge into classes and objects, and the final step as the implementation of the ontology as a fully-functional usable ontology. All the steps which are required to formulate an ontology using the approach of Methontology must be carried out with perfection in order to facilitate smooth modeling of the ontology. No ambiguities whatsoever must be present, and all the test cases must be dealt with in detail to avoid the loss of any kind of data from the created hierarchy. The ontology which is to be created must be evaluated with precision, and the analysis by the domain experts must be taken seriously, and the necessary corrections must be made to ensure that no conceptual ambiguity is present in the ontology. Finally, the evaluation report according to the metrics of the Semiotics approach must be justified, and an ontology with a suitable reuse ratio and reference ratio is obtained. The ontology is visualized using the application called WEBVOWL, and the statistics are also visualized along with the hierarchy of classes.

Conceptualization

The second step involved in ontology modeling is to convert the unstructured data which is obtained from the domain sources in step 1 to the structured form of data which has some connection between the classes and to create a hierarchy of the data with a glossary of terms which contains all the technical things used in the work a and links one part of the ontology with the other thus combining the data and compiling them into one single entity. For example, the super domain of this work is Chemical Engineering, from which a small part pertaining only to the unit operations is taken out, and the intricacy of that topic is structured strategically into an ontology. The unit operations can be briefly classified into five types which are Mass Transfer operations, Heat Transfer Operations, Thermodynamic Processes, Mechanical Processes, and Fluid flow properties. These all are employed in the chemical industry which makes it compact and enables the industry to manufacture the multitude of products that it manufactures. In other words, unit operations are the most basic things in any industry, controlling which we can control the entire plant. This makes it essential for any chemical engineer or in fact, even any mechanical engineer to have a brief and clear idea about the unit operations and the material balance equations employed in them. Figure 9.1 has all the unit operations and processes structured into a flow chart.

The superclass of the ontology is the unit operations, and all other classes are derived from it. The subclasses need to be exhaustive; in other words, entire data of the parent class must be completely represented by the subclasses and no topic must be left behind. This ensures the proper propagation of the domain knowledge and preserves the entire domain from the point of superclass to the leaflets. Many

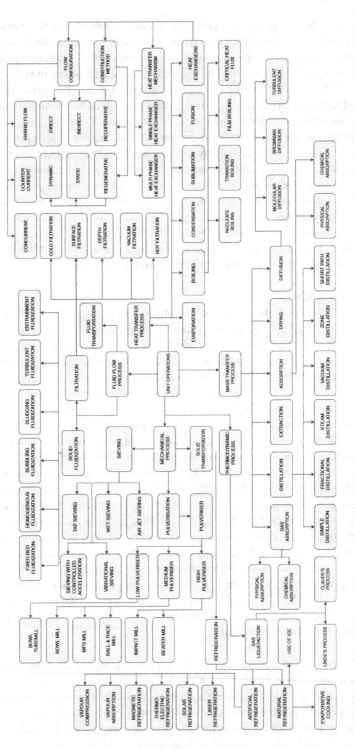

FIGURE 9.1 Domain knowledge representation.

attributes are also used to describe the qualities of the classes and the objects. The objects for the leaflets are declared by the keyword processInstance. The structural definition of a subpart of the domain is dented in the below table. Pertaining to this work, taking only the unit processes into consideration, various types of processes like "heat transfer process" and "mass transfer process" are defined as the subclasses or the child classes of the class Unit Operations. This step declares a parent-child-like relationship among the parent classes and the newly created classes as the data and the properties are inherited by the latter from the former. To denote the further branching of the newly created classes, the term "childOf" is used. For example, "Diffusion" can be classified further into "Molecular Diffusion", "Brownian Diffusion" and "Turbulent Diffusion" which are denoted here as "child" classes of the class "Diffusion". The objects of the child classes can also be applied to the parent classes due to the similar knowledge-sharing relationship among them. For example, the fundamental concept of "Brownian Diffusion" is dissimilar to that of "Molecular Diffusion" despite both sharing a common parent class i.e., Diffusion. Table 9.1 shows the branching of the classes in a hierarchical format.

9.4 IMPLEMENTATION

To accomplish the objective of the ontology, the conceptualized classes need to be implemented into a user-usable ontology that can be both machine-readable and which could be understood by the user too. This is done by the declaration of the objects to the classes and transition of all the XML-based class-related data into a functional RDF schema for the document. By defining the document type definition for the XML document, the glossary needed for the ontology is formalized, and the RDF schema of the document provides a base for the ontology to be axiomatized and finishes its requirements for being an ontology. This is done by describing all the processes in XML, initializing the objects, and forming the RDF schema of the ontology.

9.4.1 DESCRIPTION OF PROCESSES IN XML

The different unit operations should be classified in a hierarchical manner in order to develop a functional ontology. The unit operations and the unit processes are classified in a variety of ways like Thermodynamic processes are the processes that always have something to do with the study of heat i.e., Thermodynamics. Likewise, the processes which are contained in Mass Transfer superclass always have at least one component, which has a definite or indefinite mass moving from one place to the other. Hence, a logical approach is required to model them into a hierarchical ontology that will employ all of the technological data into a seamless strategic paradigm of structured knowledge. Since XML is being used in this work to classify the classes, some tags which can define the attributes of the process exhaustively are required to define the object definitely without any ambiguities. Table 9.2 depicts the description of processes for the ontology using XML.

TABLE 9.1

Hierarchical Organization of the Concepts

Unit Processes
    Fluid Flow Processes
    Mass Transfer
    ...
    Heat Transfer
        Evaporation
            [processlnstance]
        Boiling
            Nucleate Boiling
                [processinstance]
            Transition Boiling
                [processinstance]
            Critical Heat Flux Boiling
                [processinstance]
            Film Boiling
                [processinstance]
        Heat Exchangers
            Recuperative Heat Exchangers
                Direct Recuperative Heat Exchanger
                    [processinstance]
                Indirect Recuperative Heat Exchanger
                    [processinstance]
            Regenerative Heat Exchangers
                Static Regenerative Heat Exchanger
                [processInstance]
                Dynamic Regenerative Heat Exchanger
                [processinstance]
        Fusion
            [processinstance]
        Condensation
            [processinstance]
        Sublimation
            [processinstance]

9.4.2 DESCRIPTION OF OBJECTS

The main objective of the ontology is to be fully functional which can only be achieved by declaring the objects for the classes which have been declared by branching and arranging the domain knowledge. Here, in this work, the objects will be the part of the industry or the specific process where the unit operation is involved. This will give the user a brief idea of the process and also the application part of the process. An example of object declaration for a portion of the ontology is shown in Table 9.3.

TABLE 9.2

Description of Processes

```
<!-process classification -->
    <?xml version-"1.0'>
    <!DOCTYPE unitOperations SYSTEM "UPOntoDTD.dtd">
    <!-unit operations-->
    <unitOperation id="uno" title="unit operations"/>
                <definition> Any physical or chemical change in a process industry like
distillation, separation, evaporation, membrane adsorption, or decantation is termed as a unit
operation. These processes form the backbone of chemical industries.</definition>
                < MassTransferOperation id="uno/mat" title="Mass Transfer Operations/>
                <definition> Mass transfer is the study of transport phenomena induced by
chemical potential gradient which mostly focuses on concentration gradient. It is the movement of
components from one phase, or stream, or component to another. Some examples of mass transfer
operations are adsorption, absorption, distillation etc.</ definition>
                <distillationMassTransferOperation id="uno/mat/dis" title="Distillation" >
                <definition> Distillation is the process of separation of two components
based on their boiling point difference.</ definition>
                <simpleDistillationMassTransferOperation id="uno/mat/dis/sim -ind"
title="Simple Distillation">
                <definition></ definition>
                        <processInstance id="" title="" />
                </simpleDistillationMassTransferOperation>
        </distillationMassTransferOperation>
    </ MassTransferOperation >
</unitOperation >
<!--contd-->
</unit operations >
```

9.4.3   RDF/OWL CLASSIFICATION OF THE REACTION HIERARCHY

In order to finish building the ontology, the last advance is the metamorphosis of the Ontology from the conventional XML plan to the RDF structure, which enables the computer to exploit the code in a very effective manner. In order to execute this step, an RDF/OWL report has been generated, and the classes are depicted here according to the vast classifications and the categories of the unit processes and their types. Wherever possible, many references and further links are given with an aim to make the ontology sophisticated in the RDF schema and even further references in the form of comments and the extra information whichever necessary has been aptly defined in any place required. Remarks, in addition, have been utilized circumspectly in the age of this RDF record. The utilitarian assets have been depicted, and the domain of the ontology has been fixed and set to the Unit Processes. The RDF schema for UPOnto is depicted in Figure 9.2.

TABLE 9.3

Description of Objects

```
<!--object definition -->
    <?xmlversion="1.0">
    <!DOCTYPE unitOperations SYSTEM "UPOntoDTD.dtd">
    <!--unit operations-->
    <sublimationHeatTransferUnitOperation>
    <definition> <definition/>
            <processInstance id = "uno/het/sub-001"
title = "Drying of heat sensitive materials"/>
            <processInstance id = "uno/het/sub-002" title = "Purification of contaminated
solutions"/>
            <processInstance id = "uno/het/sub-003" title = "Aroma of air fresheners"/>
<! --contd-->
    </sublimationHeatTransferUnitOperation >
```

9.4.4 ONTOLOGY MODELING USING PROTÉGÉ

The theoretical ontology with all its structuring work done needs to be planted in an application called WebProtégé. This is to ensure that the ontology is generated and can be viewed as a ".owl" file and its RDF schema can be generated. In Protégé, the classes and their subclasses are defined in a hierarchical order and the relevant objects are declared. Attributes and the object properties are also defined here and the ontology comes alive. The attributes of the Ontology Web Language Tags like alias, label, comment, seeAlso are also provided wherever felt essential. The class hierarchy of the ontology as shown in WebProtégé is shown in Figure 9.3.

9.4.5 ONTOLOGY VISUALIZATION

Post modeling of the desired ontology, the subsequent stage is to envision it by conventional representation or by utilizing perception products accessible. The goal of this step is to have a visual understanding of the structuring of knowledge which at times proves to be very useful for a user who sees the ontology for the first time. In this work, to spare manual time and furthermore to imagine the ontology without any error and also as a whole, the WEBVOWL application has been utilized. So as to utilize that to imagine the modeled ontology, the demonstration has been done by Protégé. It must be uploaded as an ".owl" document and to be transferred to the page of WEBVOWL. This ontology may be required to be edited in the future owing to change of ideology or facts as new scientific theories are popping up frequently and also specific to this work, many new industrial processes will be discovered and would be used in industries which will result in further expansion of the knowledge. This step becomes inevitable as the Ontology must be dynamic and must adapt itself to the frequently changing informative world. In

```xml
<?xml version="1.0"?>
<rdf:RDF xmlns="http://webprotege.stanford.edu/project/8oOaesXM8KJmDka4uIhvoF#"
    xml:base="http://webprotege.stanford.edu/project/8oOaesXM8KJmDka4uIhvoF"
    xmlns:owl="http://www.w3.org/2002/07/owl#"
    xmlns:rdf="http://www.w3.org/1999/02/22-rdf-syntax-ns#"
    xmlns:xml="http://www.w3.org/XML/1998/namespace"
    xmlns:xsd="http://www.w3.org/2001/XMLSchema#"
    xmlns:rdfs="http://www.w3.org/2000/01/rdf-schema#">
    <owl:Ontology rdf:about="http://webprotege.stanford.edu/project/8oOaesXM8KJmDka4uIhvoF"/>
    <owl:Class rdf:ID="Unit Operations">
    <rdfs:subClassOf><owl:Class rdf:ID="Fluid Flow Processes"/> </rdfs:subClassOf>
    <rdfs:subClassOf><owl:Class rdf:ID="Heat Transfer"/> </rdfs:subClassOf>
    <rdfs:subClassOf><owl:Class rdf:ID="Mass Transfer"/> </rdfs:subClassOf>
    <rdfs:subClassOf><owl:Class rdf:ID="Thermodynamic Processes"/> </rdfs:subClassOf>
    <rdfs:subClassOf><owl:Class rdf:ID="Mechanical Processes"/> </rdfs:subClassOf>
    <rdfs:comment rdf:datatype="http://www.w3.org/2001/XMLSchema#string"> </rdfs:comment>
    </owl:Class>

    <owl:Class rdf:ID=" Heat Transfer ">
    <rdfs:subClassOf><owl:Class rdf:ID="Evaporation"/> </rdfs:subClassOf>
    <rdfs:subClassOf><owl:Class rdf:ID="Boiling "/> </rdfs:subClassOf>
    <rdfs:subClassOf><owl:Class rdf:ID=" Condensation "/> </rdfs:subClassOf>
    <rdfs:subClassOf><owl:Class rdf:ID=" Sublimation "/> </rdfs:subClassOf>
    <rdfs:subClassOf><owl:Class rdf:ID=" Fusion "/> </rdfs:subClassOf>
    <rdfs:subClassOf><owl:Class rdf:ID=" Heat Exchangers Polymerisation "/> </rdfs:subClassOf>
    <rdfs:comment rdf:datatype="http://www.w3.org/2001/XMLSchema#string"> </rdfs:comment>
    </owl:Class>

    <owl:Class rdf:ID=" Boiling ">
    <rdfs:subClassOf><owl:Class rdf:ID="Nucleate Boiling"/> </rdfs:subClassOf>
    <rdfs:subClassOf><owl:Class rdf:ID="Transition Boiling"/> </rdfs:subClassOf>
    <rdfs:subClassOf><owl:Class rdf:ID="Critical Heat Flux"/> </rdfs:subClassOf>
    <rdfs:subClassOf><owl:Class rdf:ID="Film Boiling"/> </rdfs:subClassOf>
    <rdfs:comment rdf:datatype="http://www.w3.org/2001/XMLSchema#string"> </rdfs:comment>
    </owl:Class>

    <owl:Class rdf:ID="Heat Exchangers">
    <rdfs:subClassOf><owl:Class rdf:ID="Recuperative Heat Exchangers "/> </rdfs:subClassOf>
    <rdfs:subClassOf><owl:Class rdf:ID="Regenerative Heat Exchangers "/> </rdfs:subClassOf>
    <rdfs:comment rdf:datatype="http://www.w3.org/2001/XMLSchema#string"> </rdfs:comment>
    </owl:Class>

    <owl:Class rdf:ID=" Recuperative Heat Exchangers ">
    <rdfs:subClassOf><owl:Class rdf:ID="Direct Recuperative Heat Exchangers </rdfs:subClassOf>
    <rdfs:subClassOf><owl:Class rdf:ID=" Indirect Recuperative Heat Exchangers "/> </rdfs:subClassOf>
    <rdfs:comment rdf:datatype="http://www.w3.org/2001/XMLSchema#string"> </rdfs:comment>
    </owl:Class>

    <owl:Class rdf:ID=" Regenerative Heat Exchangers ">
    <rdfs:subClassOf><owl:Class rdf:ID=" Static Regenerative Heat Exchangers "/> </rdfs:subClassOf>
    <rdfs:subClassOf><owl:Class rdf:ID="Dynamic Regenerative Heat Exchangers "/> </rdfs:subClassOf>
    <rdfs:comment rdf:datatype="http://www.w3.org/2001/XMLSchema#string"> </rdfs:comment>
    </owl:Class>

    <owl:FunctionalProperty rdf:about="#UnitProcesses">
    <rdfs:domain> Unit Processes </rdfs:domain>
    </owl:FunctionalProperty>
</rdf:RDF>
```

FIGURE 9.2 RDF schema of a part of the ontology.

the wake of transferring the effectively made ontology, the application gives the undeniable perspective on the ontological classes and subclasses with the connections among them. Since an ontology is an extremely gigantic thing to envision, a piece of the ontology perception yield produced structure WEBVOWL has been shown in Figure 9.4.

FIGURE 9.3   Class description in WebProtégé.

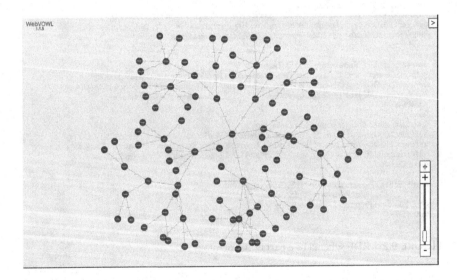

FIGURE 9.4   Ontology visualization using WebVOWL.

9.5 RESULTS AND EVALUATION

9.5.1 SEMIOTIC EVALUATION

For real-life implementation and applications, the ontologies should be evaluated according to some prescribed metrics to calculate the deviation from the required parameters to the modeled ontology values. This work employs the semiotics approach of ontology modeling in order to assess the complete quantitative and qualitative characteristics of the ontology. This step is important to customize the ontology based on the use case, and if the case requires a high or low value of any particular parameter like consistency or comprehensiveness, the ontology can be altered and made the same. The ontology can be made as rich as possible, and several required concepts and relations can be updated with the input of the domain experts.

Quantitatively, the quantity of classes, subclasses, attributes, and leaflets should be assessed. CoC refers to the count of class, CoO is the count of objects, CoSC is the count of SubClasses, DoI refers to the Depth on Inheritance, CoLC refers to the count of Leaf Classes, NoA refers to the number of Attributes for the objects which help to broaden the scope of the ontology. The reuse and reference ratio are obtained using the definitions available in the literature. Equations (9.1) and (9.2) show the formulae for calculating the reuse and reference ratios in terms of the number of classes of the ontology. Table 9.4 shows the results of the quantitative analysis of the modeled ontology.

$$\text{Reusability Ratio} = \frac{\sum \text{Number of Subclassses}}{\sum \text{Total number of classes}} \tag{9.1}$$

$$\text{Reference Ratio} = \frac{\sum \text{Referenced Elements}}{\sum \text{Reused Elements}} \tag{9.2}$$

From the above equation (9.1), the reuse ratio of UPOnto is calculated as 73.4%. Also, the reference ratio comes out to be 3% which is well within the range of the approach used for modeling of the ontology. For specified domain ontologies like this, the data and literature available in the sources are not structured properly, and hence the referencing of elements is quite low. But they have a higher tendency of being reused in other ontologies. Reuse of existing ontologies

TABLE 9.4
Quantitative Analysis of UPOnto

CoC	CoO	CoSC	DoI	CoLC	NoA	Reusability Ratio
98	21	72	8	32	13	0.734

TABLE 9.5

Qualitative Analysis of UPOnto

Metric	Very High	High	Medium	Low
Accuracy	30	26	14	0
Clarity	39	23	6	2
Comprehensiveness	26	30	13	1
Consistency	31	19	20	0
Interpretability	40	22	5	3
Lawfulness	45	19	6	0
Relevance	46	19	5	0
Richness	32	27	10	1

to form a web of knowledge can either be done via extending existing ontologies using data mining techniques or the assimilation of various already existing ontologies into one large ontology of the same or related domain. In each of them, the modeling approach differs, and one must make sure that the procedure adopted for the construction of these ontologies is suitable with the domain and its range. Large scale domains like Chemical Engineering have a wide range of applications and branches, which make it tough to form an all-inclusive web of knowledge. For those purposes where already several related ontologies are being reused, recently modeled ontologies like UPOnto with novel domain coverage and data availability are required. This used of novel classes and subclasses in an ontology makes it highly beneficial for assimilation purposes contributing to one true web of knowledge. Also, the inheritance of classes is up to level 8 in the modeled ontology which makes it highly beneficial for e-learning applications. Any exhaustive ontology should cover the entire subject matter with special cases which could be useful for several varied applications in Web 4.0. Following the quantitative evaluation, the qualitative analysis of the ontologies is carried out by a panel of 70 experts in chemical engineering and ontology modeling process. There are several metrics considered in this approach where the panel grades the ontology into qualitative scores ranging from very high to low, and the results are then summed up to check the quality of the modeled ontology via the user's perspective. The results of the qualitative analysis via the semiotics approach are tabulated in Table 9.5.

9.5.2 Retrieval of the Ontology

The retrieval of the ontology was performed on the basis of a modified firefly algorithm for the optimization of the relevance of the classes. The system specifications were Intel i5 processor with 16 GB RAM on Windows 10.

As depicted in Table 9.6, the concepts of the ontologies are arranged in a numbered order and are assigned unique id. The ids are assigned, so that searching the required classes is made faster as comparing the classes name-wise requires more

TABLE 9.6

Algorithm for Ontology Retrieval Based on Firefly

Input: The required class from the domain of the ontology to be retrieved
Output: The set of classes and concepts that satisfy the input condition

begin
Step 1: All the concepts from the ontology which are defined using XML are denoted by C_i. Each law is denoted as L_i. R_i a two tuple. $C_i = (P_i, Q_i)$ where C_i, denotes the concept i, P_i denotes the identification character of the concept, Q_i is the coefficient of relative importance assigned to the class in terms of relevance and retrieval
Step 2: Define the objective function to be optimized as f(O), $O = (O_1, O_2, \ldots, O_n)$;
Step 3: Input the classes and the concepts from the input list C., $i \in (1, n)$
Step 4: The values of Q are assumed and associated with f(O).
Step 5: A class-wise concept ranking coefficient (μ) is defined which is always a constant and if defined trivially. It is employed to increase the relevance of the classes when it resonates with the concept condition which is input by the user.
Step 6: s = 0, Con = No. of concepts to be retrieved;
    while (s = Con)
        for i = 1: n // Showing the selection of all concepts
            for j = 1: i //Selecting n concepts
                if (Q.>=Qi),
                Vary the relevance of concept with factor x using the term exp $(-\mu x)$
                Move concept i towards concept j
                Check new solutions to update the selected laws matching the input
criteria
                end if
            end for j
                Selected concepts are ranked and the most relevant item is matched
            end for i
    s=s+1;
    Post processing algorithm results and displaying
    end while
end

arithmetical computation at the beginning of the implementation process. The sum of weights of the concept in terms of relevance is denoted by the variable *s*. In the do while loop the relevance of every concept is compared with other concepts. The appropriateness of concepts is then varied based on their relevance to the input concept by the user and then it relates the laws from the input to the database of the laws in the Ontology which is stored in the RDF format. This comparison is done by using the Semantic Similarity method. The concepts that do not possess the minimum amount of similarity index requirements are neglected by the algorithm by thoroughly checking the required criteria. Even if there is one concept at least satisfying the lowest required similarity index, the concepts are relatively ranked with this as a reference because there can be many formulae and laws referring to the same quantity. Then, further scrutiny of the input data is done to find the target of the formula which then gives the exact law of the formula required. In successive iterations of the while loop, the matching set of the concepts has been found. The matching set consists of a group of concepts and laws which refer to the same quantities in relation to highly qualified candidates. The algorithm has a sequence of steps that are defined for scrutinizing the concepts and selecting the best out of the retrieved. The algorithm is clever incorporation of the already existing firefly algorithm with ontology evaluation and is even a novel approach that amalgamates the Firefly algorithm, and the semantic similarity strategies for ontology retrieval. The relevance of the concepts is compared with the input data in each of the iterations and the relevance factor is increased in the subsequent iterations because of the low sample space every second time and the proximity of the data with the desired input increases. The concepts are then ranked based on how well they match the input. This makes ontology retrieval precise and effortless.

9.5.3 Retrieval Evaluation

For a chemical engineering domain, the modeled ontology was tested with several sample classes and objects. The objective is to retrieve the optimal class for future ontology applications. If the ontology is employed in a course recommendation algorithm, it should be able to suggest a course based on the inputs received by the user. And this could only be done after the minimization of the difference between the similarity indices of the entered word and the classes present in the ontology. For this minimization algorithm, the firefly-based modified approach is used in this paper. The evaluation metrics are chosen to be Precision, Recall, Accuracy, and F Measure. These are chosen based on the requirement of the system, which is mostly classification in nature because the system should be capable of classifying the classes into relevant and not relevant according to the input and then actually retrieving the class which is the most relevant based on the similarity index specified. Precision (equation 9.3) is the number of elements that are relevant to the input class from all the retrieved elements. Hence this becomes a highly important factor that says whether the results of the experiment are inherently desirable or not. Similarly, Recall (equation 9.4) of the ontological entity is defined as the ratio of the number of relevant elements in the retrieved classes to the total number of relevant entries in the entire ontology. Recall is a measure of the ability of the ontology to give out all the desirable entities from the list provided to it, hence avoiding false negatives. F-Measure (equation 9.7) is defined as the harmonic mean of precision and recall. Accuracy (equation 9.6) is defined as the arithmetic mean of the precision and recall. Table 9.7 depicts the results of the retrieval of the system using the modified firefly algorithm.

$$\text{Precision} = \frac{\text{Relevant Retrieved}}{\text{Retrieved Elements}} \quad\quad (9.3)$$

TABLE 9.7

Performance Measures of Retrieval

Ontological Entities	Recall%	Precision%	F-measure%	Accuracy%
Sieving	93.47	95.64	94.54	94.56
Fluidization	93.81	95.77	94.78	94.79
Boiling	92.33	97.89	95.03	95.11
Heat Exchangers	94.21	95.63	94.91	94.92
Evaporation	92.83	97.91	95.30	95.37
Average	93.33	96.568	94.91	94.95

$$\text{Recall} = \frac{\text{Relevant Retrieved}}{\text{Relevant Elements}} \qquad (9.4)$$

$$\text{F-measure} = 2 \times \frac{\text{Precision} \times \text{Recall}}{\text{Precision} + \text{Recall}} \qquad (9.5)$$

$$\text{Accuracy\%} = \frac{\text{Precision} + \text{Recall}}{2} \qquad (9.6)$$

The average accuracy is 94.95% and the average F-measure of the ontology retrieval algorithm is 94.91%. The entities for ontological retrieval are Sieving, Fluidization, Boiling, Heat Exchangers, and Evaporation. When tested for 200 inputs, the ontology gives an overall average accuracy of 94.95% which is indeed commendable for domain ontologies. The hybrid Firefly algorithm of retrieval of ontologies works well with specialized domains because of the high number of interlinked entities present and the close similarity index between the ontology classes. Hence, to yield a better retrieval accuracy, this closeness in the classes should be considered and properly taken into account. Since the relevance factor is increased in each subsequent iteration, the proposed firefly algorithm captures even the relatively small differences in the similarity indices and yields better performance with higher metrics of evaluation.

With the evaluation metrics being high enough than existing models, the modeled domain ontology has a lot of future scope. The primary use of the ontology is in online knowledge archives after filling the ontology with several sources and data using data mining and extraction techniques. All chemical engineering knowledge can be structured into a massive ontology containing all the unit processes and operations from several renowned archives and literature to make it ordered and structured. This is the envisioned goal of Web 4.0 where the data is structured and linked to similar or dissimilar classes associated with it via ontological relationships. This can later be made to a self-populating ontology with algorithms designed to classify web pages into ontological classes and link it to the existing entities in the ontology. Also, the modeled ontology can be used in E-learning systems teaching Chemical Engineering to both freshers and experienced engineers. As a quick refresher, the entities of the ontology would be much more efficient than the existing naïve data structures because of predefined structures and entities falling under an automatically growing ontological model. The classes could be retrieved by one or more biologically inspired algorithms, and they could be designed into efficient knowledge-based learning systems. Also, course recommendation systems also can use ontologies to mark the keywords entered by the user and check the concepts of the ontology to recommend apt courses for them. As ontologies have entities linked to them in a parent-child fashion and are highly searchable due to the indexable nature of the entities and the accuracy of the algorithms used, they can be the best solutions for huge course recommendation systems with thousands of courses and subjects.

## 9.6	CONCLUSION

This paper was an attempt to classify all the Unit Processes which are used in the chemical industries and all other related industries into a structured strategic ontology which has been modeled using the Methontology approach of Ontology development. Methontology is an ideal methodology to model this kind of Educational Ontologies due to its dynamic nature. The ontology has been developed in the conventional XML format and then the methodology of transformation from XML to RDF has been suggested. The ontology is finally evaluated using the metrics of the semiotic approach of ontology assessment, and it is found to be a fabulous ontology for the prescribed domain. To make this ontology functional, the work is concluded by modeling the ontology using WebProtégé and visualizing it using an online Ontology Visualization tool called WebVOWL. At last, the ontology has been evaluated using the Firefly algorithm and overall accuracy of 94.95% has been achieved.

REFERENCES

1. Guo, K., Liang, Z., Tang, Y., and Chi, T. (2018). SOR: An optimized semantic ontology retrieval algorithm for heterogeneous multimedia big data. *Journal of Computational Science*, *28*, 455–465.
2. Wang, R.-L., Edwards, S., and Ives, C. (2019). Ontology-based semantic mapping of chemical toxicities. *Toxicology*, *412*, 89–100.
3. Vandana, C. P., and Chikkamannur, A. A. (2019). Semantic ontology based IoT-resource description. *International Journal of Advanced Networking and Applications*, *11*(1), 4184–4189.
4. Farazi, F., Akroyd, J., Mosbach, S., Buerger, P., Nurkowski, D., and Kraft, M. (2019). OntoKin: An ontology for chemical kinetic reaction mechanisms. *Journal of Chemical Information and Modeling*, *60*, 108–120.
5. Syamili, C., and R. V. Rekha. (2018). Developing an ontology for Greek mythology. *The Electronic Library*, *36*(1), 119–132.
6. Kim, E., Huang, K., Kononova, O., Ceder, G., and Olivetti, E. (2019). Distilling a materials synthesis ontology. *Matter*, *1*, 8–12.
7. Dooley, D. M., Griffiths, E. J., Gosal, G. S., Buttigieg, P. L., Hoehndorf, R., Lange, M. C., Schriml, L. M., Brinkman, F. S. L., and Hsiao, W. W. L. (2018). FoodOn: A harmonized food ontology to increase global food traceability, quality control and data integration. *NPJ Science of Food*, *2*(1), 23.

8. Zhou, F.-L., Li, S. C., Zhu, Y., Guo, W.-J., Shao, L.-J., Nelson, J., Simpkins, S. et al. (2019). Integrating yeast chemical genomics and mammalian cell pathway analysis. *Acta Pharmacologica Sinica, 40*, 1.

9. Cao, J., He, Y. L., and Zhu, Q. (2021). An ontology-Based procedure knowledge framework for the process industry. *The Canadian Journal of Chemical Engineering, 99*(2), 530–542.

10. Farazi, F., Salamanca, M., Mosbach, S., Akroyd, J., Eibeck, A., Aditya, L. K., ... Kraft, M. (2020). Knowledge graph approach to combustion chemistry and interoperability. *ACS Omega, 5*(29), 18342–18348.

11. Ansaldi, S., Bragatto, P., Agnello, P., and Milazzo, M. F. (2020, November). An ontology for the management of equipment ageing. In *Proceedings of the Conference ESREL2020 PSAM15* (pp. 1–5), Venice, Italy.

12. Mann, V., and Venkatasubramanian, V. (2021). Predicting chemical reaction outcomes: A grammar ontology-based transformer framework. *AIChE Journal, 67*(3), e17190.

13. Deepak, G., Kumar, A., Santhanavijayan, A., and Priyadarshini, S. J. (2021). Ontology modelling for metallurgy as a domain and retrieval using particle swarm optimization: conceptualization, modeling, and retrieval. In *Machine Learning Approaches for Improvising Modern Learning Systems* (pp. 272–301). IGI Global.

14. Deepak, G., Kumar, A., Santhanavijayan, A., Pushpa, C. N., Thriveni, J., & Venugopal, K. R. (2021). Pedagogical ontology modelling for cell biology domain with an algorithm for question generation. In *Machine Learning Approaches for Improvising Modern Learning Systems* (pp. 144–168). IGI Global.

15. Ayush Kumar, A., Deepak, G., & Santhanavijayan, A. (2021). PolyOnto: Conceptual modeling and evaluation of ontologies for polymerization reactions based on XML to OWL transfiguration. In *Advances in Automation, Signal Processing, Instrumentation, and Control* (pp. 913–925). Springer, Singapore.

16. Pushpa, C. N., Deepak, G., Kumar, A., Thriveni, J., and Venugopal, K. R. (2020, July). OntoDisco: Improving web service discovery by hybridization of ontology focused concept clustering and interface semantics. In *2020 IEEE International Conference on Electronics, Computing and Communication Technologies* (CONECCT) (pp. 1–5). IEEE.

10 Ontologies for Knowledge Representation
Tools and Techniques for Building Ontologies

Ayesha Banu

CONTENTS

DOI: 10.1201/9781003309420-10

10.1   INTRODUCTION

Semantic Web, an idea of Berners-Lee, was coined in the 1990s. As per Lee [1]
"Most of the content on the web is human readable, where the computer programs
find it difficult to manipulate meaningfully". The semantic Web is an extension of the
existing web obtained by giving meaning and structure to the pages of web content.
The semantic web develops an environment where computers and users can work
together in a better way. The fundamental idea was to design a web that can link
documents such that the semantics and facts of the documents are easily identified.
This makes it easy for both machines and users to understand the meaning or semantics of the data, which, in turn, supports interoperability across several applications.

Figure 10.1 depicts the Semantic Web layers, as given by Berners-Lee [2].
These layers express the standards of abstraction on the semantic web and every
layer takes the previous layer as its base.

I. **Uniform Resource Identifier (URI) and Unicode**: URIs lay a standard
   for referring to different entities of the web. The standard for exchanging
   symbols is called Unicode.

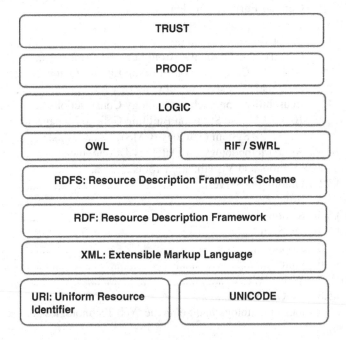

FIGURE 10.1   Layered architecture of semantic web.

II. **XML (Extensible Markup Language)**: XML allows considering the user-defined vocabulary and writing standard web documents and helps to exchange documents on the web. This language defines a certain set of rules for document encoding such that it becomes readable by both humans and machines.

III. **RDF (Resource Description Framework)**: The statements on web resources are written using RDF. This is W3C recommended standard metadata processor that supports interoperability among applications. An important feature of RDF is that it allows data merging irrespective of the underlying structure of the data. RDF also allows combining structured and semi-structured data and share across various applications. The information is made machine-understandable in this layer.

IV. **RDFS (Resource Description Framework Schema)**: RDFS helps in organizing web resources into hierarchies by using the modeling concepts such as class and subclass, properties and sub-property, domain, and range. The vocabulary definitions of OWL or SKOS are built on RDFS.

V. **Ontology vocabulary**: The complex relationships among web resources are expanded and represented by Ontology expressions. OWL (Web Ontology Language) is mainly designed for those applications which process the information content rather than just present information. This language supports great machine interpretability than compared to XML, RDF, and RDFS.

RIF is the Rule Interchange Format which is an XML language to build and express rules that can be executed by computers. SWRL is the Semantic Web Rule Language used to express both rules and logic combining OWL DL.

VI. **Logic**: A layer that extends the language to design knowledge-based applications

VII. **Proof**: This layer involves the real process for proof validation.

VIII. **Trust**: The security and trust of the semantic web are handled on the top layer using digital signatures and trusted agents [3].

10.2 ONTOLOGY REPRESENTATION LANGUAGES

Ontology describes the domain concepts and properties, relationships between these concepts. Ontologies form the basic components of the Semantic Web, and they are used to capture the knowledge of some domain of interest. Ontology is defined as a representation of concepts set within a domain and relationships between these Concepts. These concepts are elaborated using the features describing them [4].

Language is always the fundamental requirement to represent information and develop a semantic web. There must always exist some suitable language. The language used must describe the document, assist knowledge representation, support logical reasoning, and help in expressiveness. Exchange of documents must also be possible same as in Web2.0. The language must act as a medium for exchanging objects. Different types of languages are supported by the Semantic Web, which can be chosen depending on the requirement of the application.

Some of these languages like Ontology Exchange Language (XOL) are based on XML syntax that is used for authoring documents on the web [5]. Simple HTML Ontology Extension (SHOE), is an extension of HTML, developed by the University of Maryland, which incorporates machine-readable semantic knowledge in HTML and other web documents [6]. Resource Development Framework (RDF) is an XML-based language that supports web resources statement writing for Semantic Web that includes a triple known as subject, predicate, and object. The things in a statement are described by the subject. Predicate represents the relationship between subject and object. Resources may include a complete web page or selected parts of the pages [7].

All the semantics of the resources are defined by the RDF Schema (RDFS). It describes all the classes and their properties in the RDF model, and they can be arranged into hierarchies called specialization or generalization. We can also specify the domain and range for these properties. On the whole, RDFS is a fundamental building block on the semantic web for ontologies [8]. Some additional languages also exist that are a union of RDF and RDF Schema known as Ontology Inference Layer (OIL) and DAML+OIL that improve the features. Web Ontology Language (OWL) is an efficient language than RDFS to develop an ontology. It expands RDFS and builds meaningful ontologies with restrictions on the content and structure of RDF documents. OWL is known to provide better machine interpretability of web content compared to RDF and RDFS. Therefore, W3C recommends OWL as a standard language for the Semantic Web. Figure 10.2 shows a heap of all the languages supported by the semantic web [9].

Jorge Cardoso [30] surveyed ontology development languages, collecting the opinion of several respondents from academia including students, researchers, programmers, and professors with researchers forming the major percentage (21.7%). Each respondent fills in the data regarding the ontology language currently being used in their working organizations. The survey results are shown in Figure 10.3.

OWL (Web Ontology Language) and RDFs (Resource Description Framework) showed the strongest impact in the Semantic Web were More than 75% of respondents selected OWL and more than 64% opted for RDFs to develop their ontologies.

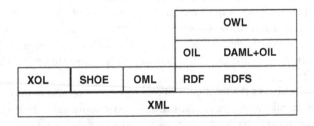

FIGURE 10.2 Languages stack in the semantic web.

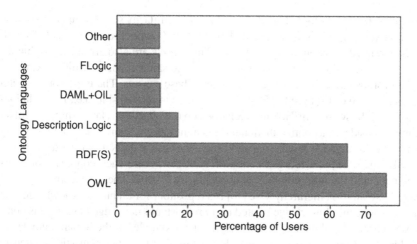

FIGURE 10.3   Survey results on ontology languages.

10.3   TYPES OF ONTOLOGIES

The first definition of the term ontology comes from the philosophical study of the existence of the nature of being. In the domain of knowledge representation and ontology engineering, this term has a different meaning. The structure describing all the domain concepts is defined as an Ontology that includes all the classes and their respective properties, attributes, and features. These properties can also be restricted [10]. The advantage behind developing any ontology is to facilitate people and web agents in understanding the ontology structure and to reuse the domain knowledge [11]. Any ontology on the web allows a system to identify the data and web resource semantics explicitly and help other systems to utilize them without any terminological problems. For example, the system can clearly distinguish the difference between the term "book", used in "reservation" or a travel agency and a "written work" in any online library [12]. Three basic elements that represent ontology knowledge include (i) Classes (or concepts) are the basic entities that are to be modeled and arranged into hierarchies. (ii) Properties (or roles or relations) correspond to the association types among the classes. (iii) Individuals (or instances) represent the class instances.

In the semantic web, the best language for representing ontologies is OWL. Irrespective of the language used for the construction of the ontology, the different ontologies available can be classified either as general-purpose ontology or as a domain-specific ontology [13].

10.3.1   GENERAL-PURPOSE ONTOLOGIES

These ontologies are not domain restricted. WordNet [14], developed by Princeton University, is a general-purpose ontology that includes nouns, verbs, adverbs, and adjectives that are organized as synonyms set called synsets. It resembles a thesaurus where the words are grouped based on their meaning. Wordnet is a natural language terms ontology that can be utilized for similarity score computations. The current version of WordNet v.3.1 includes 155,287 distinct terms and 117,659 synsets, arranged as taxonomic hierarchies. The synsets are also arranged into hierarchies corresponding to several synonyms of the same concept. Multiple kinds of relationships can also be generated between these synsets. The most popular relationships in WordNet are the Hyponym/Hypernym relationship (i.e., Is-A relationship), and the Meronym/Holonym relationship (i.e., Part-Of relationship). WordNet can be considered as both a dictionary and a thesaurus. Many researchers working on Natural Language Processing and Linguistics use WordNet as their dataset.

SENSUS [15] is a reorganization of WordNet where each concept represents one node and the total numbers of concepts are 90,000. These concepts are organized into an IS-A hierarchy. Cyc [16] is a repository for all human knowledge also as commonsense knowledge treated as an important knowledge base. Cyc is composed of terms and their respective assertions. Examples of the human knowledge included in Cyc are the facts and rules of thumb, reasoning heuristics related to objects, and events of the day-to-day life. Currently, Cyc KB consists of more than 50,000 terms, 17,000 relation types, and 7 million assertions to relate these terms.

10.3.2 DOMAIN-SPECIFIC ONTOLOGIES

These ontologies are domain-restricted.

- Unified Medical Language System (UMLS) [17] is a bio-medical domain ontology that includes a very large number of concepts related to the biomedical and health domain. It integrates and distributes all the terms, classification standards, and their associated resources to create a more efficient biomedical information system that includes Electronic Health Records.
- SNOMED [18] Systematized Nomenclature of Medicine Clinical Terms is an ontology or multi-lingual library for scientifically accepted terms in clinical healthcare making the knowledge easier to use and access. SNOMED is used by physicians and other health care providers to electronically exchange clinical health information. Currently, there are 300,000 clinical terms in SNOMED-CT.
- MeSH (Medical Subject Headings) [19] is an ontology developed by the U.S National Library of Medicine (NLM) consisting of medical and biological terms organized as a hierarchical structure. MeSH includes the subject headings appearing in MEDLINE/PubMed, the NLM Catalog, and other NLM databases. It is used for indexing, cataloging, and searching biomedical and health-related information.

10.4 TECHNIQUES FOR BUILDING ONTOLOGIES

Ontology is a hierarchical structure describing the domain concepts concerning the classes, properties, features, and attributes of the domain. Restrictions on properties can also be included in the ontology. Ontology development is practically an organized process of defining all the classes of the domain and arranging them hierarchically. The basic advantage of ontology development is to help users or web agents in understanding the ontology structure and reusing the domain knowledge. There is no particular restriction on the ontology development methodology. There are many potential ways to develop these ontologies.

All the available ontology constructions methods are classified into two segments. The first one is a knowledge-based method applicable when all the requirements are clear from the start itself. The second one is the evaluative model used when the domain requirements are not much clear from the beginning. These methods can be combined as every method has separate design ideas distinct from others. The merging can be done based on the primary goals of ontology development [20].

NF Noy and McGuinness [21] underline that there are several alternatives for ontology construction and there is no one single way. The best method depends on the application. The concepts of any ontology must represent the entities and relationships of the domain. Objects stand for nouns and relationships stand for verbs that describe the domain. Ontology building includes defining classes, arranging them as a taxonomy, defining class properties, and giving values to the properties and instances. The knowledge base can be created by defining individuals to the classes and assigning them with property values and restrictions.

10.4.1 Constructing Ontologies from Scratch

This method falls into the first group of ontology development which is used when all the requirements of the domain are clear at the beginning itself. To construct ontologies from the beginning, the method developed by Noy and McGuinness [21] is most popularly used. This method is also adopted by Grigoris and Frank [3], Berners Lee, Godel, and Turing [12]. The simple steps in the ontology development process are shown in Figure 10.4 [22].

10.4.1.1 Case Study on Building Library Ontology

Using the steps shown in Figure 10.3, a case study on building library ontology is explained in detail in [22]. This process uses the protégé ontology editor as it is a popular editor used by many researchers. After understanding the scope of the

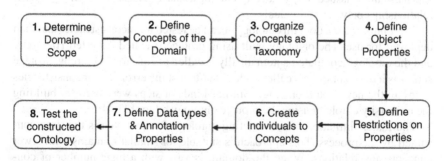

FIGURE 10.4 Steps to build ontology from scratch.

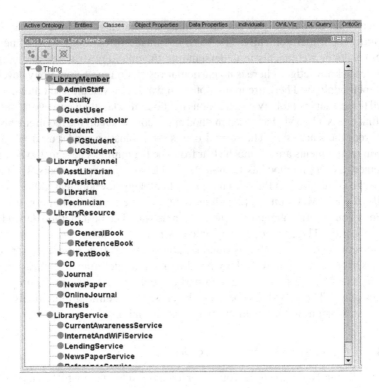

FIGURE 10.5 Concept taxonomy for library ontology.

library domain the major concepts defined and organized into a taxonomy (steps 1–3) can be shown in Figure 10.5.

Properties and corresponding restrictions can be defined on all these concepts depending on the scope of the domain. For example, consider the property BORROW which creates a relationship between LibraryMember and LibraryResource in general and specifically between Faculty or Students and Books. The restriction on the property can be like students can borrow a maximum of three books in one semester whereas for faculty it could be five books. Journals can be issued only to faculty but not students. These properties represent relationships between individuals.

After building the domain ontology, it can be tested for inconsistencies and then deployed for use. The ontology built using protégé is called Asserted Hierarchy and the ontology computed automatically is called as Inferred Hierarchy. Protégé supports a reasoner or also called a classifier to test the errors and inconsistencies if any in the newly built ontology. Similar kinds of steps were used for building the university ontology using the protégé ontology editor [23,24].

The major limitation of constructing ontology from scratch is the time consumed by the process. This approach is suitable for smaller domains with fewer concepts and relations. When the domain is vast with a huge number of concepts then we need a deep understanding of the domain before we start building the ontology. This becomes more difficult when the ontology Engineers are not domain experts. Hence, many researchers have worked on a fully automatic process of ontology construction using several intelligent agents.

10.4.2 AUTOMATIC CONSTRUCTION OF ONTOLOGIES

A fully automatic approach for constructing ontologies was proposed by Harith. Alani [25] is as shown in Figure 10.6.

- The first step searches for domain-relevant ontologies using semantic web search engines like Swoogle or Hakia. This process results in all Uniform Resource Identifier (URI's) of the domain relevant ontologies.
- The list of ontologies is ranked in the next step based on a priority basis to start with the best and relevant ontology.
- The system uses some ontology segmentation tools to extract only certain relevant parts from the extracted and ranked ontologies.

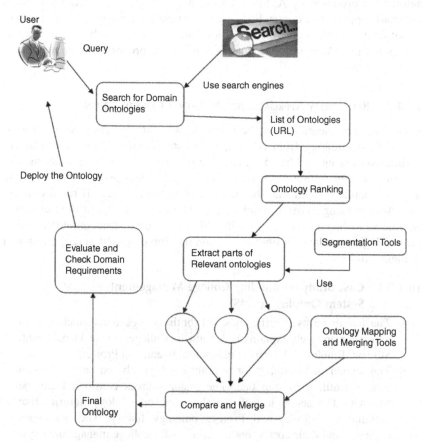

FIGURE 10.6 System for automatic ontology construction [25].

- The segments of ontologies obtained are then compared and merged by the system to form a complete domain representation. There are many tools developed for mapping and/or merging ontologies like the PROMPT Suite integrated into Protégé ontology editor and Chimeara built on Ontolingua.
- The ontology thus built by the system is evaluated to check if the resulting ontology meets the user and domain requirements or not.

The prime limitation of such a system is its successful execution due to multiple tools used at every step. The major challenge rises when the extracted parts of ontologies are mapped and merged and if it results in a large and messy ontology.

Some more automatic ontology building systems are proposed by many researchers like the Fully Automatic Enterprise Ontology Construction Using Design Patterns [26], TextOntoEx: an automatic ontology development by natural English text [27]. A method for automatic construction of domain ontology is proposed by A. Singh and P. Anand [28]. The limitation of such automatic approaches is very little user intervention during the entire process of ontology construction. A very tedious process may also be of no result if it does not meet user requirements and the entire process may be needed to start again.

10.4.3 REUSABILITY APPROACH TO ONTOLOGY CONSTRUCTION

Knowledge representation on the semantic web largely relies on ontologies. To build an ontology from scratch, gathering complete domain knowledge is time-consuming and has no guarantee of meeting users' expectations too. If a fully automatic approach is used, then it has less user intervention and the final ontology built may be messy and may not meet all requirements. Therefore, reusing existing ontologies can be both a cost and time-effective alternative to the other two approaches. Five steps to construct ontologies using the ontologies available in libraries and on the web proposed in [29] are shown in Figure 10.7.

10.4.3.1 Case Study on Building College Management System Ontology (CMS)

- The domain terms identified in step 1 for the college management system which can precisely explain the domain are College, course, Department, Student, Employee, Library, Publication, Event, and Project.
- The second step searches for relevant ontologies based on the domain terms identified in step one. The major sources considered can be SWOOGLE a semantic web search engine or ontology libraries like Ontolingua, DAML, and Protege ontology library. The ontologies retrieved and their corresponding sources for college management system ontology are listed in Table 10.1 [29].

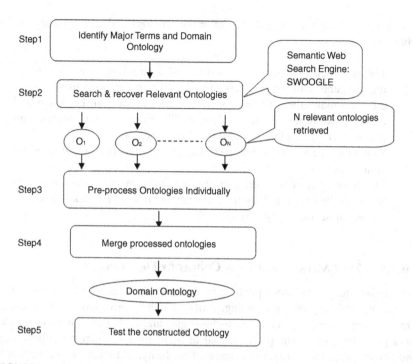

FIGURE 10.7   Reusability approach to ontology construction.

TABLE 10.1
List of Retrieved Ontologies Relevant to CMS Domain Terms

Ontology	Source	Domain Terms
Ka.owl	http://protege.cim3.net/file/pub/ontologies/ka/ka.owl	Organization, Department, Project
Univ_bench.owl	http://owl.cs.manchester.ac.uk/2008/isws-tones/ontologies/ univ_bench.owl	Employee, Student
Swrc_v0.3.owl	http://ontoware.org/swrc/swrc/swrcowl/ swrc_v0.3.owl	Event, Publication
university. Owl	http://mindswap.org/ontologies/university.owl	Course, Library
www04photo.owl	http://www.mindswap.org/2004/www04photo.owl	Event
Publication. Owl	http://ebiquity.umbc.edu/ontology/publication.owl	Publication

The ontology building method that reuse the ontology's already existing in the libraries.

- The ontologies retrieved from different sources may not meet our requirements completely. Therefore we need to pre-process these ontologies and make them fit the domain. New concepts can be added and concepts not required may be deleted. The names of the concepts can also be changed and make them suitable to the domain.
- Step four merges these processed ontologies into one single required domain ontology.
- The ontology resulting after merging may still have inconsistencies. Step five tests the final ontology to remove errors if any and make the ontology accurate and ready to use.

10.4.4 AGILE METHODOLOGY FOR ONTOLOGY DEVELOPMENT

Ontologies are getting more popular in multiple areas including semantic web search, natural language processing, artificial intelligence, bioinformatics, etc. Ontology engineering is related to technologies and tools used for building and managing ontologies. The primary goal of ontology construction methodology must be ensuring the clarity, extendibility, reusability, and reliability of the ontology. There are many challenges in the method of ontology construction because these methods are mostly applied to develop specific domain ontologies [44]. Many of these challenges are related to software engineering. If ontology development can be supported by the principles of software engineering, then the ontology can be made more reliable and adaptable [45].

The Agile Methodology for Ontology Development (AMOD) integrates the ontology engineering activities with the agile practices to fill the breach between ontology engineering and software engineering. AMOD classifies the ontology development process into three phases namely pregame, development and postgame. The phase focus on identifying the scope of the ontology to be developed and the tools and techniques, available sources. The development phase includes multiple iterative cycles called sprints. The final phase prepares the ontology and looks into its evaluation and maintenance [45].

10.5 ONTOLOGY EDITORS AND VISUALIZATION TOOLS

Several applications called ontology editors are designed that help users construct and manipulate ontologies. The ontologies can be expressed in many languages and exported using these editors. Jorge Cardoso presents a survey and comparison done on 14 different ontology editors [30]. A similar kind of comparison is done by Emhimed Alatrish on five different ontology editors [31]. Considering these two works this section focus on five ontology editors widely in use by Semantic

Web communities ontology development. Some of the ontology visualization tools are also covered in this section.

10.5.1 ONTOLINGUA SERVER

This is the first ontology editor developed in 1990 by Stanford University at Knowledge Systems Laboratory (KSL). This tool develops ontologies by using form-based applications and is considered be an easy tool as it allows reusability, i.e. parts of existing ontologies can be utilized to build a new one. Several repositories store several ontologies including many domain areas and the ontologies built using the ontolingua tool can also be included in such repositories. Ontology editor is the main module in this tool which includes many modules such as Webster, equation solver, Open Knowledge-Based Connectivity (OKBC) server, and Chimaera (an ontology merging tool), etc. This server includes translators for languages like Prolog, CORBA's, and Loom, etc. The remote editors and the local applications can browse, access, and edit ontologies from this library using the OKBC protocol. Ontolingua provides a distributed collaborative environment to browse, create, edit, modify, and use ontologies. This server supports the reusability of ontologies which is a very important information integration and developing knowledge base [32].

10.5.2 PROTÉGÉ

This is one of the free, open-source, and most widely used ontology editors developed at Stanford University supporting a large community of users for ontology construction. Experts in the medical and manufacturing domain have been using protégé for a long time for building knowledgebase systems and domain modeling. These editor extensions support ontology visualization, software engineering, project management, and many other modeling tasks. The early versions of Protégé enabled users to build only frame-based ontologies as per the Open Knowledge Base Connectivity protocol (OKBC). Later versions of protégé support several models to create, view and manage ontologies which makes it more efficient, scalable, and extendable.

The first architecture of Protégé supported only frames that were extended to OWL in 2003, which attracted researchers toward the vision of the Semantic Web. Several popular plug-ins allow ontology mapping, merging, advanced visualization, inference, etc. The OWL plug-in supports loading, save, edit and visualizing ontologies both in OWL and RDF. OntoViz tab allows presenting different graphical views of a knowledge base. Protégé's plug-in architecture can be adapted to build both simple and complex ontology-based applications. Developers can integrate the output of Protégé with rule systems or other problem solvers to construct a wide range of intelligent systems [33]. Figure 10.8 shows the survey done by Jorge Cardoso [30] on different ontology editors proving that more than 68% of the respondents use protégé for ontology development.

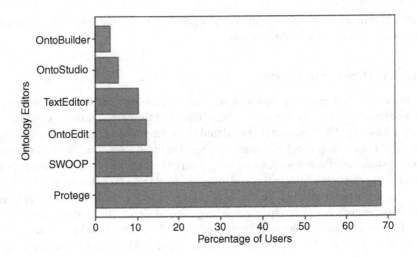

FIGURE 10.8 Survey results on ontology editors.

10.5.3 SWOOP

Swoop is an ontology editor and browser specially designed for OWL ontologies supporting validation and presentation syntax. There is also reasoning support that helps for comparing, editing, and merging ontologies. Navigation is made easy and simple with the hyperlinked capabilities in the swoop interface. There is no particular method for building ontology using a swoop, the users can reuse any external ontology by linking to the external entity, or the user can also import external ontology. We can search for several concepts across multiple ontologies stored in the Swoop knowledge base with the support of Swoop search algorithms. Swoop has a variety of mechanisms for Multiple Ontology Browsing and Editing, semantic search, Collaborative Annotation, and Multimedia Markup Extension [34].

10.5.4 ONTOEDIT

OntoEdit was developed at the AIFB Institute at the University of Karlsruhe by the Knowledge Management Group. It is an ontology editor that integrates numerous aspects of ontology engineering. It is an environment that supports ontology engineering by browsing, creating, managing, and maintaining. The environment allows the collaborative development of ontologies. This editor supports GUI to represent concept hierarchies, relations, domains, ranges, instances, and axioms. Ontoedit is multilingual where the name of each concept or relation can be specified in multiple languages. This tool supports F–Logic (Fuzzy Logic), RDF Schema, and OIL. OntoBroker is the inference engine that exploits the F-logic to represent expressive rules. Similar to Protégé, OntoEdit also depends on a plug-in architecture with three layers: GUI, OntoEdit core, and Parser. These plugins help users to extend the functionalities of onto edit [35].

10.5.5 APOLLO

This is known as a user-friendly ontology application for modeling ontology which allows users to model with fundamental concepts like classes, functions, relations, instances, etc. The internal model of Apollo is an OKBC protocol-based frame system that consists of a hierarchical organization of ontologies. This editor supports the inheritance of ontologies and imports them and accesses these ontologies. Every class can define several instances, and these instances can inherit all slots of the class. Every slot has a set of facets. The limitation of this tool is that it does not support a graph view of ontologies but is very strong in checking the consistency of ontology and storing the ontologies. Apollo is implemented in Java and a downloadable version is available at http://apollo.open.ac.uk/index.html [36].

10.5.6 ONTOVIZ

This tool helps to visualize Protégé ontologies using Graphviz which is a highly sophisticated graph visualization software developed by AT&T. The tool is highly configurable that allows selecting a particular set of classes or instances to visualize only a certain part of the ontology. The slots and slot edges can also be displayed and different colors can be specified to nodes and edges. Different closure operators can be used to visualize the vicinity of classes and sub-classes [37].

10.5.7 WEBVOWL

WebVOWL is a responsive web application used to visualize the OWL ontologies implementing the Visual Notation which is completely based on open web standards. These visualizations are generated from JSON files automatically. An exclusive converter implemented in Java called OWL2VOWL is currently in use for conversion. According to the specifications of VOWL, the ontologies are represented in the form of force-directed graph layout. The special feature of this visualization tool is its interactive mechanism that allows exploring the ontologies and customizing their visualizations [38].

10.5.8 BIOONTOVIS

This tool helps in visualizing ontologies such that it meets the different requirements of variant users from multiple application domains. BioOntoVis introduces a high-level architecture for ontology visualization that includes three steps named parsing, processing, and visual representation.

In the *parsing* step, ontology is represented as collection entities where each entity is either a class or a relation or an individual. These concepts and relations are disjoint and there exists a hierarchy between the set of concepts and among a set of relations. This step is essential to extract the ontology concepts and convert them into a JSON file to support graphical representation. The *processing* step focuses on both basic and advanced features required for visualization like color, shape, zooming, and editing. BioOntoVis supports three *visual representation* views including tree, network, and fish eye.

This tool provides a web interface using which the users can upload their ontology and choose among the types of visualizations available. This interface gives the user the metadata of the ontology and the user is allowed to make adjustments to the view as per their need [39].

10.6   APPLICATIONS OF ONTOLOGY AND SEMANTIC WEB TECHNOLOGIES

Semantic web technologies are being used in several application areas for data integration, resource classification, and discovery, improving the capabilities of domain-specific search engines, and many more using intelligent software agents to support the exchange of knowledge. Some of the applications of semantic web technologies and case studies include

a. Semantic web technologies in medicine

Generally, in medical science, the terminology and knowledge of terms and concepts are both complex and diverse and which hinders successful interdisciplinary studies. It is difficult to extract knowledge from the existing medical data due to the varied formats, schemes, and semantics they have. Interaction between health care systems has become the most challenging problem. A solution to this problem can be achieved by using ontologies and semantic web technologies [40]. Ontologies and semantic web technologies including RDF, OWL, and SPARQL provide the required semantics for the medical domain and can also serve as a tool for building robust and interoperable information systems. The primary focus of ontology in medicine is on the depiction and association of medical terms.

Doctors and medical practitioners have their specialized languages to maintain and communicate patient information and medical knowledge. Such medical terms when optimized for human processing are considered to be not clearly expressed. This increases the difficulty of medical information systems to communicate complex and detailed medical concepts. Such difficulty can be resolved by building an ontology for the medical terminology systems [40]. Currently, there are some ontology systems available in the field of medicine which include:

- **LinKBase**: is a biomedical ontology also called a knowledge base of nearly one million medical concepts that are language-independent with a formal conceptual description of the medical field.
- **MedO**: a biomedical ontology developed at the Institute of Formal Ontology and Medical Information Systems, Germany.
- **MeSH**: The Medical Subject Headings is a taxonomically -organized vocabulary of medical terms taken from the NLM Catalog, MEDLINE, and PubMed developed by the National Library of Medicine.

- **Gene ontology**: (GO) is an ontology for biological terminology created by a bioinformatics society and is considered to be the world's largest information source for functions of genes.
- **Disease ontology**: (DO) is a standard ontology developed for human disease to provide the biomedical area with reliable, reusable, and consistent descriptions of terms related to human disease their characteristics, and required medical vocabulary for the disease concepts.

b. Semantic web technologies for business

In business applications, several types of databases are used for distribution, accounting, etc. The major challenge faced by these systems is the way the business terms are understood by different systems. Even though Metadata systems are developed to help this problem, many times it becomes cumbersome to maintain the Metadata. Ontologies, address this challenge by formalizing the business rules and processes into a formal ontology. Semantic web technologies can make it possible to incorporate the same business rules to all the systems uniformly [41].

c. Semantic web technologies in IoT

IoT, the network of objects, is embedded with sensors and other software to connect and exchange data with devices and systems over the internet. These devices can range from regular household appliances to highly sophisticated types of equipment. The vision of the Internet of Things (IoT) is increasing with the number of devices being connected to the internet and being operated. The diversity of devices is becoming a challenge for current technologies to integrate their data, applications, and services more smartly. Even though the Web is a convenient platform for IoT, the Semantic Web can further enhance the capacity to understand things and improve their interoperability. A semantic web stack is also being introduced to take the benefits of semantic web technology for knowledge representation in IoT [42].

10.7 ONTOLOGIES FOR KNOWLEDGE REPRESENTATION: OPPORTUNITIES AND CHALLENGES

Knowledge representation is done in the most efficient way when ontologies take the leading role. Knowledge discovery applications are widely used where end-users write complex search requests to retrieve information. These users may not be able to grasp the semantic relationship between the data stored. Such a difficulty can be overcome by representing the knowledge and interactive queries using ontologies. Like any other technology evolving today developing semantic web, technology is also being motivated and benefitted by several opportunities including

- **Semantic web services**: Semantic web services are built around universal standards for the interchange of semantic data, which makes it easy for programmers to combine data from different sources and services without losing meaning. Semantic web services can also be used by automatic programs that run without any connection to a web browser. The semantic descriptions are registered in public registries that help the intelligent agents to migrate from one service registry to another and find required web services for the user.

- **Semantic search engines**: With the initiation of the Semantic Web, the resources on the Web are represented semantically using ontologies, and search engines can be built where queries can be executed within the context of some ontology. Swoogle is an example of a semantic web search engine for ontologies and documents saved on the web in the form of RDF and RDFS. It uses a collection of crawlers to discover the RDF and HTML documents [42].

New opportunities always come with some more new challenges and the ones faced in knowledge representation using ontologies include:

- **Content availability**: currently, the amount of data available for developing semantic web content is limited. The existing content of the web must be upgraded including the HTML pages both static and dynamic, multimedia, and web services.
- **Scalability**: A considerable effort has to be made in organizing and storing the Semantic Web content. The tasks performed must be done in a scalable manner since there may be a huge growth of data on the Semantic Web.
- **Multilingualism**: The problem of multiple languages is already prevailing in the current Web, and needs to be tackled properly in the Semantic Web. The Semantic Web must provide facilities to access information in multiple languages and allow the content access independent of the native language of users [43].

10.8 CONCLUSION

This chapter includes the basic introduction of the semantic web explaining the different layers in which the semantic web is classified. The languages which are very much essential for developing ontologies are discussed, and OWL is focused as a higher-end language to build ontologies supported by survey results. The types of ontologies are classified as general-purpose and domain-specific identifying four major methods adopted for ontology construction. The manual process starts from scratch with complete user interaction whereas the fully automatic method has very little user intervention in the entire process. A middle way also exists which is semi-automatic to take the advantage of both the methods. Software engineering principles are also adapted in the agile way of ontology construction. Later sections discuss the different editors and visualization tools available for knowledge representation using ontologies. The applications of semantic web technologies span different areas including medicine, business, and IoT. Finally, this chapter throws light on the different opportunities and challenges existing in ontological knowledge representation.

REFERENCES

1. T. Berners-Lee, J. Hendler, and O. Lassila. "The Semantic Web". Scientific American, Feature Article, May 2001.
2. https://en.wikipedia.org/wiki/Semantic_Web_Stack.
3. G. Antoniou, and F. Van Harmelen. A Semantic Web Primer. The MIT Press, Cambridge, MA, London, England, 2004.
4. V. Maniraj, and R. Sivakumar. "Ontology Languages – A Review". International Journal of Computer Theory and Engineering, vol. 2, no. 6, December 2010.
5. R. Karp, V. Chaudhri, and J. Thomere. "XOL: An XML-Based Ontology Exchange Language (version 0.4)", Aug. 1999, www.ai.sri.com/~pkarp/xol (current Jan. 2002).
6. S. Luke, and J. Heflin, "SHOE 1.01 Proposed Specification", SHOE Project, Feb. 2000. ww.cs.umd.edu/projects/plus/SHOE/spec1.01.htm (current Jan. 2002).
7. O. Lassila, and R. Webick. "Resource Description Framework (RDF) Model and Syntax Specification." W3C Recommendation, Jan. 1999, www.w3.org/TR/PR-rdf-syntax (current Jan. 2002).
8. D. Brickley, and R.V. Guha. "Resource Description Framework (RDF) Schema Specification," W3C Proposed Recommendation, Mar. 1999, www.w3.org/TR/PR-rdf-schema (current Jan. 2002).
9. A. Gómez-Pérez, and O. Corcho. "Ontology Languages for the Semantic Web". Intelligent Systems, IEEE, vol. 17, no. 1, February 2002, 54–60.
10. T. R. Gruber. "A Translation Approach to Portable Ontology Specifications". Journal of Knowledge Acquisition-Special Issue: Current Issues in Knowledge Modeling, vol. 5, no. 2, June 1993, 199–220.
11. N. Noy, and D. L. McGuinness. "Ontology Development 101: A Guide to Creating Your First Ontology". Technical Report KSL-01–05, Stanford Medical Informatics, Stanford, 2001.
12. J. G. del Rio. "Integration and Disambiguation Techniques for Semantic Heterogeneity Reduction on the Web". Ph.D. Thesis. Department of Informatics. The University of Zaragoza.
13. T. Slimani. "Description and Evaluation of Semanticsimilarity Measures Approaches". International Journal of Computer Applications, vol. 80, no. 10, October 2013, 25–33.
14. https://wordnet.princeton.edu/.
15. http://mozart.isi.edu:8003/sensus2/.
16. www.cyc.com/kb/.
17. https://www.nlm.nih.gov/research/umls/index.html.
18. https://www.snomed.org/.
19. https://www.nlm.nih.gov/mesh/meshhome.html.
20. G. Brusa, L. Caliusco, and O. Chiotti. "A Process for Building a domain Ontology: An Experience in Developing a Government Budgetary Ontology". Proceeding AOW '06 Proceedings of the Second Australasian Workshop on Advances in Ontologies, vol. 72, 7–15.
21. N. Noy, and D. L. McGuinness. 2001. "Ontology Development 101: A Guide to Creating Your First Ontology". Technical Report KSL-01–05, Stanford Medical Informatics, Stanford.

22. A. Banu, S. S. Fatima, and K. U. R. Khan. "Building OWL Ontology: LMSO-Library Management System Ontology". In: Meghanathan N., Nagamalai D., Chaki N. (eds) *Advances in Computing and Information Technology. Advances in Intelligent Systems and Computing*, vol. 178. Springer, Berlin, Heidelberg, 2013. doi:10.1007/978-3-642-31600-5_51.

23. A. Ameen, K. R. Khan, and B. P. Rani. "Construction of University Ontology", *2012 World Congress on Information and Communication Technologies*, 2012, 39–44. doi:10.1109/WICT.2012.6409047.

24. N. Malviya, N. Mishra, and S. Sahu. "Developing University Ontology using protégé OWL Tool: Process and Reasoning". *International Journal of Scientific & Engineering Research*, vol. 2, no. 9, September 2011.

25. H. Alani. "Position Paper: Ontology Construction from Online Ontologies". *WWW '06: Proceedings of the 15th International Conference on World Wide Web*, May 2006, 491–495. doi:10.1145/1135777.1135849.

26. E. Blomqvist. "Fully Automatic Construction of Enterprise Ontologies Using Design Patterns: Initial Method and First Experiences". *OTM Conferences*, 2005.

27. M. Y. Dahab, H. A. Hassan, and A. Rafea. "TextOntoEx: Automatic Ontology Construction from Natural English Text". *Expert Systems with Applications*, vol. 34, no. 2, 2008, 1474–1480. ISSN 0957-4174.

28. A. Singh, and P. Anand, "Automatic Domain Ontology Construction Mechanism". *2013 IEEE Recent Advances in Intelligent Computational Systems (RAICS)*, 2013, 304–309. doi:10.1109/RAICS.2013.6745492.

29. A. Banu, S. S. Fatima, and K. U. R. Khan. "A Re-Usability Approach to Ontology Construction". *CCSEIT '12: Proceedings of the Second International Conference on Computational Science, Engineering and Information* Technology, October 2012, 189–193. doi:10.1145/2393216.2393248.

30. J. Cardoso. "The Semantic Web Vision: Where are We?" *IEEE Intelligent Systems*, vol. 22, September/October 2007, 22–26.

31. E. Alatrish. "Comparison of Some Ontology Editors". *Management Information Systems*, vol. 8, no. 2, 2013, 018–024.

32. A. Singh, and P. Anand. "The State of Art in Ontology Development Tools". *International Journal of Advances in Computer Science and Technology*, vol. 2, no. 7, July 2013, 96–101.

33. N. F. Noy, M. Sintek, S. Decker, M. Crubezy, R. W. Fergerson, and M. A. Musen, "Creating Semantic Web Contents with Protege-2000". *IEEE Intelligent Systems*, vol. 16, no. 2, March-April 2001, 60–71. doi:10.1109/5254.920601.

34. A. Kalyanpur, B. Parsia, E. Sirin, B. C. Grau, and J. Hendler. "Swoop: A Web Ontology Editing Browser". *Journal of Web Semantics*, vol. 4, no. 2, 2006, 144–153.

35. Y. Sure, M. Erdmann, J. Angele, S. Staab, R. Studer, and D. Wenke. "OntoEdit: Collaborative Ontology Development for the Semantic Web". *The Semantic Web – ISWC 2002, First International Semantic Web Conference 2002. Proceedings.* Lecture Notes in Computer Science 2342, Springer, 2002, 221–235.

36. E. Alatrish. "Comparison of Ontology Editors". *ERAF Journal on Computing*, vol. 4, 2012, 23–38.

37. https://protegewiki.stanford.edu/wiki/OntoViz.

38. S. Lohmann, V. Link et.al. "WebVOWL: Web-Based Visualization of Ontologies". *International Conference on Knowledge Engineering and Knowledge Management.* April 2015. doi:10.1007/978-3-319-17966-7_21.

39. N. Achich et al. "BioOntoVis: An Ontology Visualization Tool." *EKAW*, 2018.

40. P. Manika et al. "Application of Ontologies and Semantic Web Technologies in the Field of Medicine." *RTA-CSIT*, 2018.

41. https://ubwp.buffalo.edu/ncor/quick-start/ontology-for-businesses/.
42. I. Szilagyi, and P. Wira. "Ontologies and Semantic Web for the Internet of Things – A Survey". *IECON 2016–42nd Annual Conference of the IEEE Industrial Electronics Society*, 2016, 6949–6954.
43. K. Munir, and M. S. Anjum. "The Use of Ontologies for Effective Knowledge Modeling and Information Retrieval". *Applied Computing and Informatics*, vol. 14, 2018, 116–126.
44. R. Benjamins et al. "Six Challenges for the Semantic Web". *KR 2002*, 2002.
45. A. S. Abdelghany, N. R. Darwish, and H. A. Hefni. "An Agile Methodology for Ontology Development". *International Journal of Intelligent Engineering and Systems*, vol. 12, no. 2, 2019. doi:10.22266/ijies2019.0430.17.

11 Data Science and Ontologies

An Exploratory Study

Prashant Kumar Sinha and
Shiva Shankar Mahato

CONTENTS

11.1 INTRODUCTION

Data are unprocessed facts that can be subsequently processed, interpreted and can be arranged to derive valuable information [1]. Data are now being produced in enormous amounts and especially in digital form. In modern times, data have been considered the most valuable asset of an organization and the field of data science (DS) has attracted a lot of attention. The studies that are being performed using DS clearly show that it has been used as a tool subject in many areas of the discipline. DS is not a buzzword now; it has emerged as a whole new subject of study. The origin of the concept of DS can be traced backed to 1960. Peter Naur termed it to be a substitute for computer science [2], whereas C. F. Jeff gave lectures on "Statistics = Data science?" [3,4] and Cleveland [5] termed it as a discipline itself, which has come true now as various undergraduate and postgraduate programs that have been developed especially for DS. Though many researchers and scientists have tried to define the term DS, there are still disagreements on the term and how

DOI: 10.1201/9781003309420-11

it has been used in the literature loosely [6,7]. Some of the major definitions of DS are (i) "an array of elementary propositions that reinforce and govern the principled extraction of information and knowledge from data" [7] (ii) "the study of the generalizable extraction of knowledge from data" [8]. (iii) "the systematic study of the building, assessing and converting of data to create meaning" [9]. (iv) "the extraction of fruitful intelligence straight from inputs through a process of revelation, or of hypothesis construction and hypothesis testing" [10]. (v) "array of doctrines, issue interpretations, algorithms, and processes for distilling the non-evident and fruitful sequences from massive datasets" [11]. These definitions clearly indicate that DS is a multifaceted and multi-disciplinary domain.

DS has now proliferated in many domains of studies like computer science [12] and library science [13] and is affecting them big time. The chief attraction towards DS is attributed to the power of the decision-making process it provides by the analysis of large amounts and a variety of data following scientific methods. Though there are a lot of pros, there are challenges in data storage, data acquisition, data wrangling, model development, validation, deployment, etc [11]. The DS process can be segregated into three parts: data pre-processing, data processing and data post-processing. These processes are composed of smaller processes such as data cleaning, data integration, data modelling, data sharing, data publishing, etc [14]. So while defining this field, it is critical that maintenance for the constant access to the expanding landscape of DS concepts and understanding of how they branch from high-level approaches to specific implementations is looked after. The concept of prior information, whether from the method or the domain, is one of the most critical and difficult problems in DS. Prior information aids in the selection of data and DS methods. Ontologies can accomplish this goal. It facilitates the structuring and conceptual representation of knowledge. It also supports the communication of explicit relationships between concepts, relationships between entities and their properties and enables machine processing of knowledge for effective information retrieval [15]. They're designed to bridge the gap between expert knowledge and computer system functionality by communicating expert knowledge in a way that computers can recognize and explore. Researchers require knowledgeable tools capable of filtering out irrelevant data and automatically categorizing relevant data, given the ever-increasing deluge of data in modern scientific domains [16]. Several ontologies are presently being advanced in portraying commodities across all the realms of science. Few fields of research have already been well described by large-scale and up-to-date taxonomies, e.g. MeSH (https://www.ncbi.nlm.nih.gov/mesh/) in Biology. Since DS is an ever-growing dynamic domain of study with many facets attached to it, ontologies can play a major role in structuring the domain as well as supporting various processes such as data integration, data annotation, etc. The primary objective of the work is to explore the role of ontologies in the DS domain and the existing data science ontologies (DSO) in the literature.

11.2 ONTOLOGIES

Gruber [17] defined ontology "as an explicit specification of a conceptualization" which was modified by Borst [18] as "a formal specification of a shared conceptualization" and then it was Studer et al. [19] explained the two definitions properly word by word where "Formal meant it should be machine-readable". "Explicit specification means it explicitly defines concepts, relations, attributes and constraints". "Shared means it is accepted by a group". "Conceptualization meant an abstract model of a phenomenon". Now ontology is termed as a knowledge artefact that has been designed to serve a purpose [20]. The ontology mainly consists of five components: concepts or classes (formalized parts of the domain), relationships (linkages between the concepts which may be taxonomic or non-taxonomic), functions (components with the objectives of estimating information from the other components), individuals or instances (representation of the domain's main artefacts) and axioms (definitions of limits, rules and logic correspondences that must be met in the relationship between ontology components) [21]. Giunchiglia et al. [22] described the ontological knowledge base as an amalgamation of Tbox a series of general explanations about what is perceived in terms of concepts, standing for collections of entities and concept properties; such descriptions establish the essential vocabulary and philosophy of the domain (e.g. persons have a date of birth) Abox a series of assertions about definite entities and the original value of their properties (e.g. the date of birth of Michelangelo Buonarroti is 6th March 1475).

To depict the ontologies in a formal manner so that it becomes machine-processable the representation languages are classified as XML-based (like ontology web language, OWL) and non-XML-based languages (Unified Modeling Language) [23]. Ontology can have different types such as domain or content ontology, meta-data ontology, general or common-sense ontology, representational/frame ontology and task/method/problem-solving ontology to serve various purposes [24]. The use of ontology is diverse which covers a wide range of topics including library and information science, knowledge management, education, medical science, information organization on the Internet and retrieval by search engines and financial institutions [25]. Since ontologies are being continuously developed in various ontology development has been an area of extensive research for the past 30 years; there have been many definitions of ontology development methodologies, but one of the most popular definitions was given by Gomez-Perez et al. [26] which defined it as "activities that concern the ontology development process, the ontology life cycle, and the methodologies, tools, and languages for building ontologies". Though the first known methodology came in 1990 with time, numerous methodologies have come into picture [27]. Few of the famous methodologies are METHONTOLOGY [28], YAMO (Yet Another Method for Ontology Construction) [29], NeOn [30], etc. The methodologies can help construct ontologies using various approaches such as manual, semi-automatic and automatic [31], and the tool support becomes very essential also while creating the ontologies. Tools like Protégé and NeOn toolkit are some of the popular tools for ontology development [32].

11.3 ROLE OF ONTOLOGIES IN DATA SCIENCE

The domain of DS comprises various aspects, but at its core is data. Different frameworks and implementations have different perspectives and mechanisms for using ontologies in DS. Ontologies can explain the data, represent the data in machine-processable format, integrate the data from heterogeneous resources, increase data quality, describe results from got DS processes, etc. Thus, ontologies can play significant roles in various aspects of DS. The review of existing literature allowed us to identify the major aspects of DS where ontologies can play a significant role, namely semantic data modelling, semantic data mining (SDM) and semantic data integration (SDI).

11.3.1 SEMANTIC DATA MODELLING

Semantic data modelling is the process of capturing the data with its appropriate explanation, with all of its underlying linkages, into a data model [33]. Ontology allows the data to be represented in a structured, meaningful manner and with logical connections. It allows various types of data such as unstructured, semi-structured and structured data, to be represented. Ontologies facilitate the creation of a network that can analyse the data and its relationships with other data in a semantic data modelling. Since ontologies represent the data in a meaningful manner, it makes them readily available for automated data processing as well in a semantic data modelling [34,35]. Ontologies provide a common understanding of information between communities and help to identify the assumptions because of its interconnectivity and interoperability. In a large organization where an extensive amount of data is generated, the ontology based semantic data modelling becomes extremely crucial for accessing and querying data. For ontologies, an important aspect is the relationships between concepts and they are suggested to work like the human brain; it provides a path for automatic reasoning which is employed easily in ontology-based semantic graph databases. Ontologies have more coherent and simple navigation as users switch from one definition to another in the ontology framework, besides the reasoning function [34,35]. Ontologies are also simple to extend because relationships and concept matching can be easily added to current ontologies. This allows the ontology-based semantic data modelling to be dynamic and strengthen with time and increase of data without affecting dependent processes and systems, even if some changes need to be made in the model that was created at the beginning [36].

Ontology-based semantic data modelling for representing and reasoning the knowledge has been developed for various purposes like to support knowledge-based document classification on disaster-resilient construction practices [37], supporting supply chain management [38], etc. A new movement that has revolutionized the DS domain is the FAIR data movement [39]. The FAIR data principle presents a framework or set of guidelines where the process of FAIRification requires the use of ontologies or vocabularies for the making the data FAIR. FAIR data will require FAIR ontologies [40]. In making ontologies

FAIR, the existing infrastructure called as ontology libraries [41] plays a major role. These libraries emphasize the depositors of ontologies to submit them in standard format, like OWL ontologies. OWL is a W3C recommended representation language for ontology-based semantic data modelling. It's built in such a way that it can reflect a wealth of information about things and their relationships. It also distinguishes between groups, properties and relationships in a detailed, consistent and meaningful manner. OWL enriches ontology modelling in semantic graph databases (a.k.a. resource description framework RDF triple stores) by describing target classes and linkage properties as well as their hierarchical plan. In such triple stores, OWL, when coupled with an OWL reasoner, supports for harmony checks (to locate some rational disparities) and satisfactory inquiries (to discover whether there are any classes that cannot have instances). It incorporates features for establishing equivalence and distinguishing instances, classes and properties. Even though different data sources define them differently, these associations help users fit definitions. They often make sure the various spellings of the same name or definition are differentiated [36].

The domain of DS has always been facing the issues of data quality. Data scientists are very concerned about the quality of data that goes into their semantic data modelling, so solutions to this problem are urgently needed. The data quality has many aspects, like data provenance, that can be improved by ontologies as it increases data provenance. The roots of a piece of data and how it is stored in a database are defined by data provenance. As a result, in order to capture the source of process data, the process flow definition must be captured (prospective provenance) [42]. Data must be semantically modelled, collected and processed for future queries in order to reap the benefits of provenance [43]. The process provenance data can be employed for analysis and process improvement after it has been taken and deposited (e.g., shorter execution time and better competence of the results). The adoption of ontology and the interpretation processes offered by it provides the diagnosis of critical intelligence for software project executive which is one desirable means to analyse processes provenance data [44]. Dalpra et al. [44] introduced a layer for the repository of software process provenance data and interpreting the data using an ontology. A W3C provenance model called PROV-O [45] was employed both for storage and investigation of these data.

11.3.2 SEMANTIC DATA INTEGRATION

Semantic data integration (SDI) is the method of taking advantage of the semantic data modelling and their relationships to reduce the data heterogeneity possibilities and ontology is the backbone structure that facilitates the process [46]. Since they present a clear and machine-comprehensible perception of a realm, ontologies have been extensively adopted in data integration operations. They've been employed in one of the three forms [47].

Single ontology approach [47]: All source schemas are precisely assigned to a shared global ontology that presents a uniform interface to the user. However, this procedure demands that all sources have approximately the same view on a domain, with the same degree of granularity [48]. A classic illustration of a system employing this procedure is SIMS [49]. This system accepts a domain-level inquiry and dynamically chooses the pertinent information sources based on their composition and availability generates a query access method that determines the procedures and their order for transforming the data and later produces semantic query optimization to curtail the overall execution time [49].

Multiple ontology approach [47]: Each data source is illustrated by its own (local) ontology separately. Instead of utilizing a universal ontology, local ontologies are mapped to each other. For this purpose, supplementary representation formalism is imperative for deciding the inter-ontology mappings. The OBSERVER technique is an illustration of this approach [50]. This is the framework that solves the main issue of vocabulary sharing. It accesses heterogeneous, distributed, and independently created data repositories using several pre-existing ontologies. Each data repository's content is represented by one or more ontology expressed using a Description Logics-based framework. Each data repository is examined at the level of the semantic concepts that are important. DL expressions based on concepts in (user) domain ontology are used to specify information requests to OBSERVER. OBSERVER uses ontological inferences to characterize the query and determine data repositories and then translates the DL expressions to the data repositories' local query languages. Mechanisms for partial translations got by integrating synonym relationships and gradual enrichment of the answers were introduced and discussed in. The replacement of a word by a combination of hyponym and hypernym relationships, which results in a shift in the query's semantics, was not considered previously.

Hybrid ontology approach [47]: A consolidation of the two preceding methods is practiced. Initially, a resident ontology is established for each source schema, which, however, is not mapped to alternative resident ontologies, but to a universal distributed ontology. Additional sources can be readily included, with no requirement for restricting existing mappings. A layered framework proposed by Cruz et al. [48] is an example of this approach. The goal was to create a unique solution to the problem of data interoperability between different databases, allowing for SDI. The problem of converting queries from one database schema into queries from another database schema was solved by leveraging their ontology relationship. The databases and ontology was modelled using RDF schema. The mappings between each database schema and the ontology were also expressed using a standard vocabulary. Though the literature provides various methods using ontologies in SDI but it's actually the lack of publicly available ontologies preferably in several popular formats, in easily discoverable places makes it difficult to create such solutions. But with the advent of ontology libraries, it is expected that researchers can be more aware of making the ontology available or accessible.

11.3.3 SEMANTIC DATA MINING

Data mining (DM) processes that structurally integrate subject information, especially formal semantics, into the task are referred to as SDM [51]. Three purposes for which ontologies have been inducted into the SDM has been summarized through literature study by Dou et al. [51] The objectives have been discussed here with specific examples from the literature as discussed by Dou et al. [51] to suffice those objectives.

a. *To fill semantic void between data, applications, DM algorithms and DM outcomes* [51]. In order to suffice these objectives, ontologies have been used post-elimination tasks and refining of the association rule mining (ARM) outcomes [52–54]. Ontologies in the semantic deep setup can be employed to illustrate data mining results, causing them simpler to distribute and reuse [51]. The process of automatically deriving organized information from text, for example, is known as information extraction (IE). The DM/text-mining outcomes emanate an array of organized information and domain insights. It is normal to use ontology to describe organized and machine-readable content. Information Extraction Using Ontologies (OBIE) [55] has extensively used this representation.

b. To present DM algorithms with a priori insight that can regulate the derivation workflow or restricts/necessitates the exploration space [51]. The explanation and reuse of previous intelligence is prominent dilemmas for SDM. As an explicit blueprint of concepts and links, ontology is a natural means to cipher the explicit semantics of previous knowledge [51]. The ciphered previous intelligence includes the power to regulate and persuade entire phases of the DM process, from pre-processing to outcome refining and depiction [51]. For example, Liu et al. [56] grab knowledge from both ontologies and data. In the work at first, a RDF hyper-graph representation was created. Ontologies were used as a priori information to skew the graph structure and to represent the separation between terms and conceptions in the graph. To illustrate both data and ontology in a suited scheme, the process converts the hyper-graph and weighted hyper-edges into a bipartite graph. For generation of semantic associations, random walks algorithm is executed with restarts being performed upon the bipartite graph.

Ontologies have domain insights encoded and as random-walk algorithms are employed it discovers the latent semantic connections that are below the data layer. Ontologies are a set of entities and predicates, thus possessing the potential to do logical interpretations and perform consistency interpretation for the predicates. Constraints are ordinarily utilized in SDM to characterize the ability to produce consistency inferences. The set of restriction induced by the ontology has the capacity to recognize conflicting data and outcomes during pre-processing phase, algorithm execution phase and the outcomes refining and creation phase. Balcan et al. [57], for example, used ontology as consistency constraints in an array of classification processes. The constraints between different classification processes are determined by the ontology. Carlson et al. [58] showed a semi-supervised IE algorithm that connects the guidance of various knowledge extractors. Using ontology as inhibitions on the collection of extractors, it gives way further conclusive results. Claudia Marinica et al. [52,53] suggested post-processing of the ARM results with the use of ontology for compatibility testing. Irrational or irreconcilable association rules are snipped and refined away through the cooperation of ontology and an interpretation engine.

c. *To present a machine interpretable way for depicting the DM workflow, starting from data pre-processing to mining results* [51]. For the structured illustration of DM outcomes ontologies are quite suitable and as they provide explicit interpretation of entities and linkages, they can cipher extensive semantics for various domains. The results of DM process originate disparate subjects and actions consistently in accordance with the depiction of ontology. For instance, IE and ARM. Specifically, in ontology-based information extraction (OBIE) [55,59], the derived information are a collection of annotated terms from the document along the relationships established in the ontology. So ontology is suitable for describing the derived information. Wimalasuriya and Dou [55] illustrated ontology as tool to explain the OBIE outcomes in a semantic affluent scheme. Ciphering OBIE outcomes in the formal ontological framework could align the DM process of separate DM actions requiring the usage of prevailing results. The inference engines developed in the domain of knowledge engineering may reduce the error possibilities that can allow DM outcomes and filter irreconcilable results. OBIE systems are better than traditional IE systems as it extracts information with better recall and accuracy. The ontology in OBIE serves as a theoretical scheme and consistency testing. It also regulates the derived knowledge in an explicit and organized manner using explicit knowledge representation [51].

11.4   DATA SCIENCE ONTOLOGIES

DSO is a knowledge artefact that structures and represents the knowledge in the domain of DS with the help of concepts, attributes and relationships that has been identified by studying the literature about DS. Ontology development is not a new trend, but the domain of DS is comparatively newer to the other domains of science like medicine, physics. Hence, the effort made to create DSO has been relatively low. Still the few efforts that have been made in the literature have been summarized.

Sean McClure [60] while working as a data scientist for Thoughts Work published the first DSO through a blog in 2014. The DSO was more so a lightweight ontology or a simple taxonomical representation of the various facets of the domain of DS. The major concepts were Model Validation, Learning Algorithms, Model Performance, Data Visualization, Databases, Production, Programming Languages, Data Clustering, Data Preparation, Statistics, Development, BusinessAccumen. The ontology was an available hierarchy as a pictorial representation in a PDF file. It does not explicitly mention any of the data properties or object properties of the ontology. Though it does not comply with the general norms of semantic community, the contribution cannot be ignored as it laid the foundation of ontology development in the domain. The ontology was further enriched, modified by Sean McClure and was made available in the same taxonomic representation but with more base concepts to cover the domain [61].

Chuprina et al. [62] aimed to equip students from various IT specializations and non IT specializations with the information necessary to create domain ontologies for their own research areas. They made a number of suggestions, one of which was to reuse existing ontologies in the construction of new ones. So the ACM Computing Classification System (CCS) by ACM [63] and Sean McClure [60] DSO were extended as a training example to improve Computer Science Skills. During this exercise, the Severo Ochoa project was created and the New Data Science Ontology (NDSO) as a part of this project was launched in order to assist instructors and learners, as well as stakeholders in industry and academia, understanding the roadmap of their personal journey in aspiring to be a data scientist. Sean McClure [60] ontology was converted to OWL a defacto format with the help of ONTOLIS environment in order to make it more usable. Subsequently Chuprina et al. [64] created a special type of learning activity called "DS-profile construction" (Data Science skills) to make the post-graduate students realize the work profiles of data scientist and how they can collaborate and contribute in their tasks. The DS-profile is an ontological synopsis of a student's history, talents and skills in DS which was created using the ONTOLIS 2.0.

Patterson et al. [65] published a DSO [66] to model DS flows semantically from original code. The concepts, in their work, formalize abstract concepts like machine learning, statistics and data computing and later map real software items to the concepts. The authors' primary advantage is that the structured formal exposition presents a means to draw semantic similarities amidst software items, such as when two sections of code represent the identical assessment steps, such as clustering. But, the semantic model developed does not reuse existing ontologies and, in its present form, is concentrated on features more intimately linked to code equivalence rather than an exhaustive interpretation of data semantics. The ontology was established in a unique ontology language called the MONoidal Ontology and Computing Language (Monocl) and has 153 DS concepts and 100 code annotations. Concepts represent the abstruse ideas of DS. They constitute the primitive types and primary functions from which more compound types and functions are established, employing the ontology language. Annotations describe instantiation by mapping the types and functions in software packages onto the types and functions of the ontology. The ontology and explanation of its units has been made open available as well [66] which can be skimmed in several different manners, like alphabetical by name, by language and by package, etc.

Shah and Subramanian [67] proposed an academic DS recommendation framework that used a DSO and a service-oriented architecture to help students find the best resources for their questions. The work's main goal was to provide students with important knowledge about books, online documentation, software resources, public code repositories and experts, tutors who can be contacted based on a student's question or previous exam scores related to. DSO was developed to define all aspects under the purview of the domain so that students can explore while querying the system. The ontology was developed in OWL file format and was created using Protégé. The developed DSO had 120 subclasses for 6 data-science classes. The major classes were DS process, data types, domain, programming tool, resourcedepth, and resourceformat.

A similar ontology called Data Science Education Ontology (DSEO) [68] is available in the ontology library [69] called Bioportal. It is a taxonomical structural view of subfields of DS for the utilization in education related applications. It is available in OWL format and has around 131 classes. The classes of the DSEO have been mapped with 38 other ontologies that are available in BioPortal as well. Sicilia et al. [70] identified various vocabularies that can be employed data pipelines. It was proposed that these ontologies could be used to semantically elucidate data transformations and semantics can be added to the current code by using sub-classing to apply meta-knowledge to unique components of the data pipeline interfaces.

It is quite evident that there is notably a less number of DSO in the literature. There are a few reasons for it as compared to some other prevalent subject like medicine; disaster, etc. the domain is relatively new. Though some of the components of DS like DM are relatively old, still a lot research is going on. Though there significant number of ontologies related to DS available like data mining ontologies, machine learning ontologies, computer science ontologies, etc., still DSO dedicated for the DS is still under progress, but these ontologies can be mapped together to develop a network of ontologies. DSO can also be connected to other domain ontologies as well because one of the important aspects of DS is domain knowledge. If these ontologies are linked to other domain ontologies, then it can help in the data analysis immensely.

11.5 RELATED WORK

Ontology exploration is at its crest now. Therefore, in a particular realm of concern, like museum, flood, medicine, etc., various kinds of ontologies are being formed to handle diverse objectives to focus on numerous investigate challenges. Therefore, an observation between such ontologies acknowledges the resemblances and diversifications among the ontologies. Attempts have been carried where ontologies have been analyzed in a particular domain. In the case of DSO, though this work is first of such attempts, previously researchers have tried to review the data mining ontologies (DMO) in the literature.

Panov et al. [71] depicted relating components that were present in the Ontology for Data Mining Entities (OntoDM) core ontology with different existing DMOs. Similarly, Arantes and Yokome [72]; Benali and Rahal [73]; Li et al. [74]; Tianxing et al. [75] also briefly described few of the existing DMO in the literature. But the narrative of the reported DMOs presented in these studies was not adequate to depict a deeper understanding of each DMO. But recently Sinha et al. [76] reviewed 35 DMOs that have been published in last 20 years based on two sets of parameters which in total amounts to 20 parameters. The effort had two fundamental purposes: (i) establishing the significant research in the domain of DMO and (ii) granularly figuring out the ongoing DMOs from various viewpoints. The initial intention was achieved by utilizing systematic literature review, which rounded off in a collection of core research comprised 35 papers that discussed the developed DMOs and established their strategic concepts within the study or as an OWL file. This study is the unique in terms of reviewing the DMOs in such a manner. It provides an extensive knowledge about the ontologies. The study can act as a one-stop point for the exploration and study of DMOs.

Along with these there are other domains also where ontologies have studied extensively to gain knowledge about them. Vilches-Blazquez et al. [77] discussed three hydrographical ontologies based on various attributes. Similarly, Zhang et al. [78] depicture the framework of four mainstream unit's ontologies. Prantner et al. [79] discussed openly available tourism ontologies, whereas Mascardi et al. [80] illustrated the popular seven upper-level ontologies. Ruiz-Iniesta and Corcho [81] reviewed three categories of ontologies that are used to describe scholarly and scientific documentation. Soares et al. [82] discussed the studies related to legal ontologies. Norris et al. [83] illustrated ontologies for human behaviour change. These are a few examples where ontological review was performed for ontologies in separate domains. Khamparia and Kaur [84] discussed medical care ontologies from a broader perspective; Messaoudi et al. [85] discussed the medical ontologies employed for the depiction of liver disease; Okikiola et al. [86] discussed the health care ontologies, whereas Yousefianzadeh and Taheri [87] discussed only the COVID-19 ontologies present in the BioPortal (https://bioportal.bioontology.org/). Liu et al. [88] discussed different disaster management vocabularies, whereas Sinha and Dutta [15] discussed only flood ontologies in their work. These are a few such examples where ontologies are being built even within a domain to serve different purposes. Thus, the application and usability of ontologies will differ from application to application.

11.6 CONCLUSION

DS and ontologies are two key fields of this modern era and when used in combination can revolutionize the field of artificial intelligence. The general notion of data scientists is that they require a more well described dataset, so that they can apply the required models to analyze the dataset. Ontologies describe data in a meaningful manner that can be utilized for various data analysis processes. By allowing any data form to be represented, including unstructured, semi-structured and structured data, ontologies enable easier data integration, simpler concepts, text mining and data-driven analytics. It also enables better semi-structured and unstructured data pre-processing, algorithm simplification, easier data access and statistical analysis [36]. Ontology is generally likewise selected to embrace issues about the most general aspects and relationships of the individuals that are available in the real world. Since ontology possess machine interpretable semantics it can perform various tasks related to DS. Ontologies can support various tasks like ARM, classification, clustering, etc. It is also expected that because of the semantic knowledge available. Users can even navigate between the concepts easily and understand their relationship because of the ontologies. The developed or the conceived semantic data model can be applied to existing dataset also to create the knowledge graph. Knowledge graphs have come to the forefront of research studies and have employed in various information retrieval systems to relate the datasets.

In this study, the role of ontologies in DS has been explored from fronts like semantic data modelling, semantic data integration and SDM using examples from the literature. The work here also summarizes some of the DSO in the literature. This part of the chapter reveals that very few DSO have been developed to suffice very few objectives like providing a taxonomical view of domain, supporting recommendation systems for education, creating a data scientist skills profile and describing the DS program code. Some of the DSO-like DS-construction ontology [64] reuse the existing ACM Computing Classification System (CCS) by ACM [63] and Sean McClure [60] DSO to create concepts. This shows the ontology development process follows best practice of reuse existing ontological resources. An ontology called DSEO was available in an ontology library called BioPortal as well and has been mapped to the other ontologies that are available in the library showing its diversified usage. There are a lot existing data mining ontologies (DMOs) in the literature available. Even the researchers have reviewed these DMOs. DS is a broader field to DM but a lot of the components of DSO can be mapped to existing DMOs. This can create a network of ontologies in the domain of DS which could represent the knowledge in a machine process able format. The researchers can use this work as a starting point when they are looking for DSOs to identify the research gaps and develop more DSOs for a variety of purposes.

REFERENCES

1. Gajbe, S. B., Tiwari, A., Gopalji, & Singh, R. K. (2021). Evaluation and analysis of data management plan tools: A parametric approach. *Information Processing & Management, 58*(3), 102480.
2. Wainer, H. *Truth or Truthiness: Distinguishing Fact from Fiction by Learning to Think Like a Data Scientist*, (Cambridge University Press: Cambridge), 2015.
3. Press, G. Data science: What's the half-life of a buzzword? Forbes, 2013. Available at: www.forbes.com/sites/gilpress/2013/08/19/data-science-whats-the-half-life-of-a-buzzword/#3e86a69c7bfd (Accessed on 15July 2020).

4. Ratner, B., *Statistical and Machine-Learning Data Mining: Techniques for Better Predictive Modeling and Analysis of Big Data*, (Chapman and Hall/CRC Press: Boca Raton, FL), 2017.
5. Cleveland, W. S. (2001). Data science: An action plan for expanding the technical areas of the field of statistics. *International Statistical Review, 69*(1), 21–26.
6. Granville, V. *Developing Analytic Talent: Becoming a Data Scientist*, (John Wiley and Sons: Hoboken, NJ), 2014.
7. Provost, F., & Fawcett, T. (2013). Data science and its relationship to big data and data-driven decision making. *Big Data, 1*(1), 51–59.
8. Dhar, V. (2013). Data science and prediction. *Communications of the ACM, 56*(12), 64–73.
9. The Data Science Association. About data science, 2017. Available at: www.datascienceassn.org/about-data-science (Accessed on July 16, 2020).
10. NIST. National institute of standards and technology (NIST) special publication 1500–1r1.NIST Big Data Interoperability Framework: Volume 1, Definitions. Version 2. NIST Big Data Public Working Group (NBD-PWG), 2018. Available at: www.nvlpubs.nist.gov/nistpubs/SpecialPublications/NIST.SP.150-1r1.pdf (Accessed on July 16, 2020).
11. Kelleher, J. D., & Tierney, B. *Data Science*, (MIT Press: Cambridge), 2018.
12. Virkus, S., & Garoufallou, E. (2019). Data science from a perspective of computer science. In: Garoufallou, E., Fallucchi, F., & William De Luca, E. (eds) *Metadata and Semantic Research. MTSR 2019. Communications in Computer and Information Science*, Vol 1057. Springer, Cham. doi:10.1007/978-3-030-36599-8_19.
13. Virkus, S., & Garoufallou, E. (2019). Data science from a library and information science perspective. *Data Technologies and Applications, 53*(4), 422–441.
14. Molina-Solana, M., Ros, M., Ruiz, M. D., Gómez-Romero, J., & Martín-Bautista, M. J. (2017). Data science for building energy management: A review. *Renewable and Sustainable Energy Reviews, 70,* 598–609.
15. Sinha, P. K., & Dutta, B. (2020). A systematic analysis of flood ontologies: A parametric approach. *Knowledge Organization, 47*(2), 138–159.
16. Hastings, J., Adams, N., Ennis, M., Hull, D., & Steinbeck, C. (2011). Chemical ontologies: What are they, what are they for and what are the challenges. *Journal of Cheminformatics, 3*(1), 1–1.
17. Gruber, T. R. (1993). A translation approach to portable ontologies. *Knowledge Acquisition, 5*(2), 199–220.
18. Borst, W. Construction of Engineering Ontologies. PhD thesis, Institute for Telematica and Information Technology, University of Twente, Enschede, The Netherlands, 1997.
19. Studer, R., Benjamins, V. R., & Fensel, D. (1998). Knowledge engineering: Principles and methods. *Data & Knowledge Engineering, 25*(1–2), 161–197.
20. Dutta, B. (2017). Examining the interrelatedness between ontologies and linked data. *Library Hi Tech, 35*(2), 312–331.
21. Reyes-Pena, C., & Tovar-Vidal, M. (2019). Ontology: Components and evaluation, a review. *Research in Computing Science, 148,* 257–265.
22. Giunchiglia, F., Dutta, B., & Maltese, V. (2014). From knowledge organization to knowledge representation. *Knowledge Organization, 41*(1), 44–56.
23. Islam, N., Abbasi, A. Z., & Shaikh, Z. A. (2010). Semantic web: Choosing the right methodologies, tools and standards. *International Conference on Information and Emerging Technologies*, IEEE, Karachi, pp. 1–5.

24. Silva, C. & Ramos, I. (2014). Ontology methodology building criteria for crowd sourcing innovation intermediaries. *International Joint Conference on Knowledge Discovery, Knowledge Engineering, and Knowledge Management*, Springer, Cham, pp. 556–570.

25. Deshpande, N. J., & Kumbhar, R. (2011). Construction and applications of ontology: Recent trends. *DESIDOC Journal of Library Information Technology, 31*(2), 84–89.

26. Gómez-Pérez, A., Fernández-López, M., & Corcho, O. (2004). Ontology tools. In: *Ontological Engineering: With Examples from the Areas of Knowledge Management, e-Commerce and the Semantic Web*, edited by Xindong Wu, Lakhmi Jain, publisher: springer, location: London, New York, pp. 293–362.

27. Gokhale, P., Deokattey, S., & Bhanumurthy, K. (2011). Ontology development methods. *DESIDOC Journal of Library & Information Technology, 31*(2), 77–83.

28. Fernández, M., et al. (1997). Methontology: From ontological art towards ontological engineering. *Proceedings of the AAAI Spring Symposium Series*, AAAI Press, Menlo Park, CA, pp. 33–40.

29. Dutta, B., Chatterjee, U., & Madalli, D. P. (2015). YAMO: Yet another methodology for large-scale faceted ontology construction, *Journal of Knowledge Management, 19*(1), 6–24.

30. Suárez-Figueroa, M. C., Gómez-Pérez, A., & Fernandez-Lopez, M. (2015). The NeOn methodology framework: A scenario-based methodology for ontology development. *Applied Ontology, 10* (2), 107–145.

31. Dutta, B., & Sinha, P. K. (2018). A bibliometric analysis of automatic and semi-automatic ontology construction processes. *Annals of Library and Information Studies, 65*(2), pp.112–121.

32. Simperl, E. & Luczak-Rösch, M. (2014). Collaborative ontology engineering: A survey". *The Knowledge Engineering Review, 29*(1), 101–131.

33. Semantic Data Models. Available at: https://www.ontotext.com/services/semantic-data-modeling/ (Accessed on April 16, 2021).

34. Ontology and Data Science. Available at: https://www.kdnuggets.com/2019/01/-ontology-data-science.html (Accessed on April 16, 2021).

35. Ontology and Data Science. Available at: https://towardsdatascience.com/ontology-and-data-science-45e916288cc5 (Accessed on April 16, 2021).

36. What are ontologies? Available at: https://www.ontotext.com/knowledgehub/fundamentals/what-are-ontologies/ (Accessed on April 16, 2021).

37. Dhakal, S., Zhang, L., & Lv, X. (2020). Ontology-based semantic modelling to support knowledge-based document classification on disaster-resilient construction practices. *International Journal of Construction Management*, 1–20.

38. Ye, Y., Yang, D., Jiang, Z., & Tong, L. (2008). Ontology-based semantic models for supply chain management. *The International Journal of Advanced Manufacturing Technology, 37*(11–12), 1250–1260.

39. Wilkinson, M. D., Dumontier, M., Aalbersberg, I. J., Appleton, G., Axton, M., Baak, A., … Mons, B. (2016). The FAIR guiding principles for scientific data management and stewardship. *Scientific data, 3*(1), 1–9.

40. Dutta, B., Jonquet, C., Magagna, B., Toulet, A. Summary report for ontology metadata task group of the vocabulary and semantic services interest group, RDA P11 – Berlin, March 2018. Available at: https://www.isibang.ac.in/~bisu/Presentation/-23-03_RDA-VSSIG_ontology_metadata_TG.pdf (Accessed March 22, 2021).

41. Naskar, D., & Dutta, B. (2016). Ontology libraries: A study from an ontofier and an ontologist perspectives. *19th International Symposium on Electronic Theses and Dissertations*, ETD, Vol. 16, pp. 1–12.

42. Buneman, P., Khanna, S., & Wang-Chiew, T. (2001, January). Why and where: A characterization of data provenance. *International conference on database theory*, Springer, Berlin, Heidelberg, pp. 316–330.
43. Marinho, A., Murta, L., Werner, C., Braganholo, V., Cruz, S. M. S. D., Ogasawara, E., & Mattoso, M. (2012). ProvManager: A provenance management system for scientific workflows. *Concurrency and Computation: Practice and Experience, 24(-13)*, 1513–1530.
44. Dalpra, H. L., Costa, G. C. B., Sirqueira, T. F. M., Braga, R. M., Campos, F., Werner, C. M. L., & David, J. M. N. (2015). Using ontology and data provenance to improve software processes. *ONTOBRAS*.
45. PROV Ontology. Available at: https://www.w3.org/TR/prov-o/ (Accessed on April 16, 2021).
46. Cruz, I. F., & Xiao, H. (2005). The role of ontologies in data integration. *Engineering Intelligent Systems for Electrical Engineering and Communications, 13(4)*, 245–261.
47. Wache, H., Voegele, T., Visser, U., Stuckenschmidt, H., Schuster, G., Neumann, H., & Hübner, S. (2001, January). Ontology-based integration of information-a survey of existing approaches. *Proceedings of the IJCAI-01 Workshop on Ontologies and Information Sharing*.
48. I. F. Cruz & H. Xiao (2003). Using a layered approach for interoperability on the semantic web. *Proceedings of the 4th International Conference on Web Information Systems Engineering* (WISE 2003), Rome, Italy, pp. 221–232.
49. Arens, Y., Knoblock, C. A., & Hsu, C. Query processing in the SIMS information mediator. *The AAAI Press*, May 1996.
50. Mena, E., Kashyap, V., Sheth, A. P., & Illarramendi, A. (1996). OBSERVER: An approach for query processing in global information systems based on interoperation across pre-existing ontologies. *Proceedings of the 1st IFCIS International Conference on Cooperative Information Systems* (CoopIS 1996), pp. 14–25.
51. Dou, D., Wang, H., & Liu, H. (2015, February). Semantic data mining: A survey of ontology-based approaches. *Proceedings of the 2015 IEEE 9th international conference on semantic computing* (IEEE ICSC 2015), pp. 244–251.
52. Mansingh, G., Osei-Bryson, K.-M., & Reichgelt, H. (2011). Using ontologies to facilitate post-processing of association rules by domain experts. *Information Sciences, 181(3)*, 419–434.
53. Marinica, C., & Guillet, F. (2010). Knowledge-based interactive postmining of association rules using ontologies. *IEEE Transactions on Knowledge and Data Engineering, 22(6)*, 784–797.
54. Marinica, C., Guillet, F., & Briand, H. (2008, December). Post-processing of discovered association rules using ontologies. *2008 IEEE International Conference on Data Mining Workshops*, IEEE, pp. 126–133.
55. Wimalasuriya, D. C., & Dou, D. (2010). Ontology-based information extraction: An introduction and a survey of current approaches. *Journal of Information Science, 36(3)*, 306–323.
56. Liu, H., Dou, D., Jin, R., LePendu, P., & Shah, N. (2013, December). Mining biomedical ontologies and data using RDF hypergraphs. *2013 12th International Conference on Machine Learning and Applications*, IEEE, Vol. 1, pp. 141–146.
57. Balcan, N., Blum, A., & Mansour, Y. (2013, May). Exploiting ontology structures and unlabeled data for learning. *International Conference on Machine Learning*, PMLR, pp. 1112–1120.

58. Carlson, A., Betteridge, J., Junior, E. R. H., & Mitchell, T. (2009, June). Coupling semi-supervised learning of categories and relations. *Proceedings of the NAACL HLT 2009 Workshop on Semi-supervised Learning for Natural Language Processing*, pp. 1–9.

59. Müller, H. M., Kenny, E. E., Sternberg, P. W., & Ashburner, M. (2004). Textpresso: An ontology-based information retrieval and extraction system for biological literature. *PLoS Biology*, *2*(11), e309.

60. Data Science Ontology. Available at: https://www.thoughtworks.com/insights/blog/-data-science-ontology (Accessed on April 16, 2021).

61. Data Science Ontology. Available at: https://collaboratescience.com/WDS/index.html (Accessed on April 16, 2021).

62. Chuprina, S., Alexandrov, V., & Alexandrov, N. (2016). Using ontology engineering methods to improve computer science and data science skills. *Procedia Computer Science*, *80*, 1780–1790.

63. ACM Computer Classification System. Available at: https://dl.acm.org/ccs (Accessed on April 16, 2021).

64. Chuprina, S., Postanogov, I., & Kostareva, T. (2017). A way how to impart data science skills to computer science students exemplified by obda-systems development. *Procedia Computer Science*, *108*, 2161–2170.

65. Patterson, E., Baldini, I., Mojsilovic, A., & Varshney, K. R. (2018). Teaching machines to understand data science code by semantic enrichment of dataflow graphs. arXiv preprint *arXiv:1807.05691*.

66. Data Science Ontology. Available at: https://www.datascienceontology.org/ (Accessed on April 16, 2021).

67. Shah, V., & Subramanian, S. (2019). Data-science recommendation system using semantic technology. *International Journal of Engineering and Advanced Technology*, *9*(1), 2592–2599.

68. Data Science Education Ontology. Available at: https://bioportal.bioontology.org/ontologies/DSEO/?p=summary(Accessed on April 16, 2021).

69. Naskar, D., & Dutta, B. (2016). Ontology libraries: A study from an ontofier and an ontologist perspectives. *19th International Symposium on Electronic Theses and Dissertations*, ETD, Vol. 16, pp. 1–12.

70. Sicilia, M. Á., García-Barriocanal, E., Sánchez-Alonso, S., Mora-Cantallops, M., & Cuadrado, J. J. (2018, October). Ontologies for data science: On its application to data pipelines. *Research Conference on Metadata and Semantics Research*, Springer, Cham, pp. 169–180.

71. Panov, P., Soldatova, L., & Džeroski, S. (2014). Ontology of core data mining entities. *Data Mining and Knowledge Discovery*, *28*(5), 1222–1265.

72. Arantes, E. A. Y. F. L. and Yokome, E. (2011). Meta-DM: An ontology for the data mining domain. *Revista de Sistemas de Informacao da FSMA*, *8*(2011), 36–45.

73. Benali, K., & Rahal, S. A. (2017). OntoDTA: Ontology-guided decision tree assistance. *Journal of Information & Knowledge Management*, *16*(03), 1–23.

74. Li, Y., Thomas, M. A., & Osei-Bryson, K. M. (2017). Ontology-based data mining model management for self-service knowledge discovery. *Information Systems Frontiers*, *19*(4), 925–943.

75. Tianxing, M., Stankova, E., Vodyaho, A., Zhukova, N., & Shichkina, Y. (2020, July). Domain-oriented multilevel ontology for adaptive data processing. *International Conference on Computational Science and Its Applications*, Springer, Cham, pp. 634–649.

76. Sinha, P. K., Gajbe, S. B., Debnath, S., Sahoo, S., Chakraborty, K., & Mahato, S. S. (2021). A review of data mining ontologies. *Data Technologies and Applications.* Vol (Ahead of Print) No. (Ahead of Print), pp. 1–33.

77. Vilches-Blázquez, L. M., Ramos, J. A., López-Pellicer, F. J., Corcho, O., & Nogueras-Iso, J. (2009). An approach to comparing different ontologies in the context of hydrographical information. *Information Fusion and Geographic Information Systems,* Springer, Berlin, Heidelberg, pp. 193–207.

78. Zhang, X., Li, K., Zhao, C., & Pan, D. (2017). A survey on units ontologies: Architecture, comparison and reuse. *Program, 51*(2), 193–213.

79. Prantner, K., Ding, Y., Luger, M., Yan, Z., & Herzog, C. (2007). Tourism ontology and semantic management system: State-of-the-arts analysis. *IADIS International Conference WWW/Internet 2007 (WWW/Internet2007),* Vila Real, Portugal, October 2007, pp. 1–5.

80. Mascardi, V., Cord, V., & Rosso, P. (2007). A comparison of upper ontologies. In: Baldoni, M., Boccalatte, A., De Paoli, F., Martelli, M., & Mascardi, V. (Eds), *8th Italian Association for Artificial Intelligence and Italian Association of Advanced Technologies Based on Object Oriented Concepts (AI*IA/TABOO) Joint Workshop "From Objects to Agents": Agents and Industry.* Technological Applications of Software Agents, Geneva, Italy, pp. 55–64.

81. Ruiz-Iniesta, A., & Corcho, O. (2014). A review of ontologies for describing scholarly and scientific documents. *4th Workshop on Semantic Publishing (SePublica) (CEUR Workshop Proceedings),* Vol. 1155. Available at: http://ceur-ws.org.

82. Soares, A. A., Martins, P. V., & da Silva, A. R. (2018). A systematic literature review of legal ontologies. *CAPSI 2018 Proceedings,* Vol. 42. Available at: https://aisel.aisnet.org/capsi2018/4.

83. Norris, E., Finnerty, A. N., Hastings, J., Stokes, G., & Michie, S. (2019). A scoping review of ontologies related to human behaviour change. *Nature Human Behaviour, 3*(2), 164–172.

84. Khamparia, A., & Pandey, B. (2017). Comprehensive analysis of semantic web reasoners and tools: A survey. *Education and Information Technologies, 22*(6), 3121–3145.

85. Messaoudi, R., Mtibaa, A., Vacavant, A., Gargouri, F., & Jaziri, F. (2020). Ontologies for liver diseases representation: A systematic literature review. *Journal of Digital Imaging, 33*(3), 563–573.

86. Okikiola, F. M., Ikotun, A. M., Adelokun, A. P., & Ishola, P. E. (2020). A systematic review of health care ontology. *Asian Journal of Research in Computer Science, 5* (1), 15–28.

87. Yousefianzadeh, O., & Taheri, A. (2020). COVID-19 ontologies and their application in medical sciences: Reviewing Bioportal. *Applied Health Information Technology, 1*(1), 30–35.

88. Liu, S., Brewster, C. and Shaw, D. (2013). Ontologies for crisis management: A review of state of the art in ontology design and usability. In: Comes, T., Fiedrich, F., Fortier, S., Geldermann, J. and Mセuller, T. (Eds), *Proceedings of the 10th International Information Systems for Crisis Response and Management Conference* (ISCRAM), Baden-Baden, Germany, 12–15 May, Karlsruhe Institute of Technology, Karlsruhe, pp. 1–10. Available at: http://cbrewster.com/papers/Liu_ISCRAM13.pdf.

12 Ontology Application to Constructing the GMDH-Based Inductive Modeling Tools

Halyna Pidnebesna and Volodymyr Stepashko

CONTENTS

DOI: 10.1201/9781003309420-12

12.1   INTRODUCTION

This chapter considers the problem of building a generalized ontological model of the subject field of inductive modeling (IM) based on the Group Method of Data Handling (GMDH) in order to apply it to develop intelligent computer modeling systems. Ontologies of this domain are used as a formal structure of knowledge representation for the development of the software tools. This provides an increase in the level of intelligence of computer modeling systems. The design methodology for the construction of ontological models of the IM area is presented. Ontological models of the main components of the GMDH-based modeling process are built.

Intelligent systems based on ontologies have shown their effectiveness in practice, but building any ontology is a difficult task. The development and creation of intelligent computing tools for solving inductive modeling tasks remains a relevant area of research making it possible to considerably enhance the performance of building models of complex systems of various natures. The basis is the results of structuring the knowledge of subject areas for constructing appropriate structures of knowledge bases, functionalities, and user interface with intelligent features.

Motivation: As of today, increasingly more attention is paid to modeling the software design processes themselves for complex computer systems. A general trend for both design processes and specifications of developed systems and domains is the application of ontological modeling methods and tools [1,2].

Ontology-Based (-Driven) Software Engineering provides three main advantages of the ontological approach: flexible models, advanced analytical capabilities, and advantages for the development of data services. A model is an ontology for describing types, entities, relationships, valid values, and semantics. It is not tied to the method of data storage which allows changes in the model much easier than in relational structures [3–5].

The amount of data increases and accordingly does the number of tasks of processing and analysis of this data, modeling of complex processes and phenomena, forecasting to support decision-making, and so on. Tools for solving these problems are constantly emerging and developed [6,7]. Creation of intelligent computer systems is a highly complex and intensive process. In addition to the creation, there are permanent problems with the need to make changes in existing systems to adapt them to new versions and capabilities of the latest information technologies. That is, the main requirements for modern tasks of software design are to simplify the process of development and modification of such tools and, therefore, reduce the design costs.

Ontological information-based semantic technologies allow building powerful applied systems for the analysis and modeling of complex objects and processes. The development of such systems is based on the outcomes of structuring knowledge aimed to studying and constructing the structure of knowledge bases and functional support. Hence, it is necessary to analyze the subject area of modeling for knowledge structuration in main stages of this process, the nature of the data, methods being used, and conditions of effectiveness of the obtained models. This is the first step in developing effective knowledge-based systems [8,9].

zThe article structure is as follows. Section 12.2 gives a brief definition of otology and its characteristics. Section 12.3 describes the field of inductive modeling, its tasks, and methods. Section 12.4 as the main part of the article is devoted to the analysis of the GMDH domain and the construction of ontological models based on the results of this analysis. Section 12.5 provides examples of the GMDH ontology implementation to the software development. The conclusions are given in Section 12.6.

12.2 ONTOLOGY AS THE KNOWLEDGE STRUCTURE OF A DOMAIN

We mean the term *ontology* as a structured set of terms as well as rules of using them when presenting a subject area, on the basis of which one can describe classes, relationships, and functions in this domain [13]. *Ontology* is the precise specification of a domain with the vocabulary of terms and their relations. It may be called a conceptual skeleton of some subject area.

Formal definition of an ontology O is given commonly as the following triplet:

$$O =< T,\ R,\ F >,$$

where T, R, and F are some finite sets: T contains terms of the subject area; R consists of relations between them; F presents interpretation functions defined for terms and relations of this ontology. That is, ontologies must first provide a glossary of domain concepts and a set of relationships among them. This facilitates their reuse and expansion by integrating existing ontologies that relate to parts of a complex subject area.

Ontologies differ in the level of generality of the presented knowledge (Figure 12.1a). *Metaontology* (higher level ontology) represents a conceptual basis for the knowledge system. *Domain ontology* generalizes concepts in a specific subject area but is separated from specific tasks [14]. *Ontology of tasks* defines objects of activity and is used to develop the functionality of applications for solving specific tasks.

The design of ontologies begins with the analysis of a subject area aimed to identifying the essences (concepts) and their properties [15]. When analyzing and presenting knowledge, it is necessary to use concepts at different levels of generality. Definition of higher-level ontologies contains a number of common problems to define concepts as their root class. The results of analyzing and structuring the process of model building from given observation data are the basis

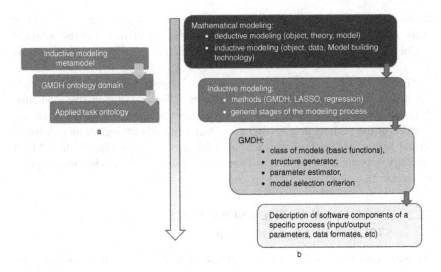

FIGURE 12.1   Ontologies of different levels of generality

for constructing the ontology of inductive modeling. Therefore, it is necessary to analyze the inductive modeling domain, the main stages of this process, the used methods, and the conditions of their effective application for constructing ontological models of the domain based on this analysis.

The automation task of intelligent computer modeling systems is to model the modeling process. Modeling is a process of investigating a real object taking into account only its specific features and conditions. We consider mathematical modeling as studying the object behavior by data-based constructing its model and analyzing it. The complexity of the modeling process is determined by the fact that in practice such tasks must be solved under conditions of uncertainty and incompleteness of information significantly affecting the results [16]. Tracking and correct consideration of these conditions is the task of intellectualizing the modeling process in computer systems. Deep consistent analysis of the whole process makes it possible to determine ontological structures of different generality levels. This becomes the basis for the construction of ontologies of various levels, from meta-ontologies to ontologies of problems.

Two main approaches to building models of an object can be distinguished [12]: one based on the study of regularities of the object functioning (*theory-driven* or *deductive* approach) and another based on the synthesis of models using the analysis and generalization of data about the object functioning (*data-driven* or *inductive* approach) (Figure 12.1b).

In particular, *deductive* modeling is a transition process from general laws of the object functioning to a specific (partial) model. Accordingly, *inductive* modeling is a transition process from specific (partial) data to a generalized model.

Hence, a constructed model can be considered as a partial or a general phenomenon depending on the used approach [12]. These approaches are also sometimes called as *theoretical* (or cognitive) and *empirical* (pragmatic) ones respectively. In this chapter, we are dealing with the inductive approach.

There are many inductive methods for finding regularities hidden in the data: regression, GMDH, associative rule detection, sequence analysis, classification, decision trees, neural networks, SVM method, genetic algorithms, LASSO, etc.

The modeling process generally consists of certain stages, and a detailed study of each of them, the definition of relationships between them allows their effective description with the maximum possible degree of formalization and adequacy. All these properties may be described by certain variables and their valid values. This information can be represented by appropriate ontologies.

12.3 GMDH-BASED INDUCTIVE MODELING AS A TECHNOLOGY OF TRANSITION FROM STATISTICAL DATA TO MATHEMATICAL MODELS

The term *inductive modeling* may be defined as a self-organizing process of evolutional transition from initial data to mathematical models reflecting some patterns of functioning objects and systems implicitly contained in the given experimental or statistical information [12].

Main theoretical results on conditions of building models of optimal complexity were described in [17] with the use of the method of critical variances. This analytical tool allowed proving that GMDH makes it possible to construct models with minimal variance of the forecasting error. GMDH enables automatically adapt the optimal model to the informativity level of available noisy data. This allows classifying GMDH as an efficient method of data mining and computational intelligence: it is aimed at automatic discovering regularities in data sets under uncertainty conditions of initial information [18–22].

It is worth noting that researchers working in the Deep Learning field [23] consider the GMDH, created in 1968 by Prof. O. Ivakhnenko, as the very first example of a deep neural network with self-organization of its structure, parameters, and depth. This opens very promising prospects for further investigations in this field.

12.3.1 Basic Principles of GMDH

Ivakhnenko has published his first article on the GMDH in 1968 in Ukrainian journal "Avtomatyka" reprinted in the USA [18]. Articles [24–26] published abroad were probably the most known sources on this method. A typical inductive modeling task reduces to automated building a model describing unknown functioning patterns of an object or process on the basis of a given data set. GMDH is an original method for inductive self-organizing construction of models from experimental data under conditions of uncertainty. To solve two optimization tasks, model structure specification (discrete task) and model parameters estimation (continues task), the following principles are applied:

- generation of models of diverse complexity;
- inductive process of constructing models with the gradual complication of structures;
- quality assessment of a model using so-called external criteria based on a dataset division into two subsets: the first for estimating parameters and the second to calculate criteria;
- non-definitive decisions when evolving models: a subset of the best model is selected at any step of the process;
- successive selection of best models according to the criteria minima.

GMDH algorithms construct in an automatic way mathematical models explicitly reconstructing unknown regularities implicitly presented in the given data on an object functioning. In the up-to-date terminology, they are original means of data mining and knowledge discovery. The effectiveness of these algorithms is confirmed when solving different applied problems of modeling complex objects and processes in various real-world areas [27–32].

12.3.2 FORMAL STATEMENT OF THE GMDH PROBLEM

GMDH is typically aimed for building models linear in parameters, first of all in the polynomial class, from a given data set $W = [Xy]$, dim $W = n \times (m+1)$, where n and m are numbers of observations and inputs (arguments) respectively, $X = [x_{ij}]$, $i = \overline{1,n}$, $j = \overline{1,m}$ is the matrix of m inputs and $y = (y_1,...,y_n)^T$ is the output vector. We build a set Φ of models of diverse complexity on the set W:

$$\hat{y}_f = f(X,\hat{\theta}_f), f \in \Phi \tag{12.1}$$

where parameters of each f are estimated as a solution to the following task:

$$\hat{\theta}_f = \arg \min_{\theta_f \in R^{s_f}} QR(X,y,\theta_f) \tag{12.2}$$

where s_f is the parameters number or the model complexity. Then the task of building the best model from data reduces generally to minimizing a given criterion $CR(\cdot)$ on the set Φ:

$$f^* = \arg \min_{f \in \Phi} CR(X,y,f) \tag{12.3}$$

In the GMDH theory, criteria in (12.2) and (12.3) are called *internal QR*($\cdot$) and *external CR*($\cdot$) ones being different: $QR(\cdot) \neq CR(\cdot)$. To satisfy this condition, the data set W is divided into two parts:

$$W = \begin{bmatrix} A \\ B \end{bmatrix} = \begin{bmatrix} X_A y_A \\ X_B y_B \end{bmatrix} \tag{12.4}$$

then the difference between internal $QR(X_A, y_A, \hat{\theta}_{Af})$ and external $CR(X_B, y_B, \hat{\theta}_{Af})$ criteria is evidently satisfied, and therefore the task (12.3) makes nontrivial sense. For instance, the *regularity* criterion is the most commonly used in GMDH algorithms being calculated as the error of a model on part B with parameters estimated on part A (see below).

Typical iteration and sorting-out GMDH algorithms are discussed in detail in [33].

12.3.3 KINDS OF PROBLEMS AND SOFTWARE IMPLEMENTATIONS OF GMDH ALGORITHMS

Within the framework of inductive modeling based on GMDH, the whole range of real-world problems is considered:

- constructing mathematical models of objects and systems;
- forecasting processes presented as time series;
- classification or building a rule for attributing an object to a particular class;
- supervised pattern recognition (identification of effective features and rules for distinguishing given classes);
- clustering (self-learning or identifying informative features, classes, and rules for distinguishing them);
- detection of informative and removal of non-informative features (feature selection), etc.

Typically, any of GMDH means is intended for solving most tasks of data mining and knowledge discovery.

There are two types of implementation: a "black box" type (*Knowledge Miner* [34], *PNN Discovery Client* [35], etc.), when the user receives only the result and has no ability to influence the modeling process, and systems in which the user is given a limited number of possible ways to solve the problem (*ASTRID* [36], *FAKE GAME* [37], *GMDH Shell* [38], etc.). In such systems, the rules for selecting options of possible algorithms are spelled out in the program.

Evidently, each particular realization of GMDH assumes fulfilling its own requirements. That is, users should be able to apply different algorithms and select their components for the most effective use in practical tasks. Furthermore, new algorithms are permanently designed; with this in mind, it is certainly reasonable to create the structure of knowledge in the inductive modeling area having properties of flexibility and versatility.

Besides, the urgent task is to develop principles for designing interactive modeling tools on the basis of knowledge gained in the GMDH field which would increase the efficiency of the process of any applied modeling system design.

12.4 STRUCTURING THE GMDH DOMAIN KNOWLEDGE TO DESIGN THE ONTOLOGY MODELS

Full implementation of the principle of self-organization requires making consistent decisions at each stage of solving a specific task, starting from acquiring statistical data. In this case, we have data for which we must first formulate a problem, i.e. what should be exactly obtained as a result of modeling, and various possible methods and algorithms that are effective for certain classes of problems. In order to choose the algorithm that is most appropriate to be used, it is necessary to determine to which class the task with the respective data belongs. In addition, to solve the modeling problem, one needs to assess the adequacy of the obtained models. The result of using GMDH can be not single best model but an ensemble, i.e. best subset of all generated models. This is an extension of the principle of non-definitive decisions to the final result of modeling.

12.4.1 DIAGRAM OF SEQUENTIAL DECISION MAKING IN THE PROCESS OF INDUCTIVE MODELING

Any data-driven process of building a mathematical model is a successive decision-making one. In general case, the sequence consists of certain fairly standard steps (Figure 12.2). The analysis of each of these stages gives grounds to form sets of relevant concepts and their properties, attributes, which form the basis of ontological models for each stage as part of general ontology of the whole modeling process.

12.4.2 DEFINITION OF BASIC CONCEPTS OF THE GMDH DOMAIN

The dictionary of the subject area of GMDH is established consisting of such concepts as model, basis, criterion, algorithm, learning and verification subsets, etc. The model concept as an entity that defines the relationship or dependence in the data is central within the system. In this case, the model belongs to some functional basis, i.e., it is determined in which class of functions the model will be built, and this basis determines the model structure. The structure and parameters of any model determine its uniqueness among other models. Analyzing the modeling process in this way, distinguishing its components, one can get a set of them as well as define sets of concepts and relationships to formally present the domain knowledge in the form of *RDF*-triplets, Table 12.1.

Data. They can be numerical or non-numerical, of physical, biological, linguistic nature, and so on. There are various methods of scaling and digitization of non-numerical data. GMDH deals exclusively with numerical data,

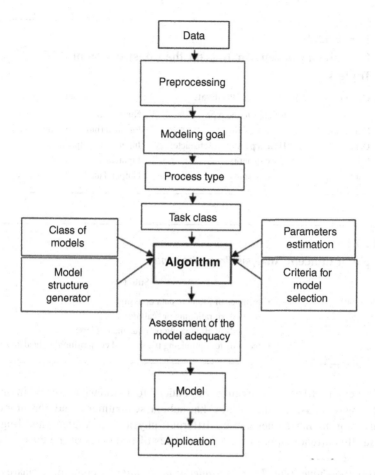

FIGURE 12.2 Diagram of main stages of successive decision-making in the process of inductive modeling

TABLE 12.1

Example of Concept Analysis for Construction of *RDF* Triplets

Object	Predicate	Subject
Model	Determines	Basis (class of models)
Basis	Determines	Model structure
Polynomial model	Has a parameter	Number of input variables
Polynomial model	Has a parameter	Degree of polynomial
Polynomial model	Has a parameter	Polynomial coefficients

TABLE 12.2

Example of Concept Analysis for the Construction of *RDF* Triplets

Object	Predicate	Subject
Data	Belongs to the type	Numerical
Non-numeric data	Have a nature	Physical (biological, linguistic)
Data	Have a parameter (characterized)	Input data (output data)
Data	Have quantity	Input data
Data	Have quantity	Output data

TABLE 12.3

Example of Concept Analysis *Preprocessing*

Class	Subclass
Preprocessing	• methods of filling in data omissions;
	• methods of rationing or scaling;
	• methods of eliminating or adjusting outliers;
	• methods for optimizing the size and composition of the data sample

and if they are different, they must be reduced to a numerical format in one or another way. Thus, the class "data" has subclasses: numeric and non-numeric. In turn, non-numeric ones can be different: physical, biological, and linguistic. But this already applies to the properties of instances of the class "data", Table 12.2.

Properties *have_type, have_parameter, have_number* combining classes and instances, so in Protege they can be described as *ObjectProperties*. The concepts *input_variables* and *output_variables* are characterized by the attribute quantity, the scope of which is a natural number specified by *DataProperties*. For any specific task of inductive modeling from a statistical data set, a special and necessary stage is data preprocessing. There is a range of problems that are solved at the stage of data preprocessing, depending on the selected class of tasks. They can be distinguished as subclasses of the class "Preliminary data processing" (Table 12.3).

To fill in the omissions, in particular, methods can be used based on modeling such as GMDH or any of the known methods of data recovery [39]. They form instances of the "fill in the omissions" class.

When the data presented in the sample have different dimensions, their normalization is applied. This is necessary for the adequate use of mathematical methods and the use of computer tools in calculations involving large and small quantities. Normalizing is an important factor influencing the accuracy of modeling.

If we talk about the purpose of modeling, then in the framework of inductive modeling, namely GMDH, a range of problems is considered which can be interpreted as ontological subclasses of the class *meta_modeling*.

There are various algorithms and their implementations for solving these classes of problems, which form a set of corresponding instances. From the point of view of the data nature, the following types of tasks should be distinguished:

- static tasks,
- time series,
- dynamic tasks.

Static tasks are those in which the dependence of variables on time is absent or exists but is insignificant: transients decay quickly and time dependences do not matter to achieve a specific goal of modeling. Time series are single or a set (vector) of time-dependent variables. If the nature of the time series is known, then polynomial, exponential, logarithmic, logistic, trigonometric, and other functions of time can be used for modeling. They can be formally considered as static models of dynamic processes.

Dynamic tasks are those where there are dynamic input–output dependencies described by different classes of differential or difference equations, including autonomous or with the presence of external influences which can be controlled or uncontrolled. If the data was measured in time, it becomes possible to automatically determine whether the process is dynamic or a time series. To determine the significance of transients, cross-correlation functions are constructed, and if they have the form of a pulse that indicates the absence of mutual influence of variables, one can talk about the static nature of the process, otherwise about its dynamic nature.

Depending on the type of process, an appropriate class of basic functions for modeling should be selected. Such analysis forms the basis for the definition of a *class_models* whose subclasses are polynomial, logarithmic, trigonometric models, and specific kinds or realizations of these functions form a set of model structures.

Any method of data-driven modeling including GMDH may be characterized by the following main elements [40]:

- class of models (basic functions),
- structure generator,
- parameter estimator,
- criterion of model selection.

In this regard, it is reasonable to analyze and compare how all these elements are implemented in available GMDH algorithms. By examining them, one can identify the main concepts and their relationships to be used when building ontological models.

Considering each method of inductive modeling (IM) in terms of these components, we can compose sets of model classes, structure generators, and methods for estimating parameters and criteria for selecting models and analyze these sets as independent scientific objects [40] to further use this knowledge for the construction of the domain ontology. As a result of knowledge structuring, principles of formation of algorithmic modules generalizing the functions of different methods are determined. Any such module can be defined by key parameters, and certain combinations of them generate particular IM methods (algorithms).

An algorithm definition. Let

- CM is a set of model classes, $CM = \{k_i\}$, $i = \overline{1, K}$;
- GS is a set of structure generators, $GS = \{g_j\}$, $j = \overline{1, G}$;
- MP is a set of methods for model parameters estimation, $MP = \{p_q\}$, $q = \overline{1, Q}$;
- CR is a set of quality criteria of models, $CR = \{r_t\}$, $t = \overline{1, R}$

Then a set L of model building methods can be presented as:

$$L = CM \times GS \times MP \times CR \qquad (12.5)$$

and any IM algorithm can be defined as an element of this set:

$$l_d = \{k_i, g_j, p_q, r_t\},$$
$$i = \overline{1, K}, j = \overline{1, G}, q = \overline{1, Q}, t = \overline{1, R}, \quad d = \overline{1, K \times G \times Q \times R} \qquad (12.6)$$

Therefore, the set $<CM, GS, OP, CR>$ can be considered as an initial basis for constructing ontological models of the GMDH-based IM domain.

Generators of structures: GMDH uses two main classes of model structure generators: iterative and sorting-out ones. Iterative algorithms are divided into multilayer and relaxation (Table 12.4). Multilayer structure generators use

---

TABLE 12.4

Classes and Subclasses of Model Structure Generators in GMDH

Classes	Subclasses
GMDH algorithms	• iterative
	• sorting-out
Iterative algorithms	• multilayer
	• relaxative
Sorting-out algorithms	• exhaustive search
	• directed search
Directed search algorithms	• with deterministic search (multistage)
	• with stochastic search
Multistage algorithms of a directed search	• sequential elimination
	• based on correlation ranking

pairwise combinations of features (arguments) of the previous layers of selection. Such algorithms are structurally similar to neural networks, so they are sometimes also called polynomial neural networks (PNN). The computational complexity of such procedures depends on the number of observations. GMDH relaxation algorithms refer to sequential approximation methods when at each step (approximation stage) the best models are complicated by adding initial arguments.

Another large class of GMDH algorithms are those of the sorting-out type, the basic principle of which is the generation of model structures that consists in exhaustive or partial (directed) search of possible candidate models [33]. Combinatorial generators with exhaustive search allow finding the global minimum of an external criterion for model selection but their use is limited in the number of arguments because the search in a reasonable time is possible approximately for $m < 30$. To eliminate this limitation, directed search algorithms are used.

A promising way to improve the algorithms is to combine different structure generators and hybridize the elements of GMDH with other means of computational intelligence. An example of such hybridization is the combinatorial-genetic algorithm *COMBI-GA* [41] which belongs to the sorting-out type but has no specific limitation in the number of input arguments. This algorithm has a stochastic model generation procedure. The multi-stage algorithm *MULTI* [42] and the correlation-rating algorithm *CORAL* [43] belong to the directed deterministic search. Having considered classes and subclasses of model structure generators in GMDH, it is possible to define their concrete instances (Table 12.5).

Methods for estimating parameters: Different methods can be used to estimate the parameters in modeling problems. Method of Least Squares (MLS) is traditionally used in the GMDH algorithms. To optimize calculations, reduce time and computational complexity, recurrent algorithms are effective, in particular, the bordering method (MLS modification) or recurrent modifications of the

---

TABLE 12.5

Classes and Instances of Model Structure Generators in GMDH

Class	Instances
Exhaustive search algorithm *COMBI*	• with binary counter • with sequential counter • with Garside counter
Algorithms of the directed search type with deterministic sorting	• multi-stage *MULTI*, • sequential elimination of irrelevant arguments, • correlation-rating algorithm *CORAL*
Algorithms of the directed search with random sorting	• combinatorial-genetic *COMBI-GA*
Multilayer algorithms	• multilayer iterative algorithm *MIA*
Relaxive algorithms	• relaxative iterative algorithm *RIA*

TABLE 12.6

Classes and Instances of Methods for Estimating Model Parameters

Class	Subclasses	Instances
Methods for parameters estimating	Non-recurrent methods	Method of Least Squares (MLS)
		Method of Least Modules (MLM)
		Maximum Likelihood Estimation (MLE)
	Recurrent methods	MLS with Bordering (MLSB),
		Recurrent Gauss method

Gauss method [44] can be used. Therefore, we can select subclasses and instances for the *parameter_valuation* methods (Table 12.6).

Model selection criteria: Model quality criteria for selecting the best/optimal model among many candidate models are usually divided into two groups [33]: those that use the entire data set (criteria of Akaike, Mallows, etc.), and those that use the set division (external ones).

External selection criteria used in GMDH are based on the division of the data set into two or more parts, and the tasks of parameter evaluation and comparison of model quality are performed on different parts. These criteria can be divided into two main groups: criteria of *accuracy* and *consistency*. Accuracy criteria express the error of the model which is calculated on different parts of the sample and are effective to avoid the model overfitting under noisy data. They include e.g. criteria of *regularity* and *stability*. The most commonly used is the *regularity* criterion:

$$AR_B(s) = AR_{B|A}(s) = \left\| y_B - \hat{y}_{B|A}(s) \right\|^2 = \left\| y_B - X_{Bs}\hat{\theta}_{As} \right\|^2 \tag{12.7}$$

where s is the model complexity and $AR_{B|A}(s)$ means an error on the checking subsample B of the model with parameters calculated on the training subsample A.

The second (consistency) group of criteria used in GMDH includes the criteria of unbiased decisions which are based on the fact that the best models differ minimally being built on A and B. The most common form of this criterion is a minimum bias of solutions (*unbiasedness* criterion):

$$CB(s) = CB_{W|A,B}(s) = \left\| \hat{y}_{W|A}(s) - \hat{y}_{W|B}(s) \right\|^2 = \left\| X_W\hat{\theta}_{As} - X_W\hat{\theta}_{Bs} \right\|^2 \tag{12.8}$$

However, the external criteria of accuracy have certain shortcomings, and above all ambiguity, i.e. lead to the ambiguous choice of the best model. Criteria of the consistency group are more noise-tolerant than other ones and have an advantage for building systems under conditions of high noise intensity or smoothing the input signal. In some cases, a combination of criteria or a sequence of several criteria is used to determine the justified choice of the optimal model [39].

TABLE 12.7

Subclasses and Instances of the Class Model_selection_criteria

Subclasses	Instances
Accuracy criteria	• regularity criterion
	• stability criterion
Consistency criteria	• unbiasedness criterion
	• variability criterion

The criteria class can be presented in the form of sets of subclasses and instances (Table 12.7).

Adequacy assessment: An obligatory step in the modeling process is the adequacy analysis of the obtained models. In GMDH, in order to assess the adequacy of the chosen model, its error is calculated on the validation/examination subset which was not used when building the models. This makes it possible to work effectively under conditions of significant incompleteness and uncertainty of a priori information about the object (process) of modeling. Depending on the problem statement, one model can be chosen among the constructed models according to the minimum of a given criterion or several best models among which the final one is determined by additional criteria. The selected group of the best models can be used as an ensemble of models, the results of which are averaged in order to obtain the most reliable forecast. Therefore, one model or a set of F (freedom of choice) models can be selected as the modeling result.

12.4.3 IDENTIFYING RELATIONSHIPS BETWEEN CONCEPTS

The next step in the analysis of the process of developing ontology is identifying relationships between entities, and the system is represented as a set of classes and objects. The result of this stage is ontological models that reflect both semantics of the selected subject area and architecture (structure) and properties of the designed software system.

After defining the set of concepts, one should specify their properties and relationships between them. Subordination genus-species relations can have the form: genus <–> species, sign <–> sign value, whole <–> part, object <–> object property, etc. They can have the following variants:

- A refers to B;
- A has the following types: A1, A2, ...;
- A is the part of B;
- A has property B;
- A is characterized by the presence of B, etc.

The property expresses one of the moments of revealing the essence of the concept in relation to others that characterizes its similarity to other concepts or difference from them (Table 12.8). Properties capture general facts about class members and their instances.

Properties can be of different types. Property functionality means that each instance of a class can be associated with this property by no more than one instance of the class. For example, a model cannot have two classes of basic function.

An example of determining the relationships for instances of combinatorial *COMBI* and combinatorial-selective *MULTI* algorithms is given in Tables 12.9 and 12.10.

Transition rules and transformation procedures constitute a set of interpretation functions. Intelligence level of a computer system means the ability to automatically (without user interference) perform certain actions and corresponding next steps. Ideally, the whole modeling process can be automatic: the result of a decision affects the automatic selection of the next step by the system. Rules of such choice are determined by the ontology which allows obtaining logical inferences. For instance, the choice procedure for an appropriate algorithm in

---

TABLE 12.8
Example of Concept Analysis to Determine Their Properties

Essence	Property
GMDH algorithm	• determines: the sequence of actions to obtain a model of optimal complexity according to the data • depends: on the division of the sample into training and checking parts
Model	• determines: relationships in the data • characterized by: parameters • has: structure • belongs to: basis (function class)
Criterion	• numerical value used for model selection • characterizes: optimality measure of the model • calculated: on the checking subsample
Basis	• determines: type of functions (class of models) • depends: on the type of an object
Models structure generator	• determines: the model structure • characterized: by parameters • has: copies
Parameter estimation	• determines: model parameters of generated structures • calculated: on the training subsample of data
Data	• numerical or non-numerical • division of the sample into training, checking (and validation) parts

---

TABLE 12.9

Example of Determining the Relationship of the Instance *COMBI*

Relation	Value
Type of models	Polynomial
Refers to the class	Sorting-out algorithms, exhaustive search
Equivalent (synonym)	Combinatorial algorithm
Characterized by parameters	X – measurement matrix
	y – input data vector
	m – number of inputs/arguments
	n – number of measurement points
	s – model complexity
Has restrictions	$m < 30$

TABLE 12.10

Example of Determining the Relationship of the Instance MULTI

Relation	Value
Type of models	Polynomial
Refers to the class	Sorting-out algorithms, directed search
Equivalent (synonym)	Combinatorial-selective algorithm
Characterized by parameters	X – measurement matrix
	y – input data vector
	m – number of inputs
	n – number of measurements
	s – model complexity
	F_s – freedom of choice
	j – index of a regressor absent in the matrix
Has restrictions	$30 < m < 100$

accordance with the purpose of modeling and the number of input arguments can be described as follows.

Let m be the number of input variables, then
If $m < 30 \Rightarrow$ Str_gen = "*COMBI*".
If $m > 30 \lor$ Target == "*feature_selection*") $\Rightarrow$ Str_gen = "*CORAL*".

This entry can be interpreted as follows. If the input number does not exceed 30, then the algorithm COMBI should be used. If this number is more than 30, and the modeling goal is feature selection, then the algorithm *CORAL* should be chosen. Interpretation of such rules makes it possible to implement the automatic selection mode of the appropriate procedure depending on the type of object or process during the tool operation.

As a result of structuring the IM subject area and determining ontological models, the representation of knowledge about a modeling task can be decomposed into a sequence of statements from complex to elementary. This makes it possible to formally present the modeling process in the form of ontologies from acquiring input data to applying an appropriate algorithm with relevant parameters.

12.4.4 DETERMINING KEY PARAMETERS OF MAIN STAGES OF THE MODELING PROCESS

As a result of the GMDH ontological analysis, the main key parameters of each stage of the model construction process from statistical data were determined [45]. For instance, the ontology of *linear model* classes defines different types of models with key parameters $\{MY, LY, MX, LX\}$, where MY, MX are numbers of input and output variables, and LY, LX are numbers of delayed values for input and output variables respectively [16]. Depending on specific values of the key parameters, most variants of linear models can be obtained which are used to model static objects, time series, and dynamic objects and processes.

For structure generators, the key parameters are $\{SI, SA, F, NI\}$, where SI and SA are the minimum and maximum complexity of generated models respectively, F is the freedom of choice, NI is the number of iterations. This can be the basis of forming a single multifunctional software module of structure generators with these key parameters summarizing a number of structure generators (including GMDH), e.g.:

- $SI = SA = m$ fits the multidimensional regression modeling;
- $SI = 1$, $SA = m$, $Fs = C^s_m$ indicates the exhaustive search by the generator *COMBI*;
- $F = 1$ means the regression procedure "inclusion";
- $1 < F < C^s_m$ corresponds to the combinatorial-selective generator *MULTI*;
- $SI = SA = 2$, $NI > 1$ represents the multilayer iterative procedure of GMDH, etc.

Ontological analysis of the subject area allows structuring knowledge, generalizing many different methods, and determining their basic parameters. This principle greatly simplifies the task of designing inductive modeling software because allows building single generalized multifunctional modules that cover a wide range of possible algorithms.

The ontology of model *selection criteria* defines key parameters that characterize the set of criteria used to build models of optimal complexity. This is a set of parameters $\{\eta_1(.), \eta_2(.), V(s), \sigma^2\}$, where $\eta_1(.), \eta_2(.)$ are multiplicative and additive penalty functions for the model complexity, $V(s)$ is a model quality measure, and σ^2 is an estimation of the unknown noise variance.

Consider a set of parameters where

- an element of the set CM of model classes is $k_i* = <linear\ regression\ models>$,
- an element of the set GS of structure generators is $g_i* = <searches:\ full\ search>$,

TABLE 12.11

Example of Key Parameters for Determining the Type of Algorithm

Parameter	RIA	COMBI
Class of models	$k_i^* = < lin\_regression>$	$k_i^* = lin\_regression>$
Structure generator	$g_i^* =< iterative\ algorithm.$ $Relaxational >$	$g_i^* =< exhaustive\ sorting\text{-}out$ $algorithm >$
Estimating method	$p_i^* =< least\text{-}squares\ method >$	$p_i^* =< least\text{-}squares\ method >$
Model selection criterion	$r_i^* =< regularity\ criterion >$	$r_i^* =< regularity\ criterion >$

- an element of the set OP of parameter estimation methods is $p_i^* ==$ <*least squares method*>,
- an element of the set CR of model quality criteria is $r_i^* =$<*regularity criterion*>.

This set of ontology parameters defines the combinatorial algorithm *COMBI*. Similarly, ontological models can be defined for other GMDH algorithms (Table 12.11). For example, if the structure element of generators GS is $g_i^* =$ <*directed search: direct search*>, it determines the multistage algorithm *MULTI*.

In the case when the structure element of generators GS is defined as $g_i^* =$< *iterative algorithm. Relaxational* >, it specifies the relaxative iterative algorithm *RIA*. In the case where the element of GS matters $g_i^* =$<*iterative algorithm: multilayer*>, it specifies a multilayer iterative algorithm *MIA* GMDH [33]. Thus, ontology allows determining both general rules of an algorithm construction (generalized ontological model, meta-ontology) and specific parameters when creating an applied program using lower-level ontological models of the task ontology.

12.5 DESIGNING GMDH TOOLS USING ONTOLOGICAL MODELS OF THE DOMAIN

12.5.1 REQUIREMENTS FOR A SET OF TOOLS

Development of modern software on the basis of standard ontologies allows re-sharing information in relevant subject areas by people and software agents as well as analysis and classification of knowledge. As of today, there are many implementations of the GMDH algorithms designed for solving various research tasks. At each step of modeling (constructing an algorithm) the consistency of the user's decision is checked. At the same time the explanation block of decisions for an automatic mode is provided and recommendations, for example, to change the algorithm or its parameters are offered. Then one can work with other methods and improve the modeling result.

Software IM tools consist typically of the following main elements:

- user interface providing interaction of a user with the system;
- implementation of a number of existing GMDH algorithms;

- ontological models of the inductive modeling domain created by the ontology editor *Protégé* as a knowledge base;
- tools for the interaction of the software and database to obtain the result needed for a user.

12.5.2 ONTOLOGICAL PRINCIPLES OF USER INTERFACE CONSTRUCTION

The ontological approach can be applied to the user interface design as an integral part of intelligent computer systems. The intelligent user interface should implement functions of communication, automatic scripting an adequate algorithm, justification of the result, and training [45]. In general, such user interface contains the system's elements and tools for affecting the interaction of a user with it. Such elements include:

- a set of user tasks being solved using the system;
- system controls;
- navigation between system units;
- means for displaying actual information and its formats;
- data input/output devices and technologies;
- interactions and transactions between the user and the computer system;
- decision-making support in a specific area;
- manual for using the software and its documentation.

Ontology-based approach to solving the task of the user interface design includes the following ideas [46]:

1. Interface development and modification should be separated from applying the interaction rules.
2. Interface information with homogeneous content should be introduced into separate components; hence, any user can select various sets of concepts, for example:
   - set of concepts describing the interaction between the user and the software;
   - set of concepts for developer regarding means used to display the interface information and implement interaction scenarios between interface and applications.
3. Interface components should be presented as declarative ontological models describing these components.

The user interface is a software element needed for adapting potential users with diverse training levels and individual preferences to a wide range of tasks being solved. To implement such kind of user-friendly interface, some universal ontological models should be used to build arrangements of the future interface, e.g.: *domain model, user model, dialog model, task model,* etc. Formation of any interface model component is specified by setting values of relevant concepts of such universal ontology which allow implementation of the adaptability feature of the IUI.

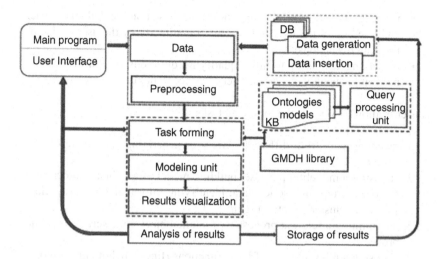

FIGURE 12.3   Flow chart of the designed software.

12.5.3 STRUCTURE OF THE SOFTWARE COMPLEX

Generally, the IM process is organized as follows. The input data are analyzed and preprocessing is performed, if necessary. Then the following elements are chosen: class of basic model functions, method for generating model structures, method of parameter estimation, and criterion for best model selection. A user can save the results in the database, generate a report, or continue modeling process by selecting other methods and parameters. The sequence of such modeling steps and relevant components are followed from the appropriate ontologies.

We implement the appropriate structure of inductive modeling software (Figure 12.3) containing the following main elements:

- realization of GMDH algorithms;
- user interface with intelligent features;
- models of the IM domain ontology created using *Protégé;*
- tools for the interaction of the software and database to obtain the results needed for a user.

12.5.4 MAIN REQUIREMENTS FOR THE INTELLIGENT USER INTERFACE

The IUI main component is the interaction management governing the following operations:

- integrating information from various channels using a data transmission model;
- creating and adjusting a real-time dialogue model, progressing the user-interface dialogue in compliance with this model based on the domain, task, and user models;

- recognizing and forming a plan of actions based on the dialogue model and a record of the user actions (entering directly by the user or due to tracking his actions);
- presenting results or forming qualifying questions to eliminate some nascent vagueness during the user-interface dialogue.

A user is allowed to solve the following tasks:

1. receiving data from different sources or generating data for test tasks (Figure 12.4);
2. choosing a modeling purpose: building model of an object; forecasting a process from time-series data; identifying informative features; classification; clustering; etc. (Figure 12.5);
3. determining a relevant process type (static object, time series, dynamic process);
4. choosing a relevant class of basic functions (linear, polynomial, autoregression, harmonic, logarithmic, exponential, difference equations, etc.);

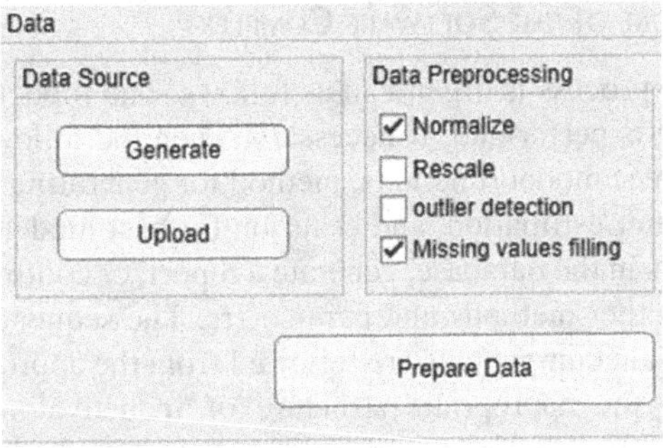

FIGURE 12.4   Interface snippet: data unit.

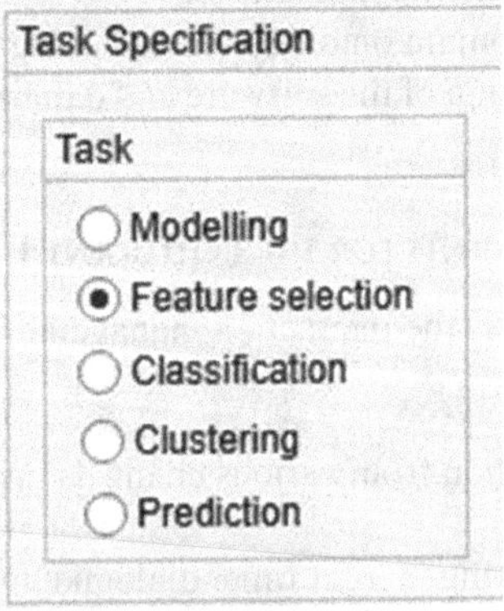

FIGURE 12.5   Fragment of the interface: the modeling goal choice.

5. deciding which external criterion is relevant to select the best model;
6. choosing a parameter estimation method;
7. deciding which method is relevant to generate model structures;
8. verifying the model adequacy by statistical criteria, accuracy on an independent validation data set, etc.

In practice, users have often different levels of experience:

- beginner: has data, knows nothing about the methods of their processing;
- experienced user: familiar with such tasks, knows methods for solving them;
- expert: knows tasks, applied methods, and specific software.

A significant IUI feature is the capability to change the system reactions pursuant to the knowledge level of a specific user. This capability is based on constructing and using the user model and properly adapting the interface. It is advisable to organize the interface and operation of the system so as to ensure the possibility of dialogue in accordance with the skills or tasks of the user, for example:

- automatic mode, designed for the unqualified user who transmits the choice to the system or in the case when the choice at this stage for the user is obvious or insignificant (Figures 12.6 and 12.7).
- menu of preferable choices: a list of the most common options used for this type of task. Designed for an experienced user who is familiar

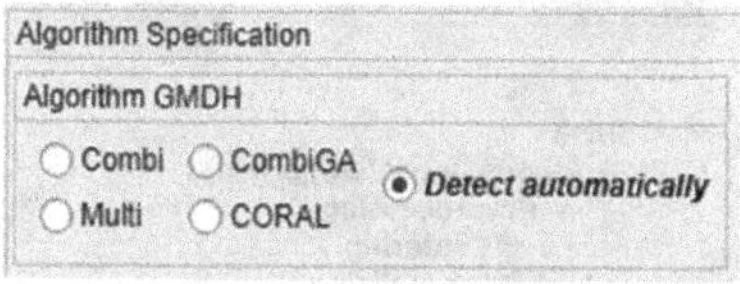

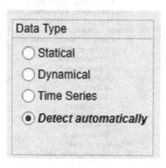

FIGURE 12.6   Menu providing automatic selection of the specific algorithm.

FIGURE 12.7   Menu providing automatic choosing the process type.

with the presented methods or already has experience working with the
system.

- complete menu: presents all possible options for choosing solutions at
  a particular stage of modeling. Designed for the expert users who solve
  specific tasks as well as are well acquainted with the presented methods
  and capabilities of the system.

In such interface, it is possible to form various standard scenarios on the basis of
a certain volume of information on the solution of similar tasks.

12.6 ONTOLOGY MODELS OF THE DOMAIN FOR DESIGNING GENERALIZED GMDH TOOL

Ontological models consist of concepts and statements about these concepts on
the basis of which classes, instances of classes, relations, and functions are built.
This can be considered as a database of a special kind with semantic information
about the IM subject area [47]: Knowledge base (KB) in the form of GMDH-based
ontological models of IM in the appropriate formats OWL/XML and RDFS.

To specify the main structural components of GMDH algorithms, meta-
ontology was created (Figure 12.8). It demonstrates the hierarchical relation of main classes to which certain low-level ontologies should be added to
describe the domain and its components. This can be considered as a generalized
GMDH-based ontological model of the IM domain. Ontologies are linked using
a URI unique to each resource.

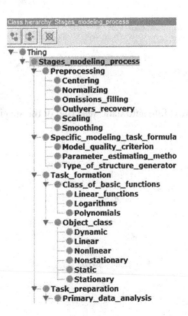

FIGURE 12.8   Meta-ontology based on GMDH.

An example of the ontological model in Figure 12.9 describes the set of available GMDH algorithms. It presents the hierarchy of classes and corresponding instances related to these classes. Another example of such components can be given as ontological models describing the set of generators of model structures (Figure 12.10) and criteria (Figure 12.11) used in GMDH algorithms. Note that in *Protégé* one can easily specify that an instance has a synonym (Same Individual as). Thus, in Figure 12.11 one of the accuracy criteria has two instances-synonyms: Regularity and AR. This is very useful if the terminology of the subject area has many equivalent definitions or for combining individual ontologies.

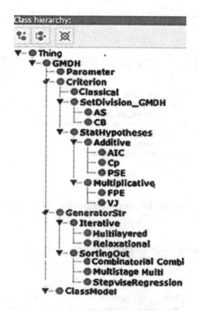

FIGURE 12.9 Ontological model of typical GMDH algorithms.

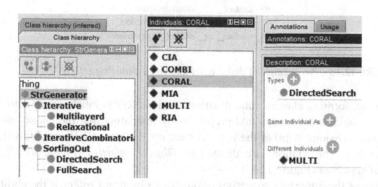

FIGURE 12.10 Ontology of model structure generators.

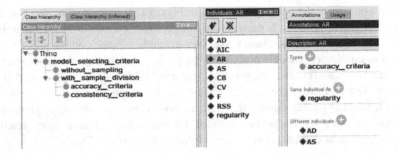

FIGURE 12.11   Ontological model of GMDH model selection criteria.

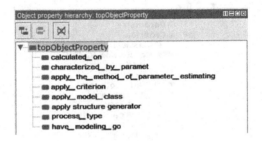

FIGURE 12.12   An example of defining relationships *ObjectProperties*.

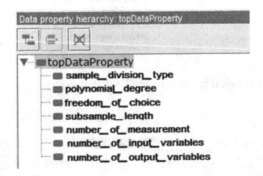

FIGURE 12.13   An example of defining relationships *DataProperties*.

When forming classes, subordination relations (class–subclass) are automatically established. One can add properties by indicating that the classes do not intersect (Disjoint with) or the instances are not different (Different Individual). Other kinds of relations are defined as *ObjectProperties* (Figure 12.12) and *DataProperties* (Figure 12.13).

One of the important properties of ontologies built-in *Protégé* is the ability to draw logical conclusions (new knowledge) from available information. For this purpose, the mechanism of logical inference, the reasoning unit (*Reasoner*), is used. In addition to *Reasoner*, there is a mechanism for constructing necessary rules on which logical conclusions are based, using a specialized language of rules SWRL (Semantic Web Rule Language). This makes it possible to define built-in functions in the form of numerical equations, simple arithmetic operations, operations with strings, time or economic functions, etc. which can be applied to specific instances.

For example, the rule for selecting the combinatorial algorithm COMBI according to the number of input variables in SWRL terms is given as follows:

```
Task(?x) ^ hasInVar(?x,?m) ^ swrlb:greaterThan(?m, 30) →
useCombi(?x).
```

This rule means: if the *task* x has the number of variables m and $m \leq 30$, then the combinatorial algorithm *COMBI* is preferable for use. To select the correlation-rating algorithm *CORAL* when solving the *feature selection* problem according to the defined purpose of modeling, the following rule is set:

```
(hasTargetMod = FeatureSelection)(?x) → useCoral(?x).
```

This builds decision trees and determines the appropriate algorithms for solving a specific modeling problem with appropriate procedures. At certain stages of the modeling process, the possibility of automatic selection of the mode is provided without user intervention.

12.7 CONCLUSION

The main stages of the inductive modeling process based on GMDH and components of GMDH algorithms were analyzed from the viewpoint of generalization of basic structural elements. To formally descript the GMDH-based inductive modeling domain, appropriate ontology models were created using the *Protégé_4*.3 ontoeditor. These ontologies form the knowledge base of the domain. Creating such a base with the use of ontologies allows for reusing information and increasing the efficiency of working with this information for data-based modeling of complex systems, forming and processing queries, and automatic obtaining logical inferences. A general framework of the GMDH-based inductive modeling tools was presented.

REFERENCES

1. Chandrasekaran B., Josephson J.R., Benjamins V.R. Ontologies: What are Ontologies, and Why Do We Need Them? *IEEE Intelligent Systems*, 14, 1999, 20–26.
2. Konys A. Knowledge Systematization for Ontology Learning Methods. 22nd International Conference on Knowledge-Based and Intelligent Information & Engineering Systems. *Procedia Computer Science*, 126, 2018, 2194–2207.

3. Dillon S., Chang E. Ontology-Based Software Engineering. *Software Engineering. ASWEC. 19th Australian Conference*, 2008. doi:10.1109/ASWEC.2008.4483185.

4. Calvaneze D. Ontologies and Databases. Tutorial. Reasoning Web Summer School, 2009. URL: http://www.inf.unibz.it/calvanese/teaching/2009-09-ReasoningWeb-school-ontologies-dbs/ReasoningWeb-2009-ontologies-dbs.pdf.

5. Belkin N.J., Brooks H.M., Daniels P.J. Knowledge Elicitation Using Discourse Analysis. *International Journal of Man-Machine Studies*, 27, 1987, 127–144.

6. Mettrey W. An Assessment of Tools for Building Large Knowledge-Based Systems. AI Magazine. 8 (4). 1987. URL: https://www.aaai.org/ojs/index.php/aimagazine/article/view/625/558.

7. Smith R.G. Knowledge-Based Systems. Concepts, Techniques. Examples. URL: http://www.reidgsmith.com/Knowledge-Based_Systems_-_Concepts_Techniques_Examples_08-May-1985.pdf.

8. Shakhovska N., Vysotska V., Chyrun L. Intelligent systems design of distance learning realization for modern youth promotion and involvement in independent scientific researches. *Advances in Intelligent Systems and Computing*. Springer, Cham. 2017, pp. 175—198. doi:10.1007/978-3-319–45991-2_12.

9. Ruy F.B., Guizzardi G., Falbo R.A., Reginato C.C., Santos V.A. From Reference Ontologies to Ontology Patterns and Back. *Data & Knowledge Engineering*, 109, 2017, 41–69.

10. Ivakhnenko A.G., Müller J.-A. Recent Developments of Self-Organizing Modeling in Prediction and Analysis of Stock Market. *Microelectronics Reliability*, 37, 1997, 1053–1072.

11. Anastasakis L., Mort N. The Development of Self-Organization Techniques in Modelling: A Review of the Group Method of Data Handling (GMDH). *Research Report-University of Sheffield Department of Automatic Control and Systems Engineering*. No. 813, The University of Sheffield, United Kingdom, 2001. 38 p.

12. Stepashko V. On the Self-organizing Induction-Based Intelligent Modeling. Advances in Intelligent Systems and Computing III. In N. Shakhovska, M.O. Medykovskyy, Editors, AISC Book Series, vol. 871, Cham: Springer, 2019, pp. 433–448.

13. Gruber T.R. A Translation Approach to Portable Ontologies. *Knowledge Acquisition*, 5(2), 1993, 199–220.

14. Kryvyi S.L. Presentation of Ontologies and Operations on Ontologies in Finite-State Machines Theory. *International Journal Information Theories and Applications*, 16(4), 2009, 349–355.

15. Gruber T. Toward Principles for the Design of Ontologies Used for Knowledge Sharing. *International Journal of Human-Computer Studies*, 43(5–6), 1995, pp. 907–928.

16. Pidnebesna H., Stepashko V. Construction of an Ontology as a Metamodel of the Inductive Modeling Subject Area. *Proceedings of the International Conference "Advanced Computer Information Technologies" ACIT–2018*, Ceske Budejovice, Czech Republic, June 1–3. pp. 137–140. ISBN 978-966-654-489-9.

17. Stepashko V.S. Method of Critical Variances as Analytical Tool of Theory of Inductive Modeling. *Journal of Automation and Information Sciences*, 40(2), 2008, 4–22.

18. Ivakhnenko A.G. Group Method of Data Handling as a Rival of Stochastic Approximation Method. *Soviet Automatic Control*, 13(3), 1968, 43–55.

19. Madala H.R., Ivakhnenko A.G. *Inductive Learning Algorithms for Complex Systems Modeling*. London, Tokyo: CRC Press Inc., 1994. 384 p.

20. Kordik P. Fully Automated Knowledge Extraction Using Group of Adaptive Model Evolution: Ph.D. Thesis, Dep. of Comp. Sci. and Computers, FEE, CTU in Prague, 2006, 150 p.

21. *Self-organizing Methods in Modeling: GMDH Type Algorithms.* Ed. S.J. Farlow. New York, Basel: Marcel Decker Inc., 1984. 350 p.
22. Voss M.S., Feng X. A New Methodology for Emergent System Identification Using Particle Swarm Optimization (PSO) and the Group Method of Data Handling (GMDH). *Proc. of the Genetic and Evolutionary Computation Conference.* Morgan Kaufmann Publishers, 2002. (9–13th of July, 2002, New-York). New-York, 2002, pp. 1227–1232.
23. Schmidhuber J. Deep Learning in Neural Networks: An Overview. *Neural Networks*, 61, 2015, 85–117.
24. Ivakhnenko A.G. Objective Computer Clasterization Based on Self-Organisation Theory. *Soviet Automatic Control*, 20(6), 1987, 1–7.
25. Ivakhnenko A.G. Heuristic Self-Organization in Problems of Automatic Control. *Automatica (IFAC)*, 3(6), 1970, 207–219.
26. Ivakhnenko A.G. Polynomial Theory of Complex Systems. *IEEE Transactions on Systems, Man, and Cybernetics*, 1(4), 1971, 364–378.
27. Ivakhnenko A.G. Inductive Sorting Method for the Forecasting of Multidimensional Random Processes and Events with the Help of Analogs Forecast Complexing. *Pattern Recogn. and Image Analysis*, 1(1), 1991, 99–108.
28. Lytvynenko V., Bidyuk P., Myrgorod V. Application of the Method and Combined Algorithm on the Basis of Immune Network and Negative Selection for Identification of Turbine Engine Surging. *Proc. of the II Intern. Conf. on Inductive Modelling ICIM-2008* (Kyiv, Sept. 15–19. 2008). Kyiv: IRTC ITS NASU, 2008, pp. 116–123.
29. Yu Z. The Investigations of Fuzzy Group Method of Data Handling with Fuzzy Inputs in the Problem of Forecasting in Financial Sphere. *Proc. of the II Intern. Conf. on Inductive Modelling ICIM-2008. (Kyiv, Sept. 15–19, 2008).* Kyiv: IRTC ITS NASU, 2008, pp. 129–133.
30. Bodyanskiy Y., Vynokurova O., Teslenko N. Cascade GMDH-Wavelet-Neuro-Fuzzy Network. *Proc. of the IV Intern. Workshop on Inductive Modelling IWIM-2011* (Kyiv-Zhukyn, July 4–11, 2011). Kyiv: IRTC ITS NASU, 2011, pp. 16–21.
31. Moroz O., Stepashko V. Data Reconstruction of Seasonal Changes of Amylolytic Microorganisms Amount in Copper Polluted Soils. *Proc. of the 13th IEEE Intern. Conf. CSIT-2018* (Lviv, Sept. 11–14, 2018), Lviv: LNPU, 2018, pp. 479–482.
32. H. Pidnebesna, V. Stepashko. Comparative Effectiveness of Some Approaches to Extracting Most Informative Factors Influencing Algae Bioproductivity. *Proceedings of the Int. Conf. "Advanced Computer Information Technologies" ACIT-2020*, Deggendorf, Germany, September 15–18, 2020, pp. 257–260.
33. Stepashko V.S. Formation and Development of Self-organizing Intelligent Technologies of Inductive Modeling. *Cybernetics and Computer Engineering*, (4), 2018, 41–61. doi:10.15407/kvt194.04.041.
34. Muller J.A., Lemke F. *Self-Organising Data Mining: An Intelligent Approach to Extract Knowledge from Data.* Dresden, Berlin: Books on Demand Verlag, 1999, 225 p.
35. Tetko I.V., Aksenova T.I., Volkovich V.V. et al. Polynomial Neural Network for Linear and Non-linear Model Selection in Quantitative-Structure Activity Relationship Studies on the Internet. *SAR and QSAR in Environ. Res.*, 11(3–4), 2000, 263–280. doi:10.1080/10629360008033235.
36. ASTRID: Problem-Oriented Program Sysytem. URL: http://www.mgua.irtc.org. ua/ua/index.php? page=astridJournal>.
37. FAKE GAME Project. URL: http://fakegame.sourceforge.net/doku.php
38. GMDH Shell: Forecasting Software for Professionals. URL: http://www.gmdhshell. com

39. Savchenko Ye. Double-Criterion Choice of an Optimal Model in GMDH Algorithms. *Proceedings of the 2nd International Conference on Inductive Modelling* (ICIM'2008). Kyiv: IRTC ITS NASU, 2008, pp. 23–27.

40. Yefimenko S., Stepashko V. Technologies of Numerical Investigation and Applying of Data-Based Modeling Methods. *Proceedings of the II International Conference on Inductive Modelling* ICIM-2008, 15–19 September 2008, Kyiv, Ukraine. Kyiv: IRTC ITS NASU, 2008, pp. 236–240.

41. Stepashko V.S. A Finite Selection Procedure for Pruning an Exhaustive Search of Models. *Soviet Automatic Control*, 16(4), 1983, 84–88.

42. Moroz O., Stepashko V. Hybrid Sorting-Out Algorithm COMBI-GA with Evolutionary Growth of Model Complexity. In: *Advances in Intelligent Systems and Computing II*, N. Shakhovska, V. Stepashko, Ed. AISC, Vol. 689. Cham: Springer, 2018, pp. 346–360.

43. Pidnebesna H. A Correlation-Based Sorting Algorithm of Inductive Modeling Using Argument Rating. *Proceedings of the IEEE 2019 14th International Scientific and Technical Conference on Computer Sciences and Information Technologies* (CSIT), September 17–20, 2019, Lviv, Ukraine. – Lviv: LNPU, 2018. Vol. 1, pp. 257–260.

44. Stepashko V., Yefimenko S. Parallel Algorithms for Solving Combinatorial Macromodelling Problems. Przegląd Elektrotechniczny (Electrical Review), ISSN 0033-2097, R. 85, NR 4, 2009, pp. 98–99.

45. Pavlov A., Pidnebesna H., Stepashko V. Ontology Application to Construct Inductive Modeling Tools with Intelligent Interface. *Control Systems and Computers*, (4), 2020, 44–55.

46. Ahmad A.-R., Basir O., Hassanein K. Adaptive User Interfaces for Intelligent E-Learning: Issues and Trends. *Proceedings of ICEB*. 2004, pp. 925–934.

47. Pidnebesna H., Stepashko V. Ontology-Based Design of Inductive Modeling Tools. *Proceedings of the Int. Conf. "Advanced Computer Information Technologies"* ACIT-2021, Deggendorf, Germany, September 18–22, 2021, pp. 731–734.

13 Exploring the Contemporary Area of Ontology Research
FAIR Ontology

Prashant Kumar Sinha

CONTENTS

13.1 INTRODUCTION

The core facet of research data management (RDM) accords with the data and with the evolution of research data repositories, RDM has picked up a load of impetus [1]. Data has consistently been prominent in research, but now the scientific community recognises its correct significance and expects it to be accessible for adoption and reuse. Thus, in this field of exploration, the proposition of FAIR Data Principles [2] has opened up new research avenues. The principles furnish guidance for scientific data management. The FAIR foundational principles are "Findability, Accessibility, Interoperability, and Reusability". These central propositions are not specifications or protocols, but somewhat simple platform/ guidelines for data publishers to embrace in order to prepare their data to be more evident and useful [2]. The intention of depositing data in a data repository should not be restricted to just making it available, but to amass data in such a form that it becomes used by diverse systems. The notion of open data and FAIR data are different, even though the theory behind such concepts is embedded in the same philosophies of data sharing and enhancing participation among researchers [3]. Open data preaches that a specific data set should be made available to all the users, whereas FAIR says the right of accessibility to specific data sets can be governed by the dataset owner. It does not compel that a data set should be open 3,4. Now, FAIR data does not reconcile to the open data principle, but for data to be reusable, it must have an explicit license, preferably machine-readable [4]. The 'I' in the FAIR principle stands for "interoperability" where the job of vocabularies or ontologies comes into the picture. It declares that data should be semantically presented so that it can be employed in different systems. Ontologies are now acknowledged as a knowledge artifact that structures and models the indicated domain of interest to represent the available knowledge in machine-processable format [5]. The engagement of the scientific community belonging to diverse backgrounds such as ontology developers, domain experts, industry experts in ontological studies has popularised the subject [6]. There have been a considerable number of increases in the literature of ontologies over the past few years.

A specific inquiry of "ontology" term in a scientific database like SCOPUS results in 136,436 records (search performed on 31/3/2021). The documents belong to diverse subject categories as well, right from computer science to social science. But it has been discovered with ontologies that are commonly troublesome to access, understand and reuse [7]. For instance, Sinha and Dutta [8] assessed flood ontologies using a parametric approach where one of the parameters dealt with the availability of ontologies in ontology libraries [9]. The study realised that among the reviewed 14 ontologies, none of them were available in the ontology libraries, only one of them was available on GitHub. So if ontology is not accessible, then understanding and reusing them becomes challenging. Keeping this very bottleneck in notice and also as the proposition of FAIR principles asserts that vocabularies or ontologies following such FAIR principles should be utilised for FAIR data; researchers have started preaching the concept of FAIR ontologies, which corroborates with FAIR principles. The present examination will encourage the scientific community involved in developing ontologies to establish that the ontologies are FAIR to strengthen their visibility and service. As this research shows a state-of-the-art work in FAIR ontologies, it will draw the attention of researchers also towards the investigations that are being performed from an ontology perspective in the upcoming domain. The primary objective of this work is to discuss the idea of FAIR ontologies, the role of ontologies in FAIRification process and the role of ontology libraries in making ontologies FAIR. The work also discusses the current research efforts made in the area of FAIR ontologies highlighting the major ideas that have been researched to pave way for the upcoming researchers.

13.2 FAIR ONTOLOGY AND FAIRIFICATION

A FAIR ontology [7] is an ontology that complies with the FAIR principles. An ontology which is:

- easily findable on the web where each of the ontology has a persistent uniform resource identifier, rich metadata with identifiers, available in searchable resources like ontology libraries [7];

- accessible to the community based on the right of accessibility given by the ontology owners with authorisation, authentication procedure put in place [7];
- can help in semantic interoperability and integration between various systems as they are presented using appropriate knowledge representation models [7];
- Described with enough information about its content, its provenance in a standard format based on the domain so that it can be reused and understood is termed as FAIR ontology [7].

FAIRification is a process that attempts to make data FAIR. A general framework for the workflow of FAIRification is segregated into three parts namely pre-FAIRification, FAIRification and post-FAIRification [10]. These three portions have seven distinct steps as depicted in Figure 13.1. These steps need not be followed in a chronological sequence and may be iterated.

The fourth step (4a) of the FAIRification workflow preaches either establishing a semantic depiction of data from scratch or reusing the prevailing semantic

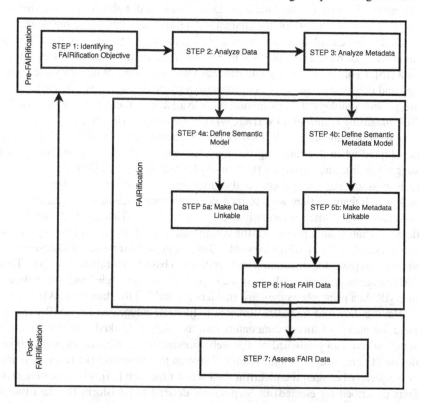

FIGURE 13.1 Generic workflow for FAIRification. (*Source*: Jacobsen, A., Kaliyaperumal, R., da Silva Santos, L. O. B., Mons, B., Schultes, E., Roos, M., & Thompson, M. (2020). A generic workflow for the data FAIRification process. *Data Intelligence*, 2(1–2), 56–65.)

models. While developing ontologies the developers can make use of contemporary methodologies like YAMO [11], NeOn [12] as it will reinforce the development of quality ontologies for use. Even though there exist various design methodologies there are no one correct methodologies to model a domain [13]. But contemporary methodologies like the ones pointed out above are developed in the modern era, so they are expected to be on par with the technological requirements of this age and have tried to surmount the glitches of previously proposed methodologies. Like YAMO implements a collection of intricate steps along with an array of general guidelines for the construction of ontologies, whereas NeOn Methodology presents a road map for building ontology networks that facilitates collaborative ontology construction, reuse of existing ontologies and likewise the dynamic evolution of ontology networks in distributed environments.

The generic workflow discusses the process of specifying semantic metadata models (4b) as well. In order to reuse the prevailing semantic model, both contents of the ontology and the ontology need to be properly illustrated. Thus, describing ontologies will make the process simpler for the users to select and employ a specific ontology for their project. The metadata vocabularies like Metadata Vocabulary for Ontology Description and Publication [14] (MOD) can be taken advantage of for characterising and annotating the ontologies. It can likewise be exploited as an explicit OWL ontology by ontology repositories/libraries to develop and offer a semantic interpretation of ontologies as linked data [15]. In order to have a straightforward understanding of ontology's technology and ecosystem, ontology documentation plays an important role [16]. There are various ontology documentation tools available in literature like *WIzard for DOCumenting Ontologies* (WIDOCO) [17], *OnToology* [18], etc. which can process such documentation. *WIDOCO* guides stakeholders through the strides to be adopted when documenting ontology, relying on familiar best practices, and suggesting missing metadata that should be introduced. *WIDOCO* also facilitates customising the generated documentation, empowering stakeholders to pick which features they wish to incorporate in their documents[17] (e.g., sections, diagrams, provenance information, etc.). Similarly, *OnToology* is a web-based device to automate a portion of the collaborative ontology development process. Provided a depository with an owl file, *OnToology* will examine it and design diagrams, complete documentation and verification based on familiar risks [18]. This FAIRification process emphasises the importance of ontologies and vocabularies in FAIR data principle as they are enablers for making the data into FAIR data.

Having data and metadata linkable is likewise advocated by FAIRification. Here the notion of linked data enters into the picture. Linked data pertains to a standardised data published on the web adhering to a collection of propositions devised to encourage the intertwining between the elements (aka resources) and consequently between the plentiful data sets on the web [15]. The semantic artifacts described (or elected) in Step 4a are desired to transform the data into a machine-readable form (Step 5a). For this method, specialised tools such as the *FAIRifier* [19] are available. It provides insights into the transmutation method and makes it repeatable by monitoring intermediate steps. This *FAIRifier* allows a post-hoc FAIRification flowchart that includes loading an available dataset, performing data wrangling process, adding FAIR attributes to the data, generating a linked data version of the data, and ultimately forcing the output to an online FAIR data infrastructure to make it usable and findable. Manually or via embedded, configurable script expressions, literal values in a set of data can be substituted with identifiers. In order to improve the dataset's interoperability, adjusting these identifiers into a relevant semantic graph framework of ontological classes and properties employing the incorporated RDF model editor is critical. A provenance route records each alteration instantly and likewise allows "undo" activities and reiterated activities on similar datasets. A FAIR data export function opens up a metadata editor to furnish knowledge about the dataset itself (https://bio.tools/FAIRifier).

Karma [20] is a data integration mechanism that empowers users to efficiently combine data from a range of data sources, comprising databases, spreadsheets, XML, JSON, delimited text files, and Web APIs. Stakeholder can integrate data using Karma by designing the data that is conforming to their preferred ontology via a graphical user interface. This automates a substantial chunk of the workflow. Karma learns to recognise the mapping of data to ontology classes and later introduces a model that associates these classes employing the ontology. Users then connect with the structure to accomplish the model that was achieved automatically. Throughout this refining, users can alter the data as required to normalise and reconstruct data expressed in various formats. After completion of the model, the integrated data can be disclosed as RDF or saved in a database. *Rightfield* [21] is a stand-alone Java application that collaborates with Microsoft documents via Apache-POI. It permits users to import existing Excel spreadsheets or construct new ones. Ontologies can be imported straight from their file systems, the web, or the *BioPortal* ontology repository. Individual units, or entire columns or rows, can be annotated with the required ontology term ranges, and an entire spreadsheet can be annotated with terms from multiple ontologies. *OntoMaton* [22] is an open source software that integrates ontology lookup and tagging functionality into a cloud-based collaborative editing environment via Google Spreadsheets and the NCBO Web services. It is an all-around, format-independent tool that might be used in conjunction with the ISA software suite. It also facilitates ontology development.

The semantic metadata representation described (or elected) in Step 4b is necessary for the transformation of the metadata into a machine-readable form (Step 5b). There is software that facilitates this conversion process for some generic metadata objects, such as the *FAIR Metadata Editor* [23], The *FAIR Metadata Editor* is a free online application that shows how to structure metadata in a FAIR-friendly manner [10]. The Center for Expanded Data Annotation and Retrieval (CEDAR) [24] is working on the establishment of global and suggestive metadata for biomedical datasets to aid data detection, data illustration and data reuse. They seize profit from rising public-based ideal templates for expressing various types of biomedical datasets and examine the value of computational procedures to aid researchers in establishing templates and filling in their contents. The initiative is building an archive of metadata to be used for determining metadata patterns that will guide predictive data entry when loading in metadata templates. The metadata repository not merely will occupy annotations specified when experimental datasets are originally formed, but likewise will absorb attaches to the printed article, consisting of secondary analyses and potential refinements or retractions of empirical perceptions [24]. By working initially with the Human Immunology Project Consortium and the developers of the *ImmPort* data repository, they are establishing and appraising an end-to-end solution to the complications of metadata authoring and management that will generalise to diverse data-management environments [24]. The FAIRification process emphasises that either semantic models for the data should be built from scratch if it's not available or the existing ones should be utilised. But in both scenarios, the ontologies should comply with the FAIR principles. These semantic linkable models provide a platform for the development of powerful global virtual knowledge graph [10]. The scientific community has acknowledged making ontologies as FAIR [25] because they understand that ontology development is a laborious, time-consuming and costly process [14]. Hence if existing ontologies are made FAIR or while developing the ontologies FAIR principles are followed, then it will not only help in making the data FAIR but also enhance the utilisation of the ontologies.

13.3 ROLE OF ONTOLOGY LIBRARIES IN FAIR ONTOLOGIES

The ontology library is characterised as an organisational network that can be accessed from any location or device in order to provide organised access to the ontologies that have been deposited in an efficient manner. The ontologies presented here are described by well-known ontology representation languages, such as RDFS and OWL [26]. Ding and Fensel [27] summarised various criteria for evaluating an ontology library system like; management (included storage, identification, and versioning of ontologies); adaptation (included searching, editing, and reasoning of ontologies); standardisation (included language, upper level ontology support). Ontology libraries are classified into three types: ontology repository, ontology registry, and ontology directory [9]. These follow the same root philosophies but might differ in some features.

Ontology repository: An ontology repository stores, manages and retrieves ontologies and linked information (e.g., ontology metadata, which includes information of the depositor of ontologies). The ontology repository's purpose is to simplify the process of downloading, searching, scanning, sharing, and handling ontologies [9,28,29]. For e.g., *BioPortal* (https://bioportal.bioontology.org/), *AgroPortal*(http://agroportal.lirmm.fr/)

Ontology registry: A registry is a data storage interface where data, information, metadata of a semantic entity, such as a person's or community's name, and the process of using data are documented. It can be described as a platform on which a set of metadata can be declared for visualising data, storing information, and gaining access to domain and scientific knowledge. Instead of containing the actual collection of ontologies, an ontology registry consists of a list of ontologies and metadata. A combination of URIs+Registry+RDF can be used to express an ontology registry [9,30]. For eg *MMI-ORR* (http://mmisw.or), *Protégé Ontology library* (https://protegewiki.stanford.edu/wiki/Protege_Ontology_Library#OWL_ontologies)

Ontology Directory: An ontology directory is a service that lists all of the ontologies that are available on a given platform. The reference to the concept of ontology-related language-based schema such as OWL, XML, and RDF Schema can be found in the Ontology Directory [31]. It includes metadata and information sources. Ontology directories are more relevant in ontologies that are related to the same domain (e.g. ODP). Crawlers for searching well-formed ontologies on the Web are included in ontology directories. Ontology directories are essentially collections of reference ontologies for a given domain (e.g., e-Government). It usually adheres to simple hierarchical structures, which are efficient in data storage and classification. An ontology directory can be thought of as a database framework that provides users with registration and service request services [9,32]. For eg *OeGov*(http://www.oegov.org/), *ONKI* (http://onki.fi/).

The objective of an ontology library is to provide a platform to the ontology developers for publishing, storing, preserving, sharing and evaluating the ontologies and to allow the ontology users to make use or reuse of particular ontology or ontologies for specific projects [9]. The library is largely in charge of ontologies deposition. An ontology library's dissemination and validation of ontologies are typically done by a community, an administrator, and/or registered users, based on the initial evaluation. Ontologies can be applied to certain libraries by unregistered users, and the ontologies are validated using automated syntax validation [9]. Few ontology libraries like *DAML Ontology Library* [33] provide a compilation of 282 libraries that have been grouped by URI, Keyword, submission date, submitting entity, and other factors. They have given email addresses from which ontologies can be submitted for correction or addition. Similarly, *Protégé Wiki library* categorises the ontologies into three categories, namely OWL ontologies, frame-based ontologies, ontologies in other formats (e.g., DAML+OIL, RDF Schema, etc.). It also includes a brief description of each of the ontologies that have been deposited, as well as the names of the contributors and links to the ontologies. This is explicitly related to FAIR's "findability," "accessibility," and "reusability" principles.

BioPortal [34], One of the most popular ontology libraries, not only offers a collection of ontologies organised by category, community, and format, but also ontology mapping and ontology recommendations. Ontology mapping is a concept that refers to the mapping of two concepts from two separate ontologies. This connection generally, but not always, indicates a degree of similarity between the two ideas. The semantics of a mapping is described by the mapping's author. Similarly, the ontology recommendation system is a structure that suggests ontology to a user based on a small section of text or a set of keywords [9]. When explaining ontologies, *BioPortal* often includes details about the various projects that are using them, as well as the number of visits each of the ontology has received. It also provides detailed information on the contents of every ontology. Similar facilities are provided by the *AgroPortal* [35]. It allows users to sign in and then submit the ontology. The portal not only provides facilities of ontology mapping, ontology recommendations and details about the project that are using *AgroPortal* resources as well. It also allows the users to provide comments for other ontologies and provides a visual representation of the data about the ontologies present in the portal. Ontology Lookup Service [36] (OLS) provides various tools such as *OxO, Zooma* and *Webulous* services. Cross-ontology mappings between concepts from different ontologies are given by *OxO*. *Zooma* is a service that facilitates mapping of data to ontologies in OLS, and *Webulous* is a tool that supports generating ontologies from spreadsheets. These facilities show how the ontologies can be "interoperable" and "reusable" which are the foundation principles of FAIR.

Keeping the FAIR principles in mind the *FAIRDOMHub* repository (https://fairdomhub.org/) has been entrusted with the responsibility of publishing FAIR (Findable, Accessible, Interoperable, and Reusable) Data, Operating Methods, and Prototypes for the Systems Biology community. It is a web-based repository for stockpiling and transmitting systems biology research assets. It enables researchers to organise, share, and publish data, prototypes, and procedures, as well as to attach them to the systems biology studies that developed them via API interfaces. By leveraging the *FAIRDOMHub*, researchers can foster collaboration with geographically dispersed collaborators, ensure the sustainability and protection of findings, and publish reproducible articles that adhere to the FAIR data stewardship guiding principles [37]. These ontology libraries also encourage the depositors of ontologies to deposit ontology in a standard format like OWL with proper descriptions of each element or contents. The libraries allow downloading of ontologies and also highlight the licensing of ontologies. Thus it can be concluded that ontology libraries provide infrastructural support for FAIR ontologies [38].

13.4   FAIR ONTOLOGIES: RESEARCH PARADIGMS

Though the FAIR ontologies concept has been well accepted by the research community for making the FAIR data, it has also led to the emergence of new research in ontologies as well. For instance, Dutta et al. [39] specified that due to the increase in the number of vocabularies or ontologies the task of selecting and using ontologies has become cumbersome. So, there is a requirement to elucidate ontologies with metadata vocabularies like ontology metadata vocabulary [40] (OMV). A survey was conducted on the uses of metadata vocabularies in various ontology libraries and repositories. The findings were reported like; developers of ontologies use different metadata vocabularies and OMV which was published in 2005 is not widely used; there was a strong overlap between various metadata vocabularies but still they do not depend on one another; the analysed ontology libraries use metadata elements but every time they do not use standard metadata vocabularies. Keeping these findings in mind a new version of Metadata Vocabulary for Ontology Description and Publication, termed as MOD 1.2, an extension of the previous MOD work [14] presented in 2015 was developed. This vocabulary can describe the ontologies in a detailed manner which will increase an ontology's "findability", "accessibility" and "reusability". Guizzardi [41] work dealt with more fundamental knowledge about ontologies. The importance of semantic interoperability was discussed and characterisation for it was proposed. It also discussed the role of formal ontology theories to provide a base for the construction of representation languages, as well as methodological and computational tools that support ontology engineering in the context of FAIR.

Ren et al. [42] proposed a unified data service model that described the ocean data resources and supported FAIR data services. At first, a semantic model named Ocean Environmental Data Ontology (OEDO) representing the diverse heterogeneous ocean data resources was proposed. The idea here was to offer interoperability data services. So once the ontology was developed then with the support of extended Quick Service Query List (QSQL) data structure the domain concepts were further extended to optimise QSQL by the lexical database WordNet. Once these two infrastructures were built the ocean data service publishing method termed Data Ontology and List-based Publishing (DOLP) was put forth in order to enhance the data discovery and data access. Nuniger et al. [43] proposed an ontological approach to connect the archaeological digital data and modelling projects as they wanted to establish a link between the actual data and the models. The solution was built towards solving the problem of creating useful metadata for the obtained data in proper context. It was done so that the same archaeological digital data can be used by different studies in which they are relevant. Thus truly making the data ``interoperable". These kinds of ontologies are absolutely impetuous as they facilitate FAIRness for both the resources. Similarly, Gillespie et al. [44] also proposed an ontology called Neuron Phenotype Ontology (NPO). It established a semantically-enhanced FAIR data model for describing the complex cellular phenotypes created by neuroscientists operating on both individual and large-scale brain projects. It enables the development of machine-generated taxonomies and provides a machine-configurable naming convention. Brewster et al. [45] on a technical level, elucidated a mechanism for applying ontology-based authentication protocols to structured data. The method essentially used the applied ontology's structure to handle data access by assigning semantic or topological restrictions to users based on individual or group policies. The project had a direct influence on data access authentication and use. Becker et al. [46] emphasised the importance of findability (F), accessibility (A), interoperability (I), and reusability (R) of research data as critical and recognised factors for efficient data re-use, for example, in data-driven science. However, it was observed that there are currently no common standards or tools for publishing data in accordance with these FAIR data principles in the field of plasma technology. To address this issue, this contribution described the development of a modular metadata model for representing subject- and method-specific metadata in the field of plasma technology. The model is based on the Plasma-MDS core metadata schema (https://arxiv.org/abs/1907.07744). Ontologies are used to connect and semantically describe the metadata modules. The developed tools and services are accessible via a knowledge graph for plasma technology and the data platform (https://www.inptdat.de/). They are intended to be reviewed and expanded upon by the community of low-temperature plasma scientists in order to establish a common framework for open science and research data management in accordance with the FAIR principle.

Poveda-Villalón et al. [47] drew attention towards the more pressing subject on FAIR ontologies i.e. the process on how to publish ontologies abiding by the FAIR principles. The work reviewed the technical and social needs required to define a roadmap for generating and publishing FAIR ontologies on the Web. They also investigated four initiatives for ontology publication by aligning them in a common framework for comparison. In a similar line of work about publishing FAIR ontologies on the web Garijo and Poveda-Villalón [7] elucidated the basic implementation guidelines and recommendations for making ontologies FAIR. It discussed that ontologies can be made findable through metadata registries and annotations; can be made accessible through good practices in URI design and content negotiation, can be made interoperable by displaying how to make ontologies available in various standard serialisations format; and made reusable by a proper description of the metadata and the proper guidelines for the diagram for an appropriate understanding of the ontologies on the web while following the linked data principles. As a proof of concept, they also demonstrated the process of employing the recommendation using ontology. The guidelines mentioned in this work can act as a beginning point for the developers of ontology for making FAIR ontologies. Cox et al. [48] presented ten simple rules that allow the conversion of a legacy vocabulary into a list of terms available in a print-based glossary or in a table not accessible using web standards—into a FAIR vocabulary.

One more ontology research area that has come to the forefront is the assessment of the FAIRness of ontology. In this regard, Oliveira et al. [49] presented a systematic mapping study of state-of-the-art core reference security ontologies, published in the last two decades. The FAIRness of the foundational ontologies was investigated and it was observed that they are not being utilised very often (interoperability) and the security core ontologies are not available publicly as well (findability). These facts about the security ontology suggest that it does not allow them to be analysed and thus making their use or reuse difficult. Similarly, Mazimwe et al. [50] examined the degree to which allocation and reuse of emergency management knowledge in the realm are in order with FAIR recommendations. They accomplish this through a methodical exploration and inspection of publications in the emergency management sphere based on a predefined inclusion and exclusion criteria. Basereh et al. [51] worked on analysing the relationship of FAIR ontologies with the accountability, clarity of ontology-dependent artificial intelligence systems. The works also investigated the governance-related issues in ontology quality evaluation metrics by assessing their relationship with FAIR principles and FAcct (Fairness, Accountability, Transparency) governance features. Based on the use case the work recommends a strong correlation between the FAIR principles and FAccT AI. The work also recommends that FAIRness should also be assessed for ontologies and can be inducted as an evaluation criterion for reviewing the quality of ontologies. Amdouni et al. [52] acknowledged the importance of ontology repositories for making semantic resources FAIR. They proposed a generic seven-step ontology FAIRness assessment methodology. The methodology was employed on the ontologies in AgroPortal for their automatic assessment and the work is still under progress.

Based on the above discussion it can be seen that various areas of ontology research have been explored keeping FAIR ontology as the central idea. It starts with developing FAIR ontologies that develop a systematic, formal framework for defining and correlating the different concepts. It enables one to understand the different connections between the entities with the relationships that have been employed. The ontologies facilitate the metadata to locate and access meaningful data about the particular concepts which will in turn help in interoperability [42]. Then metadata vocabularies like MOD 1.2 have been developed for elucidating the ontologies with the help of a collection of metadata elements. These vocabularies can help to describe the ontology in ontology libraries which in turn will help in discovering, classifying and selecting the appropriate ontologies from ontological libraries [14]. This facilitates in making the ontology comply with FAIR principles. A set of recommendations has also been proposed for making FAIR ontologies that are based on the experiences of ontology engineering [47]. There exist a lot of ontologies in literature and a set of ideas has also emerged to assess them for FAIRness. Keeping this FAIRness point in perspective, general methodologies are in development and being applied to assess ontologies automatically in ontology repositories like Agroportal [52]. Figure 13.2 shows the FAIR ontology research areas in literature.

13.5 CONCLUSION

Ontologies enable domain interpretation, knowledge reuse, content aggregation, publication, semantic assimilation of data and isolation of practical intelligence and domain theory. In recent times, ontology-running structures have gained prominence because of the qualities they afford. In general, ontologies provide a standardised mutual representation of domain information and thus it is expected that data would be easier to find if it is defined using ontologies. The level of usability can be improved by using ontologies to describe metadata. Interoperability can be facilitated by using the same terminology as defined by ontologies. Finally, since ontologies are shared and standardised, they can be reused. FAIR ontologies allow us to fulfil all these aspects. The idea of FAIRness allows the ontology to be used more in the community which in turn increases attention towards creating quality FAIR ontologies. The development of ontologies and making ontologies FAIR are two processes but when these two aspects are performed together it enhances the chances of use of ontologies. In this work, a small introduction to FAIR ontology and the exact use of ontologies in the FAIRification process has been explained. Some tools like *FAIRifier, Karma, OntoMaton, FAIR Metadata*

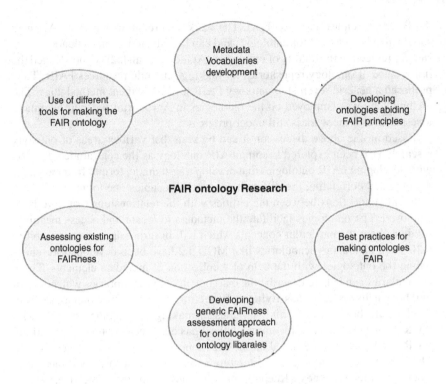

FIGURE 13.2   FAIR ontology research areas in literature.

Editor, WIDOCO which can provide support in making ontologies FAIR have been discussed as well. The work here draws attention towards the already existing infrastructural support of FAIR ontologies i.e. ontology libraries. The facilities that are provided at ontology libraries allow making the ontologies FAIR. Few specific ontology libraries like *BioPortal, FAIRDOMHub,* Ontology Lookup Service and specific facilities which help to make the ontologies FAIR are discussed as well. The article also presents a state of the art literature where various areas of ontology research have been explored keeping FAIR ontology as the core idea like best practices for making ontologies FAIR, developing ontologies abiding FAIR guidelines, development of generic FAIRness assessment methodology in ontology libraries, etc.

Though the research in the field of FAIR ontologies has picked up pace in the last few years, but there are still a lot of challenges. The first challenge is the availability of ontologies. Even though there are ontologies repositories and a lot of ontologies are being developed but mostly ontologies that are available in the repositories are biological/medical science related ontologies. Sinha et al. [53] recently conducted a review of data mining ontologies. The work reviewed 35 data mining ontologies out of which only 8 were openly available and among them, 3 were also available in ontology repositories. Similar to this Mazimwe et al. [50] was investigating disaster ontologies among 69 ontologies only 17 were findable. A lot of ontologies are made available through GitHub as well. But these ontologies should also be deposited in designated ontology repositories this will certainly increase their chance of being reused and the ontology libraries are the available infrastructure for making them FAIR. The next challenge is lack of research studies. A Google Scholar search of "FAIR ontologies" retrieves only 38 results and for "FAIR Semantics" retrieves 187 results as of 26/10/2021. The number of studies is comparatively less but in the last 2 years research in the field has significantly picked up pace. The literature studies on ontological research demonstrate that several ontologies are built even within a single domain of interest like tourism, disaster, medicine to fulfil different objectives [53]. Ontologies have the potential of creating unified data models with real datasets. This particular capability has given rise to the concept of knowledge graph. According to Schrader [54] (2020) ontology+data=knowledge graph. So if we have FAIR data+FAIR Ontology we get FAIR Knowledge Graph as well (FAIR data+FAIR Ontology= FAIR Knowledge Graph). Thus the power of FAIR ontologies is immense.

REFERENCES

1. Patel, D. (2016). Research data management: A conceptual framework. *Library Review, 65*(4/5), 226–241.
2. Wilkinson, M. D., Dumontier, M., Aalbersberg, I. J., Appleton, G., Axton, M., Baak, A., ... Mons, B. (2016). The FAIR Guiding Principles for scientific data management and stewardship. *Scientific Data, 3*(1), 1–9.
3. OGOOV. Available at: https://www.ogoov.com/en/blog/open-data-and-fair-data-differences-and-similarities/.
4. GO FAIR. Available at: https://www.go-fair.org/resources/faq/ask-question-difference-fair-data-open-data/.
5. Dutta, B. (2017). Examining the interrelatedness between ontologies and linked data. *Library Hi Tech, 35*(2), 312–331.
6. Dutta, B., & Sinha, P.K. (2018). A bibliometric analysis of automatic and semi-automatic ontology construction processes. *Annals of Library and Information Studies, 65*(2), 112–121.
7. Garijo, D., & Poveda-Villalón, M. (2020). Best practices for implementing FAIR vocabularies and ontologies on the web. *arXiv preprint arXiv:2003.13084*.
8. Sinha, P. K., & Dutta, B. (2020). A systematic analysis of flood ontologies: A parametric approach. *Knowledge Organization, 47*(2), 138–159.
9. Naskar, D., & Dutta, B. (2016). Ontology libraries: A study from an ontofier and an ontologist perspectives. *In 19th international symposium on electronic theses and dissertations*, ETD (Vol. 16, pp. 1–12).
10. Jacobsen, A., Kaliyaperumal, R., da Silva Santos, L. O. B., Mons, B., Schultes, E., Roos, M., & Thompson, M. (2020). A generic workflow for the data FAIRification process. *Data Intelligence, 2*(1–2), 56–65.
11. Dutta, B., Chatterjee, U. & Madalli, D.P. (2015). YAMO: yet another methodology for large-scale faceted ontology construction. *Journal of Knowledge Management, 19*(1), 6–24.
12. Suárez-Figueroa, M. C., Gómez-Pérez, A., & Fernandez-Lopez, M. (2015). The NeOn Methodology framework: A scenario-based methodology for ontology development. *Applied Ontology, 10*(2), 107–145.

13. Noy, N. F., & McGuinness, D. L. (2001). Ontology development 101: A guide to creating your first ontology. Technical Report SMI-2001–0880, Stanford: Stanford knowledge systems laboratory, Stanford University).

14. Dutta, B., Nandini, D., & Shahi, G. K. (2015, September). MOD: Metadata for ontology description and publication. *International Conference on Dublin Core and Metadata Applications* (pp. 1–9).

15. "Linked data". Available at: www.w3.org/standards/semanticweb/data (Accessed March 22, 2021).

16. Ontology Documentation. Available at https://ontio.github.io/documentation/tutorial_for_developer_en.html (Accessed March 22, 2021).

17. Garijo, D. (2017, October). WIDOCO: A wizard for documenting ontologies. *International Semantic Web Conference* (pp. 94–102). Springer, Cham.

18. Alobaid, A., Garijo, D., Poveda-Villalón, M., Santana-Perez, I., & Corcho, O. (2015, July). OnToology, a tool for collaborative development of ontologies. In *ICBO*.

19. FAIRifier. Available at https://github.com/FAIRDataTeam/FAIRifier (Accessed March 22, 2021).

20. KARMA. Available at https://usc-isi-i2.github.io/karma/ (Accessed March 22, 2021).

21. Rightfield. Available at https://rightfield.org.uk/ (Accessed March 22, 2021).

22. Maguire, E., González-Beltrán, A., Whetzel, P. L., Sansone, S. A., & Rocca-Serra, P. (2013). OntoMaton: A bioportal powered ontology widget for Google Spreadsheets. *Bioinformatics*, 29(4), 525–527.

23. Thompson, M., Burger, K., Kaliyaperumal, R., Roos, M., & da Silva Santos, L. O. B. (2020). Making FAIR easy with FAIR tools: From creolization to convergence. *Data Intelligence*, 2(1–2), 87–95.

24. Musen, M. A., Bean, C. A., Cheung, K. H., Dumontier, M., Durante, K. A., Gevaert, O., ... CEDAR team. (2015). The center for expanded data annotation and retrieval. *Journal of the American Medical Informatics Association*, 22(6), 1148–1152.

25. Janowicz, K., Hitzler, P., Adams, B., Kolas, D., & Vardeman II, C. (2014). Five stars of linked data vocabulary use. *Semantic Web*, 5(3), 173–176.

26. Noy, N. F., Griffith, N., & Musen, M. A. (2008, October). Collecting community-based mappings in an ontology repository. *International Semantic Web Conference* (pp. 371–386). Springer, Berlin, Heidelberg.

27. Ding, Y., & Fensel, D. (2001, July). Ontology library systems: The key to successful ontology reuse. *SWWS* (pp. 93–112).

28. http://ontologforum.org/index.php/OpenOntologyRepository_Scope.

29. d'Aquin, M., & Noy, N. F. (2012). Where to publish and find ontologies? A survey of ontology libraries. *Journal of Web Semantics*, 11, 96–111.

30. Stock, K. M., Atkinson, R., Higgins, C., Small, M., Woolf, A., Millard, K., & Arctur, D. (2010). A semantic registry using a Feature Type Catalogue instead of ontologies to support spatial data infrastructures. *International Journal of Geographical Information Science*, 24(2), 231–252.

31. Tudorache, T., Nyulas, C., Noy, N. F., & Musen, M. A. (2013). WebProtégé: A collaborative ontology editor and knowledge acquisition tool for the web. *Semantic web*, 4(1), 89–99.

32. Li, G., Lv, S., & Chang, P. (2009, December). ODS: Ontology directory services for ontology integration in semantic web services. *2009 International Conference on Research Challenges in Computer Science* (pp. 32–36). IEEE.

33. DAML Ontology Library. Available at http://www.daml.org/ontologies/ (Accessed March 22, 2021).

34. BioPortal, Available at https://bioportal.bioontology.org/ (Accessed March 22, 2021).

35. AgroPortal. Available at http://agroportal.lirmm.fr/ (Accessed March 22, 2021).
36. Ontology Lookup Service, Available at https://www.ebi.ac.uk/ols/index (Accessed March 22, 2021).
37. Wolstencroft, K., Krebs, O., Snoep, J. L., Stanford, N. J., Bacall, F., Golebiewski, M., ... Goble, C. (2017). FAIRDOMHub: A repository and collaboration environment for sharing systems biology research. *Nucleic Acids Research*, *45*(D1), D404–D407.
38. Dutta, B., Jonquet, C., Magagna, B., Toulet, A. Summary report for ontology metadata task group of the vocabulary and semantic services interest group, RDA P11 – Berlin, March 2018. Available at https://www.isibang.ac.in/~bisu/Presentation/23-03_RDA-VSSIG_ontology_metadata_TG.pdf(accessed March 22, 2021)Journal>.
39. Dutta, B., Toulet, A., Emonet, V., & Jonquet, C. (2017, November). New generation metadata vocabulary for ontology description and publication. *Research Conference on Metadata and Semantics Research* (pp. 173–185). Springer, Cham.
40. OMV. Available at http://mayor2.dia.fi.upm.es/oeg-upm/index.php/en/downloads/-75-omv/index.html.
41. Guizzardi, G. (2020). Ontology, ontologies and the "I" of FAIR. *Data Intelligence*, *2*(1–2), 181–191.
42. Ren, X., Li, X., Deng, K., Ren, K., Zhou, A., & Song, J. (2020, November). Bringing semantics to support ocean FAIR data services with ontologies. *2020 IEEE International Conference on Services Computing (SCC)* (pp. 30–37). IEEE.
43. Nuninger, L., Opitz, R., Verhagen, P., Libourel, T., Laplaige, C., Leturcq, S.,.... & Rodier, X. (2020). Developing FAIR ontological pathways: Linking evidence of movement in lidar to models of human behaviour. *Journal of Computer Applications in Archaeology*, *3*(1), 63–75.
44. Gillespie, T. H., Tripathy, S., Sy, M. F., Martone, M. E., & Hill, S. L. (2020). The neuron phenotype ontology: A FAIR approach to proposing and classifying neuronal types. *bioRxiv*.
45. Brewster, C., Nouwt, B., Raaijmakers, S., & Verhoosel, J. (2020). Ontology-based access control for FAIR data. *Data Intelligence*, *2*(1–2), 66–77.
46. Becker, M. M., Franke, S., Loffhagen, D., Vilardell Scholten, L., Hoppe, F., Tietz, T., & Sack, H. (2020). Metadata schema and ontologies for FAIR research data in plasma technology. *APS Annual Gaseous Electronics Meeting Abstracts* (pp. LT1–007).
47. Poveda-Villalón, M., Espinoza-Arias, P., Garijo, D., & Corcho, O. (2020, September). Coming to terms with FAIR ontologies. *International Conference on Knowledge Engineering and Knowledge Management* (pp. 255–270). Springer, Cham.
48. Cox, S. J., Gonzalez-Beltran, A. N., Magagna, B., & Marinescu, M. C. (2021). Ten simple rules for making a vocabulary FAIR. *PLOS Computational Biology*, *17*(6), e1009041.
49. Oliveira, Í., Fumagalli, M., Prince Sales, T., & Guizzardi, G. (2021, May). How FAIR are security core ontologies? A systematic mapping study. *International Conference on Research Challenges in Information Science* (pp. 107–123). Springer, Cham.
50. Mazimwe, A., Hammouda, I., & Gidudu, A. (2021). Implementation of FAIR principles for ontologies in the disaster domain: A systematic literature review. *ISPRS International Journal of Geo-Information*, *10*(5), 324.
51. Basereh, M., Caputo, A., & Brennan, R. (2021, September). FAIR ontologies for transparent and accountable AI: A hospital adverse incidents vocabulary case study. *2021 Third International Conference on Transdisciplinary AI (TransAI)* (pp. 92–97). IEEE.

52. Ontology_FAIRness_assessment. Available at https://www.researchgate.net/profile/
    Emna-Amdouni-2/publication/340654621_Ontology_FAIRness_assessment_
    the_case_of_AgroPortal/links/5e9719e64585150839dea0b1/Ontology-FAIRness-
    assessment-the-case-of-AgroPortal.pdf (Accessed March 22, 2021).
53. Sinha, P. K., Gajbe, S. B., Debnath, S., Sahoo, S., Chakraborty, K., & Mahato, S. S.
    (2021). A review of data mining ontologies. *Data Technologies and Applications.*
    Vol (Ahead of Print) No. (Ahead of Print). 1–33.
54. Schrader, B. (2020). What's the difference between an ontology and a knowledge
    graph? Available at https://enterprise-knowledge.com/wp-content/uploads/2020/01/-
    Ontologies-vs.-Knowledge-Graphs.pdf.

14 Analysis of Ontology-Based Semantic Association Rule Mining

G. Jeyakodi and P. Shanthi Bala

CONTENTS

14.1 INTRODUCTION

Covid-19 pandemic builds the world towards digital transformation and all the organizations generate a huge amount of data every day, among which hospitals and social media are playing a major role. The data interpretation, data replication, multidimensional data, and the complications in integrating data from multiple sources make data analysis a challenging one. The analysis helps to identify the potentially important and hidden information. Various tools based on artificial intelligence, statistics, natural language processing, and data mining are available to manage and analyze the data [1]. Data mining helps to identify and remove undesirable and replicated data in an efficient manner. There are many techniques for knowledge discovery in which association rule mining helps to detect relationships from large databases [2].

DOI: 10.1201/9781003309420-14

Association rules are generally expressed in the form of 'if-then' statements in order to represent the relationship occurrence in the transactional data and can be evaluated using the support and confidence metrics [2]. The association rules help to discover the correlations and co-occurrences among the data to improve the decision-making in business, medical, bioinformatics, geographical information systems, fraud detection, trend analysis, census data, etc. [3]. The algorithms such as apriori which is based on breadth-first search, Eclat - equivalence class transformation that is based on depth-first search, and 'FP (Frequent Pattern) - growth' are available for deriving association rules [4]. The algorithm complexity depends on the data which always consists of noise and inconsistent values, and its performance depends on the data representation [2]. Also, these algorithms generate a greater number of rules that consist of redundant, semantically wrong, uncurious, and previously known relations.

The major challenge in the association rule mining algorithm is the unsuitability in interpreting semantically organized data that describe the clear meaning of the concepts [5]. Semantic data mining enhances the performance of association rule mining, classification, clustering, and information extraction by integrating the background knowledge [6]. RDFS (Resource Description Framework Schema) and OWL (Web Ontology Language) are the World Wide Web Consortium (W3C) recommended representation frameworks for semantic knowledge representation [7]. The knowledge can be represented formally or informally. The formal knowledge represents semantic association implicitly while the informal represents the semantic association explicitly. The text corpora is an example of unstructured or informal knowledge sources and the dictionaries, thesauri, lexical databases, encyclopedias, and ontologies are examples of formal knowledge sources [8].

Ontologies are suitable for semantic association mining as they have a semantically rich taxonomic structure that consists of concepts and their relations. They are described by the ontology specification languages such as RDFS, OWL, OIL, and DAML+OIL. OIL (Ontology Interchange Language) is used to combine frame-based model primitives with semantics and reasoning services provided by description logic. DAML (DARPA Agent Markup Language) is a semantic markup language specifically designed for the web which is based on RDF and DAML+OIL. Ontologies help to generalize the relations identified by the semantic pattern mining algorithm and improve the quality by removing the redundant patterns [9]. Hence, ontologies are adapted in semantic association mining algorithms to extract the deep semantic association patterns and to reduce the number of shallow associations. The recommendation system for metadata, ranking of semantic medical association rules, and information extraction system is the few real-time applications of the ontology-based semantic association rule mining approach [10–12].

This chapter explores association rule mining and its applications in various domains, types of lexical relations, ontology representation languages, and the framework of ontology-based semantic association rule mining. Further, the advantages and issues in ontology-based semantic association rule mining have been discussed.

14.2 ASSOCIATION RULE MINING

Analyzing trends from the huge transactional data is challenging and important in business, medical, banking, academics, and so on. Trend analysis helps to take decisions that necessitate the understanding and analysis of complex data. The association rule mining helps to identify the trends by discovering the relations among data. The relations between the biomedical entities are helpful to understand the relationship between the diseases and their corresponding symptoms including their treatment. The trend breakthrough in customer's loan may help the bank managers to decide whether the loan can be sanctioned or not. In business, the relations among the items have been used in marketing to promote product sales. The association mining algorithms are used to generate all the possible relations among the entities and the interesting relations can be derived using support and confidence metrics. The association rules are expressed in the form of antecedent and consequent which consists of the items list. For example, the rule depicted in Figure 14.1 shows that 5% of antivirus software is bought with the purchase of a computer and printer, and 60% of the customers who purchased antivirus software also bought a computer and printer.

14.2.1 INTERESTINGNESS MEASURE

Interestingness measures are used to identify the important association rules. The support and confidence measures are commonly used. The support expresses how frequently an association rule occurred in all the transactions and it helps to eliminate the rules which do not have much importance. The confidence defines the strong co-occurrence relation of the item sets in the association rule and measures the reliability of the generated rule. The lift is another performance measure that is used to find the ratio between confidence and expected confidence [13,14]. The equations (14.1–14.3) are used to calculate support, confidence, and lift. The other measures such as all-confidence, collective strength, conviction, coverage, and leverage are available for measuring the rule interestingness [15]. The conviction-based algorithm and pruning can be used for better prediction. The support and confidence measures are limited in producing the semantics of associations [16].

$$\text{Support}(\{X\} \rightarrow \{Y\}) = \frac{\text{Transactions containing both } X \text{ and } Y}{\text{Total number of transactions}} \quad (14.1)$$

$$\text{Confidence}(\{X\} \rightarrow \{Y\}) = \frac{\text{Transactions containing both } X \text{ and } Y}{\text{Transactions containing } X} \quad (14.2)$$

{Computer, Printer} ----▶ {Antivirus Software} [support=5%, confidence=60%]
Antecent   Consequent

FIGURE 14.1 Example of association rule.

$$\text{Lift}(\{X\} \rightarrow \{Y\})$$

$$= \frac{(\text{Transactions containing both } X \text{ and } Y) / (\text{Transactions containing } X)}{\text{Fraction of Transactions containing } Y} \quad (14.3)$$

where $\{X\}$ is the Itemsets in Antecent

   $\{Y\}$ is the Itemsets in Consequent

14.2.2 RULES GENERATION

Association rules can be generated from text documents, Comma Separated Values (CSV) files, and Attribute Relation File Format (ARFF) files by data mining tools. Natural language processing (NLP) is used to extract the related entities from the text documents. Data mining open-source tools such as Weka, Orange, Tanagra, Rapidminer, Knime, Frida, etc., are available for generating and visualizing association rules. Among, Waikato Environment for Knowledge Analysis (Weka) is simple and it supports R, scikit-learn, and DeepLearning4j and also can run in any platform [17,18].

The association rule generation includes the steps like data loading and exploring, pre-processing, model building, and visualizing the rules that are depicted in Figure 14.2. The data pre-processing or data cleaning is important to eliminate the inconsistent data that are liable for invalid rule generation. It handles missing and null values, removes irrelevant attributes, and performs data transformation such as scaling, ranging for effective association rule mining. Among various association mining algorithms, SETM, Artificial Immune System (AIS), Equivalence Class Clustering and bottom-up Lattice Traversal (Eclat), Frequent Pattern (FP) growth, the apriori algorithm have been widely used in many real-time applications [19]. The minimum support and confidence parameters are used to reduce the number of generated association rules. For example, Figure 14.3 shows the association rules on the diabetes dataset.

The data greater than 50 Megabytes cannot be loaded in weka [20]. The libraries in python or R language provide a mechanism to handle huge data and generate association rules. Figure 14.4 shows the python coding for association rule generation by apriori algorithm implementation. In general, the apriori algorithm generates rules that are positively related to each other. However, the negative rules are the rules that are not related with each other but also have significant importance in health care and crime detection. Sikha Bagui et al., 2019 implemented an apriori algorithm in Hadoop's distributed and parallel MapReduce

Association Rule Generation

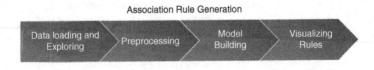

FIGURE 14.2   Steps in association rule generation.

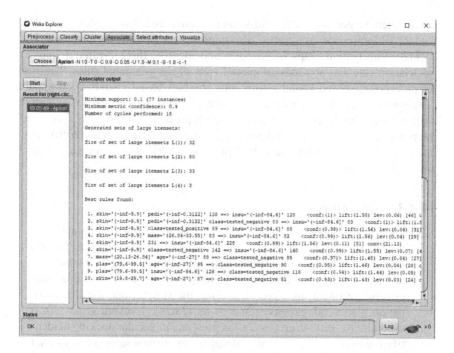

FIGURE 14.3 Association rules generated on diabetes data.

```
#Association Rules Generation in Python

# Importing Libraries
import numpy as np
import pandas as pd
import matplotlib.pyplot as plt
from apyori import apriori

#Loading and Exploring data
# Read data from E drive and store as data frame
data_store = pd.read_csv(E:\\association_data.csv')
data_store.head()  # displays the data
data_store.columns # displays the columns

#Data preprocessing: Converting the data frame into list
records = []
for i in range(0, 10000):
    records.append([str(data_store.values[i,j]) for j in range(0, 20)])

#Build model using apriori algorithm
#min_lift filters the rules, min_length specifies the number of items in the rule
rules = apriori(records, min_support=0.0045, min_confidence=0.2,  min_lift=3,
min_length=3)
association_results = list(rules)

#Exploring the rules
print(len(association_rules))     # display the number of associations generated
print(association_rules[0])       # display the first rule
print(association_rules.head())   # display the top rules
```

FIGURE 14.4 Apriori algorithm implementation in Python.

environment to generate both positive and negative association rules in big data. Hadoop's MapReduce architecture also improves the runtime performance [21].

The visualization helps to envision and identify the top-ranked frequent association rule and provides an overall view of the generated rules. The scatter plot, matrix-based visualization, and graphics-based visualization are the common visualization techniques. The scatter plot represents each rule with its support and confidence. The matrix-based representation helps to specify all the possible item sets in antecedents and consequents. Cheng et al. (2016) developed an interactive technique called InterVisAR for association rules search and visualization [22].

14.2.3 Applications

Biomedical: In biomedical, the apriori algorithm is used to generate association rules for disease-gene-drug, tissue-gene, disease-risk factors, disease symptoms, etc. The various problems solved in biomedical are predicting the unidentified disease, nature of the disease, drug side effects, treatment duration, identifying the risk factors, the chance of patient's survival, expenses, and so on [23,24]. Jeyakodi et al. (2021) used disease-gene-drug association rules and molecular docking to identify the novel repurpose drugs (drugs for multiple diseases) that can act as a knowledge base of pharmacogenomics. The text mining and Unified Medical Language System (UMLS) thesaurus have been used to extract and map biomedical entities from PubMed articles. The apriori algorithm has been used to generate association rules [25].

Ankit Agrawal et al. (2011) used a data mining hotspot association rule mining algorithm to analyze the lung cancer data to predict the patient's survival outcome. The findings may help to avoid situations that reduce the survival rate and encourage the circumstances to increase the survival rate [26]. Domadiya et al. (2018) analyze the heart disease data and identified the problems in mining association rules from electronic health records without violating patient's privacy and developed a privacy-preserving distributed association rule mining (PPDARM) approach [27].

Business: Association rule mining is used for business intelligence (BI) to identify the correlation between the items that can either improve the business performance or understand the business happenings. Business intelligence can be applied for various business purposes such as measurement, monitoring business progress, analytics, enterprise reporting, the collaboration platform for data sharing, electronic data interchange, and knowledge management. In addition, BI can also be used to alert if the measures exceed the thresholds and anomalies also highlighted [28]. Daniele et al. (2019) proposed a data-driven association rule mining approach to monitoring the coffee machine by correlating the parameters such as coffee ground size, ground amount, and water pressure in the real-world espresso data. This approach helps to spotlight the espresso quality of the coffee-making industries [29].

Education: Association rule mining in education assists the management in taking decisions for improving the quality of education [30]. The educational data analysis discovers the factors that affect the academics and helps to redesign the curriculum, changing teaching and assessment methodologies. Varun Kumar et al. (2012) used association rules for assessing student data using the data mining tool called Tanagra [31]. The association rule mining is also effective in improving the learning efficiency of the cognitive tutor. It also helps to evaluate and improve the student's performance skills [32]. The feature selection algorithm such as ReliefF can reduce the number of association rules generated from huge educational data and help to improve the higher education policies [33]. The key application of association rules in education is to build a job prediction model to find a suitable job based on the student's ability. An association mining algorithm can be used to correlate student parameters (marks, programming skill, aptitude, etc.,) and the nature of the job (IT, bank, teaching, or business) to identify their suitable job. Rojanavasu (2019) applied an association rule mining algorithm for the admission process and to predict student's job [34].

Others: Association rule mining in software defect prediction helps to reduce the time and cost of software testing and assists the project managers to allocate the resources. Shao et al. (2020) observed that imbalanced data generates a large number of non-defective association rules and ignored defective rules. The correlation weighted class association rule mining handles the imbalance data, assigns feature weight, and also optimizes the ranking based on the weighted support [35]. The association rule mining on traffic data helps to monitor the road traffic by predicting the accidents and to alert the drivers. The ELECRE TRI, a multi-criteria decision-making approach has been used to manage the redundant rules for improving road safety [36]. Yucel et al. (2002) preserved privacy association rules by hiding the sensitive rules by either reducing the support or confidence. The online secret information of private or government data such as email, mobile number, account number, and login credentials can also be secured [37].

14.3 SEMANTIC ASSOCIATION RULE MINING (SARM)

The growth of heterogeneous data increases the challenges in analysis and data mining. The semantic data known as background knowledge is used to analyze heterogeneous data consisting of objects and relationships among the objects that help to interact with the data automatically and reduce the redundant or inconsistent data [5]. The relation between the two concepts may be either classical or non-classical. The classical relations are easily identifiable such as 'pearl is a

TABLE 14.1

Types of Lexical Relations

Relation	Description	Example
Hypernymy	Words that are superior	Mosquito is a hypernym of insect
Hyponymy	Words that are specialized and generic	The hyponym of mosquito are *anopheles culicifacies, anopheles stephensi*
Synonymy	Words that have the same or almost the same meaning	Intelligent, bright
Antonymy	Words that have opposite meaning	Doctor vs Patient
Monosemy	Single sense words, generalize meanings	Cousin can generalize as Son of father's brother, daughter of mother's brother, etc
Polysemy	The word has multiple related senses.	John is a good man - Moral judgment Biju is a good artist. Judgment of skill
Homonymy	Words have the same pronunciation sound, same spelling but different meanings.	A vector in mathematics – Quantity A vector in biology – Disease-causing insect
Meronymy	Part-whole relation	The keyboard is a meronymy of computer
Holonymy	Whole - part relation	A computer is a holonymy of keyboard

gemstone' and the complex relations like 'tamarind is sour' that are not easily identifiable are non-classical relations.

The semantic associations represent asymmetrical, direct or indirect, and non-classical relations. The semantics among the entities/concepts are liable for exploring appropriate semantic associations and can be assessed by semantic similarity, similarity relatedness, and semantic distance. The lexical relations such as synonymy, hypernymy, and hyponymy share similar compositions. Semantic relatedness is used to relate the association among the concepts which are not of similar nature. For example, insecticide and resistance do not have similar nature but they have strong cause-effect relations. The semantic relatedness can be applied to lexical relations such as antonymy, meronymy, and holonymy [38]. The lexical relations are the semantic relations that are used to explore and analyze how the words are semantically related to each other [39]. Table 14.1 shows the different types of lexical relations with examples.

14.3.1 SEMANTIC PROXIES

The semantic proxies are the background knowledge sources in which the semantic concepts are expressed implicitly or explicitly for computing the measure of the semantic associations. The semantic concepts from structured, unstructured, and semi-structured sources can be graphically represented as a knowledge graph. It may be directed or undirected semantic network that are used to store contextual and conceptual information which consists of patterns or taxonomies. Patterns are the meaningful structure that is used to recognize semantic relations. Taxonomies are the hierarchical structure that defines subtype or supertype relations. The semantic proxies are classified as formal or informal that is based on the type of semantic representations. The informal proxies are unstructured and have implicit representation and the formal proxies contain explicit structural representation.

Informal proxies: Informal proxies are unstructured knowledge sources. The text corpus is a large collection of documents from a single or multiple knowledge sources for analyzing the words, phrases, and languages. The linguistics, lexicographers, scientists, humanities, and experts used text corpus for computational linguistics, speech recognition, spell check, grammar correction, machine translation, natural language processing, philologies, etc. The different types of corpus are:

- **Monolingual corpus**: Monolingual identify the frequent patterns or new trends. The Sketch Engine is an example of a monolingual corpus.
- **Comparable corpus**: This is a set of two or more monolingual corpora. Examples are CHILDES corpora, Wikipedia corpora, and Araneum corpora.
- **Diachronic corpus**: This corpus consists of documents collected from different periods used for trend analysis. They are historical. The ARCHER corpus, a century of prose corpus, and the corpus of newsbooks are some of the examples of the diachronic corpus.
- **Synchronic corpus**: This corpus consists of text from the same period either present or a particular period in the past. They are descriptive.
- **Static corpus**: This corpus is called a reference and complete corpus. The brown corpus is an example of a static corpus.
- **Monitor corpus**: This corpus is used to monitor the language changes and is updated regularly. The timestamped corpus is an example of a monitor corpus.

The Google Books corpus, Google's N-Gram Corpus, Google Web-Based English, International Corpus of Learner English are some of the examples of a text corpus.

Formal proxies: Formal proxies are multilingual, general-purpose structural knowledge sources. The dictionaries and thesauri are used earlier for semantic association computation but nowadays, online dictionaries, encyclopedias, and domain-specific ontologies have been used due to the huge amount of web data. The WordNet, Wikipedia, and Wiktionary are multilingual formal proxies [8,40]. Table 14.2 shows the summary of different formal proxies used for semantic association computations.

TABLE 14.2

Formal Proxies for Semantic Association Computations

Sl. No	Formal Proxy	Description	Examples
1	Dictionary	• Monolingual or Multilingual knowledge sources • Alphabetical list of words organized as a taxonomy • Used to find meaning or definition of a word	Wiktionary
2	Wiktionary	• Freely available online dictionary • Multilingual • Called as thesaurus • Defines and explicitly encode lexical-semantic relations	ENGLAWI - Human-to Machine-Readable Wiktionary
3	Thesaurus	• Semantically similar words in a specific context	Roget's thesaurus
4	Lexical database	• Lexical resources • Associated with software to access its content such as synonyms, relations, and lexical categorizations • Verbs, nouns, and adjectives are organized as synsets (set of synonyms)	WordNet
5	Encyclopedia	• Information of particular or many fields • Compared to dictionaries, it provides more definitions and part-of-speech details • Focus on factual information • Represent relation labels explicitly • Either general-purpose or domain-specific	General – Britanica, Wikipedia Domain – Medline Plus, MeSH
6	Wikipedia	• Provide detailed information on topic as an article • Group's information into categories, list and info boxes • Provide hyperlinks to connect semantically related articles • Excellent topical coverage compared to WordNet • Used as a knowledge source for ontology learning, information extraction, text wikification and so on	–
7	Ontology	• Relations and concepts are expressed explicitly as a taxonomy structure • Provide semantically rich and high-quality content • Used as a knowledge base to extract semantic associations	Disease Ontology, Gene Ontology, Vaccine Ontology

14.3.2 SEMANTIC ASSOCIATION APPROACHES

Semantic association relations can be derived from the text as well as the web by using formal representation languages. The semantic web provides a large platform for semantic data on accumulating knowledge from different sources by sharing and interlinking multiple domains' structural data. The technological growth made more semantic data to be stored on the web, a rich source of hidden patterns. The semantic web's heterogeneous data with semantic properties are expressed as OWL or RDFS in a subject, predicate, or object representation. The description logic, a formal knowledge representation language is used to represent information into OWL and RDFS. An ontology represents the semantic web knowledge in the form of classes, relations, and properties, and plays a vital role in extracting patterns from semantic web data. The traditional association rule mining technique cannot apply to semantic data to extract relations. The combination of association rule mining with ontology as background knowledge can effectively extract semantic relations from semantic data [41]. The online software tools FRED can be used to convert the text into semantically connected ontologies in RDFS or OWL format [42].

Inductive Logic Programming (ILP) is one of the eminent techniques for mining semantic associations from RDFS and OWL documents. The limitation of ILP is that it works based on closed-world assumptions and is not suited for incomplete evidence. Luis Galarraga developed AMIE (Association Rule Mining under Incomplete Evidence), a multi-threaded approach to support open-world assumptions and to handle a huge knowledge base. The limitation of these approaches are considering only instance-level data and ignoring schema-level knowledge. But semantic web contains knowledge on both levels. The Semantic Web Association Rule Mining (SWARM) approach can overcome this limitation by automatically mining the semantic association rules from RDFS without the support of domain experts [41,43]. A multi–Modal Semantic Association Rule approach has been proposed by Ruhan He et al. (2010) to associate keywords and visual features of web images [44]. Ait-mlouk et al. (2020) developed a platform WINFRA for semantic association rule generation using Natural Language Processing (NLP) and data mining techniques. WINFRA can also generate knowledge graphs along with semantic knowledge extraction [45].

Resource Description Framework Schema (RDFS): RDFS is the knowledge representation data model used to describe the classes and properties of the ontologies. RDFS is a semantic web language built on Uniform Resource Identifier (URI) and Extensible Markup Language (XML). It is formal semantics and property-centric, which allows to describe the resources by defining their properties. RDFS consists of set of classes, properties and utility properties to define vocabularies in the simple triple format <subject> <predicate> <object> to express relationship. The directed graph can be constructed that consists of an elliptical node with a label to represent a resource, a rectangular node to represent

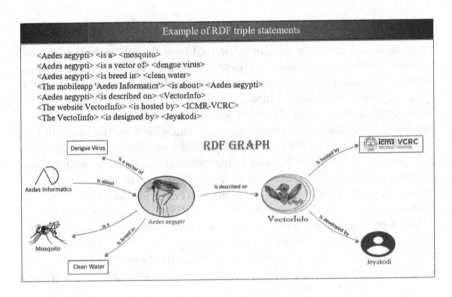

FIGURE 14.5   RDF triple statements and RDF graph.

literal values, and an arc to represent a property of those resources [46].
Figure 14.5 depicts the sample RDF triple statements and its RDF graph.
The semantic web data can be converted into RDF by the freely avail-
able online tool EasyRDF. The sample RDF is shown in Box 14.1.

Web Ontology Language (OWL): The web ontology language is a data
modeling language, a part of RDFS used to define semantic web ontologies for describing web resource properties. OWL differs from RDFS by
providing rich vocabulary, logical consistency, and constraints. It also
consists of operators to describe concepts, boolean operators such as
intersection, union, and complement, and explicit quantifiers to define
properties and relationships. OWL defines the transitivity, domain, and
range characteristics have the ability to reason automatically and also
uses external ontologies [47]. Box 14.2 depicts the OWL representation
of union operation.

14.4   ONTOLOGY-BASED SEMANTIC ASSOCIATION RULE MINING (OSARM)

Ontology helps to share and reuse the semantically represented knowledge. The
National Center for Biomedical Ontology (NCBO) provides a bioportal for reusable biomedical ontologies [48]. The ontology-based approach uses ontology
knowledge to reduce the semantic knowledge gap and improve the accuracy of
data pre-processing and association rule mining. Semantic knowledge also helps
to understand the relations. In general, the association rule mining algorithm is

BOX 14.1   RESOURCE DESCRIPTION FORMAT FOR WIKIPEDIA SEMANTIC DATA MODEL WEBSITE

```
<?xml version="1.0" encoding="utf-8" ?>
<rdf:RDF xmlns:rdf="http://www.
w3.org/1999/02/22-rdf-syntax-ns#"
   xmlns:og=http://ogp.me/ns#
   xmlns:xhv="http://www.w3.org/1999/xhtml/vocab#">
<rdf:Description rdf:about="https://en.wikipedia.
org/wiki/Semantic_data_model">
<og:image xml:lang="en">https://upload.wikimedia.org/
wikipedia/commons/thumb/5/58/A2_4_Semantic_Data_Models.
svg/1200px-A2_4_Semantic_Data_Models.svg.png</og:image>
<og:title xml:lang="en">Semantic data model
- Wikipedia</og:title>
   <og:type xml:lang="en">website</og:type>
   <xhv:license rdf:resource="https://creativecommons.
org/licenses/by-sa/3.0/"/>
   <xhv:license rdf:resource="https://en.wikipedia.
org/wiki/Wikipedia:Text_of_Creative_Commons_
Attribution-ShareAlike_3.0_Unported_License"/>
</rdf:Description>
</rdf:RDF>
```

BOX 14.2   OWL REPRESENTATION OF UNION OPERATION

```
<owl:Class rdf:ID="AedesBreedingPlaces">
<rdfs:subClassOf>
<owl:Class rdf:about="#AedesBreeding"/>
</rdfs:subClassOf>
<owl:equivalentClass>
<owl:Class>
<owl:unionOf rdf:parseType="Collection">
<owl:Class rdf:about="#Coconut_Shell"/>
<owl:Class rdf:about="#Tank"/>
<owl:Class rdf:about="#Tyre"/>
<owl:Class rdf:about="#MoneyPlant"/>
</owl:unionOf>
</owl:Class>
</owl:equivalentClass>
</owl:Class>
```

generic and not suitable for identifying semantics, however, ontology helps to resolve it. Ontologies can also be used to improve the association rule mining by pruning and eliminating the inconsistent rules. The ontology can be represented as hypergraphs that are used to generate semantic association rules using the random walk model. Ontologies help to discover concept association rules instead of instance-level association rules [9].

14.4.1 FRAMEWORK

The framework of ontology-based semantic association rule mining includes data collection, ontology development, ontology into triple encoding transformation, and semantic association rule generation based on the semantic association rule mining algorithm as depicted in Figure 14.6.

Data collection and preprocessing: Data is collected from text documents or the web. The preprocessing includes stop word removal, POS tagging (part of speech), filtration, stemming, and lemmatization. The filtration removes the highly relevant words like articles, pronouns, determiners, prepositions and conjunctions, common adverbs, and non-informative verbs. The stemming and lemmatization remove the prefix and suffix and return the primitive word using WordNet. Finally, the TF-IDF (Term Frequency, Inverse Document Frequency) identifies the highly distinguished words by applying higher weights. Online toolkits such as RapidMiner, OpenNLP, AllenNLP, CoreNLP, etc., are available for data preprocessing.

Ontology construction: Ontology development constructs the ontology by representing the domain knowledge into formal representation such as OWL or RDF graph. The quality of the developed ontology depends on the involvement of the domain expert. The Uschold and King ontology, Toronto Virtual Enterprise Method, Methontology, KACTUS, and Lexicon-based ontology development are the different approaches available for ontology construction. Horrocks ontology development method

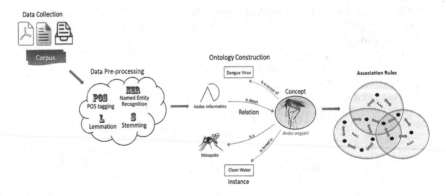

FIGURE 14.6   The framework of ontology-based association rule mining.

is simple and useful in developing and modifying OWL ontologies. Since there is no standard procedure for ontology construction, most of the applications follow the Noy and McGuinness manual ontology development technique [49] as shown in Table 14.3. The computational tools such as Text2Onto and OntoGain are also available for constructing automatic domain ontologies.

Ontology Encoding: Ontology encoding encodes the semantic ontology by considering the graph structure, lexical information, and logical constraints. Table 14.4 depicts RDF triples and Figure 14.7 shows its RDF graph representation. The semantic query language SPARQL is used for extracting the semantic triplet information from the linked open data or RDF datasets. In addition to the SPARQL query, the regular expressions can also be used for querying or matching the paths in the RDF graph [50].

Semantic Association Rules Generation: Association rules are generated from the semantic data using an association mining algorithm. The apriori algorithm is widely used for generating related rules. The algorithm

TABLE 14.3

Steps in Ontology Development

Step	Description	Purpose
1	Determine the domain and scope of the ontology	To identify the relevancy and its scope
2	Considering existing ontologies	To identify the existing ontology that can be extended or modified instead of created from scratch.
3	Extracting Concepts and Terms	To identify the concepts and terms required for the ontology
4	Defining Classes and Hierarchy	To identify the classes and their relation
5	Defining Class properties (slots)	To identify the attributes to define the class internal structure (taxonomy)
6	Slot facets	To describe the attribute value type, range, cardinality, and so on
7	Instances	To create the individual instances of the class

TABLE 14.4

Example of RDF Triples

Subject	Predicate	Object
<Mosquito>	<hasName>	Aedes aegypti
<Mosquito>	<hasResistant>	<Insecticide>
<Insecticide>	<hasName>	DDT
<Insecticide>	<hasName>	Deltamethrin

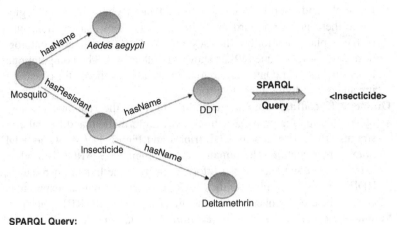

SPARQL Query:
SELECT ?x WHERE {?x <hasName> "DDT"; ?x <hasName> "Deltamethrin"}

FIGURE 14.7.   RDF graph and result for SPARQL query.

**BOX 14.3   SEMANTIC ASSOCIATION
RULES FOR SPECIES CLASS**

Semantic Association Rules for Species class
   {Bacteria}: (genus, Bacillus) => (family, Bacillacease) support = 0.2,
confidence = 1.0
   {Mammal}: (genus, Balaenottera) => (family, Rorqual) support=0.61,
confidence=0.99

---

uses ontology as the background knowledge for generating association rules. The algorithm with ontology support can generate a smaller number of effective association rules, provide meaningful decisions, and also reduces the execution time. The domain ontology also helps to evaluate the association rules by the semantic distance measure [51]. Barati et al. (2017) derived semantic association rules from six different classes of DBPedia ontology using (SWARM) approach [41]. The semantic association rules derived from the Species class of DBPedia ontology are depicted in Box 14.3.

14.4.2 APPLICATIONS

The ontology-based semantic association rule mining concept is adapted in many applications. Sumathi et al. (2020) developed a recommendation system for query recommendation applications in health care using ontology and association rules. The fuzzy weighted iterative concept has been used to mine positive and negative association rules. The healthcare ontologies such as disease ontology, drug ontology, and lab ontology are used as background knowledge [52]. Viktoratos et al. (2018) developed a hybrid approach for a rule-based context-aware recommendation system to overcome its cold start problem. The combination of association rules, probability metrics, and user's context has been used. The recommendation system recommends the appropriate rules to the user based on the user's context [53]. Lakshmi et al. (2019) identified the associations among the diseases using a weighted association rule mining algorithm. The protein-protein interactions data, disease pathway information, Gene Ontology (GO), and Unified Medical Language System (UMLS) medical ontologies have been used as background knowledge. The generated association rules have been compared with clinical data [54].

Sornalakshmi et al. (2020) used a hybrid method to predict the trends in the patient's medical condition and behavior. An enhanced apriori algorithm and the context ontology have been used for association rule mining. The prediction model used sequential minimal optimization regression and achieved better accuracy and reduces the execution time [55]. Heritage et al. (2018) identified the disease-drug associations to indicate the side effect of the drugs if the patient has multiple diseases. The GO, Kyoto Encyclopedia of Genes and Genomes database (KEGG), and disease ontology have been used for knowledge. The predicted associations are useful for optimizing the medical prescription [56].

Skorupa Parolin et al. (2019) developed a system to automatically extract verbal patterns from political news articles using Conflict and Mediation Event Observations ontology, word-to-vec or word embeddings, and apriori association rule mining algorithm [57]. Kulkarni (2017) developed a probabilistic Intent-Action Ontology (IAO) and Tone Matching algorithm (TMA) for predicting the probable series of events that occurred in the different timelines. IAO helps to determine the relevance and TMA helps to analyze the emotions. This prediction helps to solve real-life applications such as criminal cases, story completion, and so on [58].

Cao et al. (2020) developed a risk management system to detect and prevent railway accidents using railway accident risk ontology and context ontology [59]. Kafkas et al. (2018) mined the associations among the disease and pathogens from text using National Center for Biotechnology Information (NCBI) Taxonomy and Human Disease Ontology helps to provide background knowledge on pathogens and infectious diseases [60]. Peral et al. (2018) defined an ontology-oriented architecture to handle heterogeneous data such as sensors and social networks to mine patient details such as diagnosis, and treatment for telemedicine usage [61].

Mohammed Thamer et al. (2020) compared the medical rules generated by traditional apriori algorithms and ontology-based apriori algorithms using the performance metrics such as number of frequent items, number of rules, computation time, and so on. The performance of the ontology-based apriori algorithm is superior and generated fewer association rules [49]. In semantic association rule mining, the quality of the association rules depends on the quality of the ontology.

More concepts can be captured by extending the ontology or integrating it with other related ontologies [1].

The integration of operational definitions known as semantics that are expressed as semantic web rules with a descriptive knowledge base that helps to perform decision making and enhance the knowledge base expressivity. Runumi Deve et al. (2020) formalized dengue disease for supporting clinical and diagnostic reasoning from the patient's case sheet. The operational definitions have been incorporated with semantic web rule language to enhance the expressive capability of the dengue knowledge base [62]. Semantic ontology helps to generate domain-specific image captions for providing natural language descriptions [63]. The machine learning concepts can provide dynamic semantics to enrich concepts [64].

Big data environment produces association rules in high quantity that may increase the complexity of semantic analysis and understanding [65]. Engy Yehia et al. (2019) used patient clinical data ontology to extract semantic relations using the rule-based relation extraction method. This approach could not identify some clinical concepts and abbreviations. And also suffered in inferring new knowledge [66]. Liu Xiaoxue et al. (2019) extracted hierarchical and non-hierarchical semantic relations among the crop pest entities such as plant diseases to pesticides using a knowledge graph. Neural network model based on syntactic tree can achieve better results in relation extraction that shows syntactic information is useful for relation extraction [67].

14.5 DISCUSSION

The association rules generated by the association rule mining algorithm need to be enhanced due to the discovery of a large number of semantically non-interesting association rules. The pruning technique, support and confidence factors and subjective measures such as unexpectedness and actionability overcome the issues and thus reduces the greater number of generated rules. The semantic association rule mining enhances the association rules exactness by providing background knowledge for a better understanding of semantic data. SARM simplifies the difficulty in extracting association rules from complex and heterogeneous data with the help of schema-level knowledge. The efficiency of the traditional association mining algorithm is enhanced by the ontology-based semantic association rule mining by extracting the constraint-based multi-level association rules.

In general, Noy and McGuiness methods are commonly used for constructing domain-specific ontology manually. The ontology characteristics play a key role in semantic association rule generation. The domain-specific ontology with semantic association rule mining algorithm not only reduces the number of generated association rules, in addition, it generates more semantically related rules when compared to the association rule mining approach. Logical rules can also be applied to ontology-based semantic associations to predict additional interesting patterns. Since Text2Onto tool relies on the single-word term for concept extraction, OntoGain produces rich semantic rules that rely on the multi-word term extraction technique.

Even though, ontology-based semantic association rule approach discovers significant and exact rules, extracting rules from huge data is problematic and could not extract deep non-taxonomic relations. The ontology combination with deep learning can generate dynamic semantic association rules. Deep learning is observed as the recently emerging technique for handling vast heterogeneous unstructured data. The ontology combination with deep learning, NLP, and ARM may identify and extract new deep association rules.

14.6 CONCLUSION

This chapter explored the significance of ontologies in extracting semantic relations using the association rule mining technique. The growth of semantic data and the various proxies used to discover semantic associations are discussed. The various semantic data representation approaches are also explained. The issues in using association mining algorithms for extracting relations in semantic data and how the ontologies may help to extract semantic relations are also enlightened. The ontology-based semantic association rule mining framework is described. In addition, the usage of ontology-based semantic association rules in various applications is explored. Finally, the advantages and issues in ontology-based semantic association rule mining are also discussed. This chapter helps to provide an insight into ontologies and semantic association rule mining for effective knowledge discovery.

REFERENCES

1. Afolabi, I., Sowunmi, O., & Daramola, O. (2017). Semantic association rule mining in text using domain ontology. *International Journal of Metadata, Semantics and Ontologies, 12*(1), 28–34. doi:10.1504/IJMSO.2017.087646.
2. Kale, K. A. (2016). Review on mining association rule from semantic data. *International Journal of Computer Science and Information Technologies, 7*(3), 1328–1331.
3. Rajak, A., & Gupta, M. K. (2008). Association rule mining: Applications in various areas. *International Conference on Data Management*, January 2008, 3–7. https://pdfs.semanticscholar.org/e5db/a964a791763d7cd9c9d8979f2c00604e7b9a.pdf.
4. Hipp, J., Güntzer, U., & Nakhaeizadeh, G. (2000). Algorithms for association rule mining: A general survey and comparison. *ACM SIGKDD Explorations Newsletter, 2*(1), 58–64. doi:10.1145/360402.360421.
5. Dou, D., Wang, H., & Liu, H. (2015). Semantic data mining: A survey of ontology-based approaches. *Proceedings of the 2015 IEEE 9th International Conference on Semantic Computing*, IEEE ICSC 2015, 244–251. doi:10.1109/ICOSC.2015.7050814.
6. Sirichanya, C., & Kraisak, K. (2021). Semantic data mining in the information age: A systematic review. *International Journal of Intelligent Systems, 36*(8), 3880–3916. doi:10.1002/int.22443.
7. Jabeen, S., Gao, X., & Andreae, P. (2020). Semantic association computation: A comprehensive survey. *Artificial Intelligence Review, 53*(6), 3849–3899. doi:10.1007/s10462-019-09781-w.

8. Champin, P. A. (2007). Representing data as resources in RDF and OWL. *ICDT Workshop on Emerging Research Opportunities in Web Data Management* (EROW 2007), Barcelona, Spain. pp.1–9. ⟨hal-01501729⟩.

9. He, C., & Zhang, Y. F. (2010). Research on semantic association pattern mining model based on ontology. *ICACTE 2010-2010 3rd International Conference on Advanced Computer Theory and Engineering, Proceedings, 1,* 497–501. doi:10.1109/ICACTE.2010.5578967.

10. Martínez-Romero, M., O'Connor, M. J., Egyedi, A. L., Willrett, D., Hardi, J., Graybeal, J., & Musen, M. A. (2019). Using association rule mining and ontologies to generate metadata recommendations from multiple biomedical databases. *Database, 2019*(1), 1–25. doi:10.1093/database/baz059.

11. Idoudi, R., Ettabaa, K. S., Solaiman, B., & Hamrouni, K. (2016). Ontology knowledge mining based association rules ranking. *Procedia Computer Science,* 96(September), 345–354. doi:10.1016/j.procs.2016.08.147.

12. Song, M., Song, I. Y., Hu, X., & Allen, R. B. (2007). Integration of association rules and ontologies for semantic query expansion. *Data and Knowledge Engineering,* 63(1), 63–75. doi:10.1016/j.datak.2006.10.010.

13. Zhang, C., & Zhang, S. (2002). *Association Rule Mining, Models and Algorithms. Lecture Notes in Computer Science.* Springer, Berlin, Heidelberg. doi:10.1007/3-540-46027-6.

14. Garg, A. (2018). Complete guide to association rules (1/2). Available at: https://towardsdatascience.com/association-rules-2-aa9a77241654.

15. Tan, P. N., Kumar, V., & Srivastava, J. (2004). Selecting the right objective measure for association analysis. *Information Systems,* 29(4), 293–313. doi:10.1016/S0306-4379(03)00072-3.

16. Adamo, J. M. (2001) Beyond support-confidence framework. In: *Data Mining for Association Rules and Sequential Patterns.* Springer, New York, 151–184. doi:10.1007/978-1-4613-0085-4_9.

17. Nafie Ali, F. M., & Mohamed Hamed, A. A. (2018). Usage Apriori and clustering algorithms in WEKA tools to mining dataset of traffic accidents. *Journal of Information and Telecommunication, 2*(3), 231–245. doi:10.1080/24751839.2018.1448205.

18. Hall, M., Frank, E., Holmes, G., Pfahringer, B., Reutemann, P., & Witten, I. H. (2009). The WEKA data mining software: An update. *SIGKDD Explorations, 11*(1), 10–18.

19. Saxena, A., & Rajpoot, V. (2021). A comparative analysis of association rule mining algorithms. *IOP Conference Series: Materials Science and Engineering, 1099*(1), 012032. doi:10.1088/1757-899x/1099/1/012032.

20. Kotak, P., & Modi, H. (2020). Enhancing the data mining tool WEKA. *Proceedings of the 2020 International Conference on Computing, Communication and Security,* ICCCS 2020, 4–9. doi:10.1109/ICCCS49678.2020.9276870.

21. Bagui, S., & Dhar, P. C. (2019). Positive and negative association rule mining in Hadoop's MapReduce environment. *Journal of Big Data,* 6(1). doi:10.1186/s40537-019-0238-8.

22. Cheng, C. W., Sha, Y., & Wang, M. D. (2016). Intervisar: An interactive visualization for association rule search. *ACM-BCB 2016–7th ACM Conference on Bioinformatics, Computational Biology, and Health Informatics,* 175–184. doi:10.1145/2975167.2975185.

23. Naidenova, X., Ganapolsky, V., Yakovlev, A., & Martirova, T. (2011). Application association rule mining in medical biological investigations: A survey. *Knowledge Management, 1*(6), 25–33.

24. Borah, A., & Nath, B. (2018). Identifying risk factors for adverse diseases using dynamic rare association rule mining. *Expert Systems with Applications*, 113, 233–263. doi:10.1016/j.eswa.2018.07.010.

25. Gopal, J., Prakash Sinnarasan, V. S., & Venkatesan, A. (2021). Identification of repurpose drugs by computational analysis of disease–gene–drug associations. *Journal of Computational Biology*, 28(10), 975–984. doi:10.1089/cmb.2020.0356.

26. Agrawal, A., & Choudhary, A. (2011). Identifying hotspots in lung cancer data using association rule mining. *Proceedings - IEEE International Conference on Data Mining*, ICDM, 995–1002. doi:10.1109/ICDMW.2011.93.

27. Domadiya, N., & Rao, U. P. (2018). Privacy-preserving association rule mining for horizontally partitioned healthcare data: A case study on the heart diseases. *Sadhana – Academy Proceedings in Engineering Sciences*, 43(8), 1–9. doi:10.1007/s12046-018-0916-9.

28. Jha, R. (2014). Association rules mining for business intelligence. *International Journal of Scientific and Research Publications*, 4(5), ISSN 2250-3153.

29. Apiletti, D., & Pastor, E. (2020). Correlating espresso quality with coffee-machine parameters by means of association rule mining. *Electronics (Switzerland)*, 9(1), 1–19. doi:10.3390/electronics9010100.

30. Venkatesan, R. K. K. G. S., Sundararajan, M., & Arulselvi, S. (2015). *An Applying Association Rule Mining in Education Management Systems*, 3771–3777. doi:10.15680/IJIRSET.2015.0405199.

31. Kumar, V., & Chadha, A. (2012). Mining association rules in students assessment data. *International Journal of Computer Science Issues*, 9(5), 211–216.

32. Jha, J. (2017). Effectiveness of association rule mining for learning efficiency analysis. *International Journal of Engineering Research and Technology (IJERT)*, 5(01), 1–7.

33. Hashim, A. S., Hamoud, A. K., & Awadh, W. A. (2018). Analyzing students' answers using association rule mining based on feature selection. *Journal of Southwest Jiaotong University*, 53(5), 1–16.

34. Rojanavasu, P. (2019). Educational Data Analytics using Association Rule Mining and Classification. *Joint International Conference on Digital Arts, Media and Technology with ECTI Northern Section Conference on Electrical, Electronics, Computer and Telecommunications Engineering* (ECTI DAMT-NCON), 142–145. doi:10.1109/ECTI-NCON.2019.8692274.

35. Shao, Y., Liu, B., Wang, S., & Li, G. (2020). Software defect prediction based on correlation weighted class association rule mining. *Knowledge-Based Systems*, 196, 105742. doi:10.1016/j.knosys.2020.105742.

36. Ait-Mlouk, A., Gharnati, F., & Agouti, T. (2017). An improved approach for association rule mining using a multi-criteria decision support system: A case study in road safety. *European Transport Research Review*, 9(3). doi:10.1007/s12544-017-0257-5.

37. Saygin, Y., Verykios, V.S., & Elmagarmid, A.K. (2002). Privacy preserving association rule mining. *Proceedings Twelfth International Workshop on Research Issues in Data Engineering: Engineering E-Commerce/E-Business Systems* RIDE-2EC 2002, 151–158. doi:10.1109/RIDE.2002.995109.

38. Morris, J., & Hirst, G. (2004). Non-classical lexical semantic relations. *Proceedings of the Computational Lexical Semantics Workshop at HLT-NAACL 2004*, Association for Computational Linguistics. 46–51. https://aclanthology.org/W04-2607.

39. Malik, M. (2017). The significance of the use of lexical relations in English language. *International Journal of Advanced Research*, 5(4), 944–947. doi:10.21474/ijar01/3900.

40. Gentile, A. L., Basile, P., Iaquinta, L., & Semeraro, G. (2008). Lexical and semantic resources for NLP: From words to meanings. *Lecture Notes in Computer Science (Including Subseries Lecture Notes in Artificial Intelligence and Lecture Notes in Bioinformatics), 5179 LNAI* (PART 3), 277–284. doi:10.1007/978-3-540-85567-5_35.

41. Barati, M., Bai, Q., & Liu, Q. (2017). Mining semantic association rules from RDF data. *Knowledge-Based Systems, 133,* 183–196. doi:10.1016/j.knosys.2017.07.009.

42. Lopez, V., Schlobach, S., Eds, J. V., & Hutchison, D. (2013). The semantic web - ESWC 2013 satellite events, revised selected papers. *Lecture Notes in Computer Science (including subseries Lecture Notes in Artificial Intelligence and Lecture Notes in Bioinformatics): Vol. 7955 LNCS* (Issue May).

43. Galárraga, L., Teflioudi, C., Hose, K., & Suchanek, F. M. (2013). AMIE: Association rule mining under incomplete evidence in ontological knowledge bases. *International World Wide Web Conference Committee (IW3C2).* WWW 2013, May 13–17, 2013, 413–422, Rio de Janeiro, Brazil. ACM 978-1-4503-2035-1/13/05.

44. He, R., Xiong, N., Yang, L. T., & Park, J. H. (2011). Using multi-modal semantic association rules to fuse keywords and visual features automatically for web image retrieval. *Information Fusion, 12*(3), 223–230. doi:10.1016/j.inffus.2010.02.001.

45. Ait-mlouk, A., Vu, X., & Jiang, L. (2020). WINFRA: A web-based platform for semantic data retrieval and data analytics. *Mathematics, 8*(11). doi:10.3390/math8112090.

46. McBride, B. (2004) The resource description framework (RDF) and its vocabulary description language RDFS. In: Staab, S., Studer, R. (eds) *Handbook on Ontologies. International Handbooks on Information Systems.* Springer, Berlin, Heidelberg. doi:10.1007/978-3-540-24750-0_3.

47. Bechhofer, S. (2009) OWL: Web ontology language. In: Liu L., ÖZSU M.T. (eds) *Encyclopedia of Database Systems.* Springer, Boston, MA. doi:10.1007/978-0-387-39940-9_1073.

48. Whetzel, PL., Noy, NF., Shah, NH., Alexander, PR., Nyulas, C., Tudorache, T., & Musen, M.A. (2011). BioPortal: Enhanced functionality via new web services from the National Center for Biomedical Ontology to access and use ontologies in software applications. *Nucleic Acids Research, 39*(Web Server issue), W541–W545. Epub 2011 Jun 14.

49. Thamer, M., El-Sappagh, S., & El-Shishtawy, T. (2020). A Semantic approach for extracting medical association rules. *International Journal of Intelligent Engineering and Systems, 13*(3), 280–292. doi:10.22266/IJIES2020.0630.26.

50. Lee, J., Pham, M.-D., Lee, J., Han, W.-S., Cho, H., Yu, H., & Lee, J.-H. (2011). Processing SPARQL queries with regular expressions in RDF databases. *BMC Bioinformatics, 12*(S2), S6. doi:10.1186/1471-2105-12-s2-s6.

51. Benali, K. (2021). Using association rules for ontology enrichment. *CEUR Workshop Proceedings, 2904,* 105–119.

52. Sumathi, G., & Akilandeswari, J. (2020). Improved fuzzy weighted-iterative association rule based ontology postprocessing in data mining for query recommendation applications. *Computational Intelligence, 36*(2), 773–782. doi:10.1111/coin.12269.

53. Viktoratos, I., Tsadiras, A., & Bassiliades, N. (2018). Combining community-based knowledge with association rule mining to alleviate the cold start problem in context-aware recommender systems. *Expert Systems with Applications, 101,* 78–90. doi:10.1016/j.eswa.2018.01.044.

54. Lakshmi, K. S., & Vadivu, G. (2019). A novel approach for disease comorbidity prediction using weighted association rule mining. *Journal of Ambient Intelligence and Humanized Computing.* doi:10.1007/s12652-019-01217-1.

55. Sornalakshmi, M., Balamurali, S., Venkatesulu, M., Navaneetha Krishnan, M., Ramasamy, L. K., Kadry, S., Manogaran, G., Hsu, C. H., & Muthu, B. A. (2020). Hybrid method for mining rules based on enhanced Apriori algorithm with sequential minimal optimization in healthcare industry. *Neural Computing and Applications*, 2. doi:10.1007/s00521-020-04862-2.

56. Heritage, J., McDonald, S., & McGarry, K. (2018). Integrating association rules mined from health-care data with ontological information for automated knowledge generation. *Advances in Intelligent Systems and Computing*, *650*, 3–16. doi:10.1007/978-3-319-66939-7_1.

57. Skorupa Parolin, E., Salam, S., Khan, L., Brandt, P., & Holmes, J. (2019). Automated verbal-pattern extraction from political news articles using CAMEO event coding ontology. *Proceedings -5th IEEE International Conference on Big Data Security on Cloud, BigDataSecurity 2019, 5th IEEE International Conference on High Performance and Smart Computing, HPSC 2019 and 4th IEEE International Conference on Intelligent Data and Security, IDS 2019*, 258–266. doi:10.1109/BigDataSecurity-HPSC-IDS.2019.00056.

58. Kulkarni, H. (2017). Intelligent context based prediction using probabilistic intent-action ontology and tone matching algorithm. *International Conference on Advances in Computing, Communications and Informatics*, ICACCI 2017, 2017-January, 656–662. doi:10.1109/ICACCI.2017.8125916.

59. Cao, T., Mu, W., Gou, J., & Peng, L. (2020). A study of risk relevance reasoning based on a context ontology of railway accidents. *Risk Analysis*, *40*(8), 1589–1611. doi:10.1111/risa.13506.

60. Kafkas, Ş., & Hoehndorf, R. (2018). Ontology based mining of pathogen-disease associations from literature. *Journal of Biomedical Semantics*, *10*(1), 15. doi:10.1186/s13326-019-0208-2.

61. Peral, J., Ferrandez, A., Gil, D., Munoz-Terol, R., & Mora, H. (2018). An ontology-oriented architecture for dealing with heterogeneous data applied to telemedicine systems. *IEEE Access*, *6*(c), 41118–41138. doi:10.1109/ACCESS.2018.2857499.

62. Devi, R., Mehrotra, D., Zghal, H. B., & Besbes, G. (2020). SWRL reasoning on ontology-based clinical dengue knowledge base. *International Journal of Metadata, Semantics and Ontologies*, 14(1), 39–53. doi:10.1504/IJMSO.2020.107795.

63. Han, S. H., & Choi, H. J. (2020). Domain-specific image caption generator with semantic ontology. *Proceedings -2020 IEEE International Conference on Big Data and Smart Computing, BigComp* 2020, 526–530. doi:10.1109/BigComp48618.2020.00-12.

64. Rahman, H., & Hussain, M. I. (2020). A light-weight dynamic ontology for Internet of Things using machine learning technique. *ICT Express*, 7(3), 355–360. doi:10.1016/j.icte.2020.12.002.

65. Sajjad, R., Bajwa, I. S., & Kazmi, R. (2019). Handling semantic complexity of big data using machine learning and RDF ontology model. *Symmetry*, 11(3). doi:10.3390/sym11030309.

66. Engy Yehia, Hussein Boshnak, & Sayed AbdelGaber. (2019). Ontology-based clinical information extraction from physician free-text notes. *Journal of Biomedical Informatics*, *98*, 103276. doi:10.1016/j.jbi.2019.103276.

67. Xiaoxue, L., Xuesong, B., Longhe, W., Bingyuan, R., Shuhan, L., & Lin, L. (2019). Review and trend analysis of knowledge graphs for crop pest and diseases. *IEEE Access*, 7, 62251–62264. doi:10.1109/ACCESS.2019.2915987.

15 Visualizing Chat-Bot Knowledge Graph Using RDF

*Noman Islam, Darakhshan Syed,
Mariz Zafar, and Asif Raza*

CONTENTS

15.1 INTRODUCTION

15.1.1 SEMANTIC WEB

Credential and life-long training possibilities and services are getting increasingly diverse. In Germany, there seem to be currently over 20,000 study courses available by around 450 qualifying centres in the educational area alone [40]. This can be solved by using Artificial Intelligence as this particular scenario requires lots of planning. Software agents (SAs) are symbolic artificial intelligence systems that can explore the internet and make decisions on their own using ontologies. Booking travels, scheduling doctor appointments, and building eligibility routes, or patterns of qualifying behaviours that lead to a specific qualification goal, are examples of SA-oriented tasks. Berners-Lee coined the term "semantic web" to describe this data network. Furthermore, the field's early industry engagement, which was strong from the start, was focused on using Semantic Web principles for knowledge implementation and coordination. One may claim that the discipline is about developing efficient (low-cost) methodologies and techniques for information sharing, exploration, integration, and exploitation and that the World Wide Web is or is not a data communication vehicle in this framework. This knowledge of the topic relates specifically to databases, or the part of data science that deals with data processing and management.

Semantic web enables the integration of web contents by defining the information available over the internet with the help of ontology languages [11]. An ontology is the shared specification of the information of a domain in a formal language. Over the years, a number of researches have been done that defines the knowledge in the form of ontology in various languages such as RDF, DAML, OWL, etc. [10]. Each of these three viewpoints may have significance, and the discipline arises at the intersection of these, with ontologies, linked data, and knowledge graphs serving as core ideas for the research area, with standards comprising specialized exchange formats that consolidate the field on a compositional level. For instance, Islam and Laeeq define the ontology for Islamic prayers in [12]. An ontology can be written with the help of an ontology editing tool such as Protégé, TODE, Web ODE, and SWOOP [13,14]. An ontology for Salat is developed by defining essential concepts and then several use cases were presented. This chapter is about the challenges involved in encoding a chat-bot system knowledge using ontology and then integrating it with Linked Open Data (LOD). Ontology development requires consideration of various aspects such as the language to be used for ontology development, the editing tool facilitating ontology development, ontology development methodology, software agents, and semantic web services [10]. There has been a significant amount of contradictory Textual Medical Knowledge, which plays a significant role in health information systems, as a result of the emergence of healthcare information. Existing work on linking and exploiting textual medical knowledge focuses mostly on establishing direct linkages and pays little emphasis to make computers accurately read and retrieve knowledge. Researchers [43] are investigating a unique strategy for organizing and integrating this data into Knowledge graphs. Finally, it provides a system for retrieving high-precision knowledge from knowledge graphs efficiently. The approach aims to create a high-quality, exhaustive, and sustainable Medical Knowledge Graph.

The Semantic Web's grandiose ambition is to establish an online extender centred on providing information machine easily understandable, i.e., giving web information with well-defined context (Berners-Lee, 2001). The goal is to allow systems, software agents, online services, and other devices to interpret and use web data in new and more intelligent ways [36] for example, enabling software agents to do tasks that would otherwise need human interaction, or easing the integration of semantically linked data into new content constructions. While the Semantic Web has been under constant and rapid growth for years, a portion of the research on the Semantic Web and its related technologies has focused on its practical application in the educational domain.

The Semantic Web is a vast area of research and applications that draws on a wide range of fields inside and outside of machine learning [41]. According to one perspective, the field is centred on attaining the long-term goal of building The Semantic Web (as a resource) and all the resources and procedures required for its creation, alteration, and use. The Semantic Web is usually thought of as an addition to the current World Wide Web that incorporates machine-understandable facts and features that make use of it. The metadata is typically in the form of ontologies or, at the very minimum, a technical language with logical semantics that permits argumentation over the interpretation of the information on the Semantic Web, or, in other words, the formal description will be investigated further down. The Semantic Web field is seen as having a substantial overlap with the field of Artificial Intelligence because of this, as well as the knowledge that intelligent/ software agents will use the knowledge.

An ontology can be written with the help of an ontology editing tool such as Protégé, TODE, Web ODE, and SWOOP [13,14]. These tools provide support with the help of a visual interface, drag-and-drop support, and provision for an ontology development methodology. Ontologies for the semantic Web might be thought of as formal terminology, or a complicated and formal set of concepts. Ontologies like these are often used to assist and facilitate semantic web-based knowledge organization and advanced analytics of information through integration. There are now a variety of ontology editors available for developing OWL ontologies, with Protégé being one of the most well-known. The semantic web leverages the Resource Description Framework for data modelling in addition to semantics. The item, premise, and topic triples are used to model data.

This paper is about the challenges involved in developing a chat-bot system knowledge using ontology and then integrating it with linked open data (LOD). Standardized protocols such as the hypertext transfer protocol, resource description framework, and unified resource identifier are used to create linked data i.e. HTTP, RDF, and URI, respectively.

15.1.2 CHATBOTS

A chatbot is a programming tool that uses written and audio to connect with its customers on a given topic or in a specified domain in a casual [37], courteous style. Chatbots have been used for a variety of reasons in a broad variety of domains, including advertising, customer care, technical assistance, training, and entertainment [6]. Personal digital assistants like Apple's Siri, Amazon's Alexa, Microsoft's Cortana, and Google's Assistant are at the cutting edge of automatic speech recognition and machine learning algorithms. These digital assistants employ machine-learning techniques to help their users become even more productive by managing some of the day-to-day activities of traditional assistants or secretaries (such as email categorization, emphasizing the most important material and conversations). A plethora of easier and more knowledgeable text-based chatbots supplement priority features like raising support tickets to provide comments, distributing content for publishing sites, buying a hotel accommodation, making a dinner reservation, and so on. To answer questions by users, text-based chatbots often follow a series of pre-defined rules or processes [38]. These guidelines or processes allow them to reply efficiently to queries within a certain area, but they are ineffective in responding to inquiries whose behaviour does not match the rules on which the chatbot was taught.

Chatbots have been increasingly popular in a variety of industries in recent years, including health care, advertising, academia, assistance programs, native culture, amusement, and many others [32]. Because various chatbots have previously been constructed, they can broadly be divided into two groups i.e. retrieval-based and generative-model chatbots [22]. Among the various approaches, research can classify those that use templates from those that use semantic-like search techniques. Unger et al. [23] offer a framework for RDF data that falls under the first classification. The authors adapt a method for parsing user queries into a SPARQL framework that closely resembles the question's structural properties. Researchers [34] presented a description of the knowledge graph in order to encourage collaboration and a shared vision for future graph-based knowledge studies. Graph learning is used in certain template-based ways to map user requests to SPARQL queries [24]. In the next classification, numerous techniques, such as those of [25,26], that focus on semantic search-like strategies were proposed. These algorithms essentially execute entity linking on characteristics and entities, finally getting the top k-ranked elements that reflect the query's purpose.

15.1.3 Paper's Contribution

In the context of connected data, the number one mission of chat-bot structures is to get applicable records from more than one information base via way of means of the usage of herbal language information and semantic net technologies. This motive is addressed in this chapter via way of means of remodelling herbal language right into a SPARQL query. The effects of a talk between a chatbot and a person are stored as RDF triples, i.e. subject, predicate, and object. Linking this information to the LOD cloud will increase the reaction first-rate of the chat bot and may additionally assist in getting the character of the person with the aid of using reading the statistics gathered. The extensive adoption of the RDF information model, additionally due to the LOD initiative, has made to be had large, connected datasets which have the capability to deliver worthwhile knowledge. Accessing, comparing, and know-how those datasets as published, though, calls for great schooling and expertise inside the discipline of the semantic web, making those treasured reassets of records inaccessible to a much broader audience.

Developing a chatbot using linked data presents a number of issues, including comprehending user queries, supporting numerous knowledge bases, and language support. The purpose of this article is to design and construct an architecture for an engaging graphical user interface. The study also presents a machine learning strategy for recognizing user intents and generating SPARQL queries relying on intent classification and natural language understanding.

The studies afford a brand new social community dataset (myPersonality) that may be delivered to present-day understanding bases to assist chatbots to reply analytical inquiries. The device can be elevated with a brand new area on request, is versatile, has several understanding bases, is bilingual, and lets in for smooth assignment implementation and execution over a huge variety of topics. Numerous evaluation and alertness instances inside the chatbot display the way it helps collaborative semantic facts in the direction of awesome actual utility regions and show off the proposed method for an understanding graph-pushed chatbot [1].

The rest of the study is laid out as follows. Afterward, you'll find the literature review. Next, there is a suggested approach. Thereafter, the implementation details are presented for your consideration. There will be a report of the results at the end. Suggestions and limits of the planned work conclude the paper.

15.2 RELATED WORK

Many efforts are being made in recent years to develop systems that allow for the viewing and investigation of ontology-based data. A few of those structures rely upon procedures that permit them to restrict the number of facts displayed via way of means of imparting aggregated, filtered, or summarizing get admission to datasets, even as others begin exploring the dataset primarily based totally on consumer sports such key-word searches and queries.

The underlying generation is crucial for the system's long-time period viability, in addition to the characterization of the input's needs, the datasets that can be visualized, and the visualization types available. The next paragraphs provide an overview of these strategies, their advantages and disadvantages, and the datasets they can accommodate. This review will give the reader a thorough grasp of the issues associated with the representation of huge, linked datasets, as well as a classification of the strategies established to address them and an analysis of the systems available and their capabilities [2].

Domain-particular expertise graphs can constitute complicated area expertise in an established layout and feature accomplished excellent achievement in sensible applications. Recently, expertise graphs were broadly utilized in recommender structures due to their cap potential to combine diverse advice fashions and cope with information sparseness and cold-begin problems. In one of the papers, the authors proposed a technique to extract film-associated records from Linked Open Data (LOD) and assemble the expertise graph of film area. The

Neo4j graph database which is characterized with the aid of using a pleasant person interface and a brief inquiry, is used to visualize the expertise graph [3].

In [4], the researchers discuss a particular project, the circumstances that led to it, the design development, and the initial outcomes. It investigates the terminology used in the linked open data cloud and compares it to terminology used in the universal decimal categorization and the basic ideas classification.

Machine learning (ML) is the prevailing application of AI, rooted in the notion that ought to provide computers with a way to go through and process data and allow them to identify (or learn) patterns and acquire knowledge from the data on their own. In the context of chat bot, machine learning comes into play not just in understanding user input but also in finding the accurate information being requested by the user. Most NLP techniques and programming libraries use pre-trained machine learning models for tasks such as part of speech tagging, dependency parsing, and co-reference resolution. Lastly, machine learning also helps in Natural Language Generation for forming coherent sentences that contain the user-requested information in a human-understandable language. The paragraphs below now review machine learning-based approaches. There are also papers that target primarily the development of textual documents of sets of triples [28,29], produce textual abstracts from knowledge base triples [30], and use neural networks to construct natural language interpretations from knowledge bases [31].

In several domains, conversational agents have the potential to facilitate behaviour modification and well-being. Users can utilize both text and speech to communicate with conversational agents. As a result, researchers emphasize the importance of understanding how the architecture of these pathways facilitates behaviour change. Because chatbots are expected to respond to user queries, various research projects have focused on constructing query-answering systems [20]. The concerned reader is directed to [21] for a comprehensive overview of the subject. Researchers [17] created a chat bot for the workplace to enhance workers' activity tracking and self-learning through reflection in order to continue responding to the query.

Ait-Mlouk et al. [18] take on the problems of user inquiries and knowledge from multiple support by first designing and developing an architecture for an interacting user interface. Then, in order to comprehend user intents and construct SPARQL queries, this research provides a machine learning approach based on intent categorization and natural language understanding. Another semantics-based technique is described in the FRASI [27] research study, which aims to construct a question-answering framework to help consumers in finding information about products of interest. The decision was made to use pattern-matching chatbot technologies to create a Question-Answering (QA) platform.

Lopez et al. utilize natural language interpretation, conversation management, emotional situation control, and natural language creation in conjunction with an ontology that collects the agent's knowledge base and specifies a semantic language that the units use to communicate with one another in [5]. In 71.8% of the cases, the generated emotional conduct is the anticipated one, even though the diverse tiers of the policies must be multiplied in order that 100% of every emotional characteristic can be completed at some point of a whole dialogue (a few attributes did now no longer attain the 65% for the seven dialogues).

In [8], the dialogues are scripted earlier which shows a rule-primarily based totally machine. The authors additionally point out the usage of a custom hierarchical transition network-primarily based totally scripting language, and a visible speak layout tool. Results imply that sanatorium sufferers with low fitness literacy observed the machine clean to apply and mentioned excessive ranges of 20 satisfaction. The solution kinds ought to be associated with query kinds for smooth identity of solutions from the expertise base. The look for a solution with inside the expertise base the use of records retrieval methodologies consequences in a fixed of associated solutions. The expertise has been advanced primarily based totally on the area records amassed from many online assets and saved in Dbpedia, Wikipedia, etc. that may be accessed as offline dumps as well. The gadget turned into evaluation via way of means of a set of ten college students of various ages. The findings are as follows: Intelligent Answer furnished 60% times, Interesting Answer 80% and Wrong Answer 20%. The Watson Conversation Service turned into used to construct this chat-bot right into a scientific selection help gadget, and the Bluemix infrastructure turned into used to create and teach it.

The Watson Conversation API aided text mining and natural language processing (NLP) utilizing deep learning. When first-level characteristics are utilized to assess the incidence of distinct preventive pathways, 74.65% of the Area under ROC Curve (AUC) is used to evaluate the occurrence of different preventative pathways. When disease-specific features are included, the Chatbot achieves a more precise preventative route evaluation in 86.78% of cases. In terms of intervention adherence, preliminary findings from 15 patients show good results. A corpus of research questions that would likely be asked by a clinician and replies that could be given by the virtual patient (VP) was used in another chatbot [8], which used a language model in a speech recognizer to use a corpus of questions that would likely be requested by a clinician and reactions received that could be given by the virtual patient (VP). The natural language group at the University Of Southern California Institute for Creative Technologies [9] created a statistical text classifiers technique, which was used in this study. The response of the system is packaged into a Functional Markup Language (FML) message structure once it has been selected. Apart from the text, the overall response load for the VP does not provide any other additional details.

Semantic web protocols have reported to be efficient in a variety of situations. In addition, a study paper [39] emphasizes the need of using pattern recognition algorithms. Restoration and extraction of data and formalization are carried out manually due to stringent demands on data precision and reliability. However, this is not always possible for all collected content, as the archive of the free theatre exemplifies. As a result, prospective candidates must be identified within a large search space, i.e. material that has not yet been defined but is likely worth documenting. As a result, applying automated pattern recognition is a potential strategy.

A model for offering transparency to computational processes is presented in a research article [42]. For the high risk and high effect context of defence intelligence investigation to produce effective analysis, the given study includes key factors not addressed in later studies. The suggested framework's main concepts are also discussed, with an interactive application being used as an example. A conversational agent for knowledge inquiry is included in this software. In a critical decision-making hypothetical scenario, the agent displays shared human-machine logic. There are also some fresh findings from interviews with a small number of analysts, as well as directions for future analysis.

R2D2, an intelligent chatbot based on semantic web applications and enabling an intelligent supervised natural language interaction for acquiring information available in DBpedia, is presented by Haridimos et al. [15]. The chatbot receives structured data, allowing users to ask questions in the form of triple-patterns, which are then responded to by the actual engine. While typing, an auto-complete feature assists users in constructing triple patterns by recommending resources from the DBpedia. The related SPARQL queries are generated automatically based on user interaction. The requests are sent to the appropriate DBpedia SPARQL interface, and R2D2 receives the results, which are then enhanced with maps and visualizations before being shown to the user. The results of the usability testing demonstrate the benefits and utility of the research product. To promote dialogue modelling and generation, Wang et al. [16] suggest augmenting neural network-based generative infrastructure with knowledge embedding and knowledge attentive reader to include external textual knowledge into the conversation model. The goal of the research technique is to test the model with the Ubuntu dataset using both automated and human performance metrics. The findings of the experiments were also shared. It also shows how attention operates in a discussion setting, allowing the usefulness of the knowledge attentive reader technique to be verified.

Vegesna et al. [26] present an ontology-based chatbot for the e-commerce domain in another way. External data is obtained using e-commerce APIs (e.g., eBay API), while data from the source of data is retrieved employing ontology templates and Jape principles. Ronzhin et al. [33] discuss their experience building and using the Netherlands' first government accountability knowledge graph. The researchers elucidate the benefit of having such a graph and illustrate its usage in the context of effective data browsing, multi-criterion evaluation for town development, and the construction of location-aware chat bots, based on the created demos. It offers the necessary knowledge dimensions for linking facts that would otherwise have little in similar. The Open Geographic Consortium's development of the GeoSPARQL standard [35] offered basic components for standardizing geospatial semantics in data and enabling geospatial argumentation in SPARQL in this situation.

According to a research article [44], a hybrid chatbot system incorporating Q&A and knowledgebase methodologies should be developed. For responding to questions and response pairs, machine learning approaches are applied. A meta graph model is used to process the pool of knowledge, which necessitates the combination of two methodologies. However, there is still room to increase the quality of Q&A computation. To boost efficiency, the knowledgebase computing module needs to include more question replying functionalities.

A chatbot system that collects and analyses data in real-time to build an autonomously scalable knowledge network and uses it as the base data was proposed in a research paper [19]. To increase performance, a fine-tuned BERT-based relation extraction model will be used for auto-growing graphs. It evaluates the performance and reliability of a knowledge graph by creating a chatbot that learns human rational thinking using an auto-growing knowledge base as a graph.

15.3 PROPOSED WORK

Because chatbots are expected to respond to user queries, various research projects have focused on constructing query-answering systems [20]. The concerned reader is directed to [21] for a comprehensive overview of the subject. The targeted work suggested in this research is based on a modular technique (Figure 15.1). The proposed technique utilizes the supremacy of semantic web approaches, graph of available knowledge and machine learning as well. The suggested chatbot can do a variety of functions (e.g., analytical inquiries, FAQs, etc.), aggregating data from numerous sources (KBs, online services), but this knowledge is in the form of triples, which we are generating to represent in a way that users can readily comprehend how and why chatbot delivers a certain response with semantic relatedness on it. Users can communicate with the service through a chit-chat system or voice-based communications to improve usability. The suggested chatbot is built with the Flask framework and can run in an independent or distributed manner to optimize information-gathering response times. The user input and inquiries are anonymously saved in a records for future learning and enhancement. The next subsections will provide further information on the suggested chatbot's functionalities and working mechanisms.

15.3.1 QUERY PREPROCESSING & NLP

Natural Language Processing (NLP) is a phrase that describes a device's potential to take in what is spoken to it, break it down, interpret its significance, take necessary action, and react in a language that the user understands [7] (see Figure15.1).

Natural language understanding (NLL) is a subgroup of PNL that handles the closest element of the problem; however, it's also vital to know how and where to better manage unorganized sources and transfer them into a structured component that a computer can comprehend and act on. While people are adept at handling faulty pronunciations, word substitutions, contractions, and colloquialism, systems are less skilled at dealing with surprising entries. The typical NLP pipeline (see Figure 15.2) that covers sentence splitting, tokenization, lemmatization,

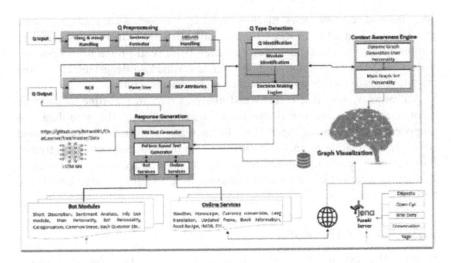

FIGURE 15.1 Proposed modular architecture for chat-bot.

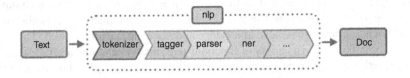

FIGURE 15.2 NLP pipeline that covers sentence splitting, tokenization, lemmatization, and part of speech tagging is applied to each user.

and part of speech tagging is applied to each user query so 34 that the NLU part can be applied. In addition to the above-mentioned NLP tasks, dependency parsing (see Figure 15.3), has been used for converting the unstructured natural language query into structured data based on the grammar and inter-dependencies between words.

The NLP pipeline, shown in Figure 15.2, detects the nouns in the sentence and select entity types are also detected by the NLP library we're using called Spacy. However, a lot of domain-specific entities that contain multiple words are not identified by Spacy as the language models are trained on generic information and may not contain some highly specific medical terminology. These are used by proposed architecture as a knowledge source. The below algorithm explains the step-by-step process of identifying multi-word concepts and entities using ontologies.

Algorithm 15.1: Identifying Multi-Word Concepts in the Sentence

INPUT:

Sentence (*S*)
OUTPUT:
List of strings
FUNCTION:
Break *S* into bigrams and trigrams
For each bigram and trigram
Check if an entry exists in domain-specific ontology/Wordnet
If yes, it is a concept so add it to the list to be returned

15.3.2 QUERY TYPE DETECTION

This module identifies the query type by using two different approaches. In the first approach, a few variations have been defined along with machine learning techniques. The other one is brain objects which define the abstract level of query items that will map with the ontology triples like info box. Once the user's type has been identified by semantic relations, it calls the respective handler function to service the request. The handler function is just a plain Python function in the codebase that may route the request further for more processing or handle it directly (Figure 15.3).

15.3.3 BOT MODULES

This chatbot handles online services like Weather, horoscope, currency conversion, lang translation, etc. and for that, we have defined pattern-based Responses by rec. input from third-party services, on the other hand, we have a few other modules like a short description, and sentiment handling info box, in these module responses would take place from ontology. We also implemented a generative approach LSTM for handling all general-purpose queries.

15.3.4 CONTEXT AWARENESS ENGINE

There are two types of dynamic graphs generated in which one represents the user presently on the fly and the other one is bot personality. Every query is

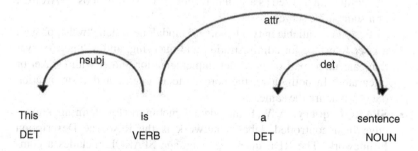

FIGURE 15.3 Example of dependency parsing.

dynamically stored into the bot brains in the form of triples. We are providing a dashboard where the admin can visualize the graph.

Algorithm 15.2: Ontology-Based Context Awareness

INPUT:

sentence (S)

OUTPUT:

Enriched sentence

FUNCTION:

1. Expand all contractions in S (e.g. isn't -> is not)
2. Run the NLP pipeline on S
3. Make a list of all the nouns and noun chunks identified in the NLP pipeline
4. Get a list of multi-word concepts - See Algorithm 15.1
5. For each term in the noun list:
   a. Check if a concept exists in any medical ontology on Dbpedia
      i. If yes, get its synonyms from medical ontologies
      ii. If no, check if the term is found in WordNet
         1. If found, get its synonyms from wordnet
   b. Replace the occurrence of the term in S with the term and its enriched context

15.4   IMPLEMENTATION DETAILS

15.4.1   Tools/Technologies Used

We will first discuss the technologies used in this project.

- **Apache Jena Fuseki**: Apache Jena is a proprietary and free source Java platform for constructing Semantic Web and Linked Applications. It is made up of several APIs that work together to process Resource Description Framework data. Jena Fuseki is an Apache Jena SPARQL server that can be used as a system resource, a Java web apps (WAR file), or a standalone server.

  Fuseki is available in two flavors: a standalone system "webapp" with a User Interface for administration and querying, and a "main" server that can run as part of a broader implementation, either with Docker or integrated. In both cases, the core protocol engine and customization data format are the same.

- **SPARQL query**: A Web metadata benchmark for defining knowledge in a controlled, labeled network is the Resource Description Framework. The RDF querying language SPARQL includes a comprehensive syntax and ontology definition. Whether the data resides as

RDF or accessed as RDF through processing, SPARQL can be used to perform searches over a wide variety of sources. Both necessary and optional graph patterns, as well as their conjunctions and disjunctions, are query able using SPARQL. SPARQL also supports extensive value verification and query limitations based on the source's RDF graph. As a consequence of SPARQL queries, you can get either results sets or RDF networks.

- **Python**: Python is a powerful programming language which is general-purpose with a high level of abstraction. It is an open-source software, cross-platform programming language. Python can be used in different domains such as Web development, Machine Learning, Data Analysis, Scripting, Game development, etc. We have used python to perform Network Analysis of RDF graphs and visualization of knowledge graphs. Python is becoming a widely used programming language among smart professionals due to its numerous advantages, which make it ideal for machine learning and data mining research inititiatives. Python's large number of machine learning-specific frameworks and libraries make the design process easier and faster.
- **NLP (Stanford NLP java library)**: In Java, CoreNLP is used to process human language. Token and sentence boundaries, parts of speech, named entities, numeric and time values, dependence, and constituency parses, reference, sentiment, quotation inferences, and relations are all possible with CoreNLP. The pipeline is the heart of CoreNLP. Pipelines take in text data in raw form, process it via a sequence of NLP transcribers, and output a final set of remarks.
- **NER module**: A Named Entity Recognizer is implemented in Java by the Stanford NER. Named Entity Recognition (NER) detects and labels sequence data of words in the text that are the names of specific people, businesses, or genes and protein synthesis. It includes well-designed feature extractors for Named Entity Recognition as well as a plethora of features extraction design options (see Figure 15.4).

15.4.2 STEPS OF IMPLEMENTATION

a. Fuseki server setup, loading graph data.

The first step is to load a dataset in the TDB datastore. We downloaded and extracted the datasets from DBpedia downloads.

b. Using SPARQL Endpoint to access linked open data.

RDF data base expresses a complex relationship between the entities such as people, things, places, etc. The databases are called graph base as they are structured as a network or graph where nodes are connected to edges showing the relationship between each entity. RDF data represents information in three parts, the triple comprises subject, predicate, and object. SPARQL helps us in translating heavily linked data into the normalized, tabular data with rows and columns.

NER Module Details

- This module consists of three types of functions
  - NER (type-2) detection from Dbpedia:

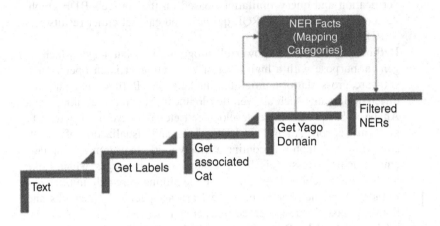

FIGURE 15.4   Steps of implementation.

c. Dbpedia.

The Linked Open Data Cloud is created by DBpedia and made available to the community. Data is available in strict accordance with the "Linked Open Data" standards, which include the following specifications:

1. All the entities are identified using URIs.
2. Entities are described using RDF Language-based triples where the subjects and predicates are identified by URIs, while objects may be identified using either an HTTP URI or a Literal.
3. Entity descriptions are being published to HTTP networks (the World Wide Web) using RDF documents.
4. We have taken DBpedia as the source of getting the RDF data for visualization. We will be using URIs in our python code to perform network analysis and visualization on different entities. Following are example URLs:

   url = 'http://dbpedia.org/resource/Yasir_Khan'
   url = 'http://dbpedia.org/resource/Maria_Wasti'
   url = 'http://dbpedia.org/resource/Ayesha_Khan'
   url = 'http://dbpedia.org/resource/Kamran_Akmal'

d. Network Analysis & Visualization

In the next step, we have performed a network analysis of RDF performing easily with Python rdflib and network. The analysis of the network and visualization of graph was done in 4 steps.

i. Loading an arbitrary RDF graph into rdflib. Following is the sample script:

```
import rdflib
fromrdflib.extras.external_graph_libsimportrdflib_to_networkx_
multidigraph
url='http://dbpedia.org/resource/Kamran_Akmal'
g=rdflib.Graph()
result=g.parse(url, format='turtle')
G=rdflib_to_networkx_multidigraph(result)
```

ii. Next Step is to convert the rdflib Graph into a networkx Graph.

```
# rdflib's Conversion.
# Graphs to network(x) graphs
G = rdflib_to_networkx_graph(g)
PRINT("networkx Graph loaded successfully with length {}".format(len(G)))
```

This step tells us the length of the network as well as the output. For example network graph for Kamran Akmal is of length 481. It will show the following output:

networkx Graph loaded successfully with length 481

The last step is to get a community evaluation record via way of means of strolling networkx's algorithms on that information structure.

e. Sample Analysis Report

This step will give below analysis (Figures 15.5–15.7).

1. General network metrics (network size, pendants, density)

EIGENVECTOR CENTRALITY
==============================

The mean network eigenvector centrality is 0.0333253513068688, with stdev 0.031151895404202243

The maximum node is http://dbpedia.org/resource/Kamran_Akmal, with value 0.7151193516122513

The minimum node is http://dbpedia.org/resource/List_of_inter-national_cricket_five-wicket_hauls_by_Anil_Kumble, with value 0.03190494713956625

(0.7151193516122513, 0.03190494713956625) (1, 480)

BETWEENNESS CENTRALITY
==============================

The mean betwenness centrality is 0.002079002079002079, with stdev 0.04559607525875532

The maximum node is http://dbpedia.org/resource/Kamran_Akmal, with value 1.0

The minimum node is http://dbpedia.org/resource/List_of_interna-tional_cricket_five-wicket_hauls_by_Anil_Kumble, with value 0.0

(1.0, 0.0) (1, 480)

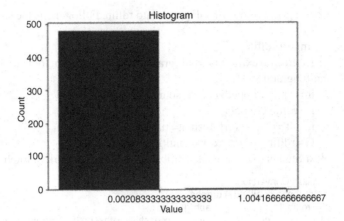

FIGURE 15.5 Degree centrality.

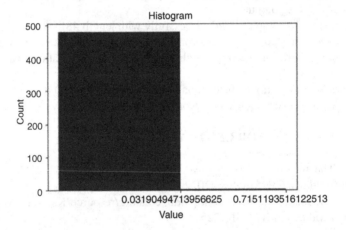

FIGURE 15.6 Eigen vector centrality.

CONNECTED COMPONENTS

There are 1 related components in the graph.
There are 481 nodes in interlinked module 0
2. Clustering metrics (connected components, clustering)
   Figures 15.8–15.10 shows the various graphs of clustering.
   It is concluded that network gives multiple visualizations to present
our network graphs.

CLUSTERS

There are 481 groupings in the networked graph.

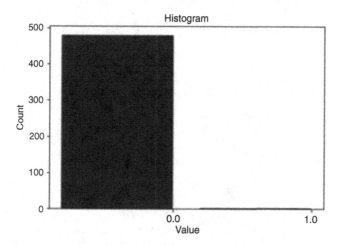

FIGURE 15.7 Betweeness centrality.

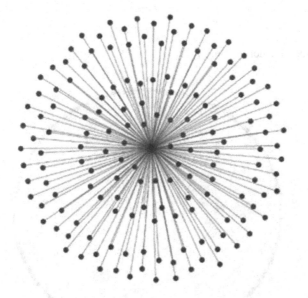

FIGURE 15.8 Spring format.

Clusters 0 has a total of 90 connections.
Clusters 1 has a total of 40 connections.
Clusters 2 has a total of 46 connections.
Clusters 3 has a total of 35 connections.
Clusters 4 has a total of 59 connections.
Clusters 5 has a total of 40 connections.
Clusters 6 has a total of 57 connections.
Clusters 7 has a total of 59 connections.

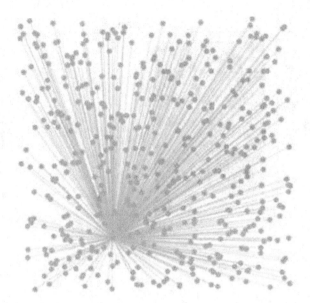

FIGURE 15.9 Random graph.

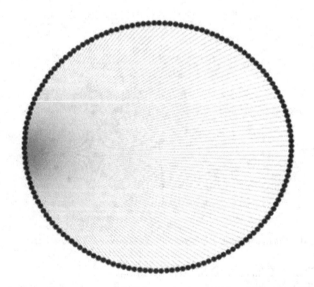

FIGURE 15.10 Shell format.

Clusters 8 has a total of 46 connections.
Clusters 9 has a total of 43 connections.
Clusters 10 has a total of 48 connections.
Clusters 11 has a total of 53 connections.

Clusters 12 has a total of 91 connections.
Clusters 13 has a total of 53 connections.
Clusters 14 has a total of 263 connections.
Clusters 15 has a total of 38 connections.
Clusters 16 has a total of 35 connections.
Clusters 17 has a total of 38 connections.
Clusters 18 has a total of 44 connections.
Clusters 19 has a total of 48 connections.
Clusters 20 has a total of 40 connections.
Clusters 21 has a total of 4 connections.

15.5 RESULTS

We performed Network Analysis and Visualization on multiple entities. Below is the Network Analysis and Network Graphs (Figures 15.11–15.15):

We have implemented context based chatbot that generated brains objects in the form of RDF graph, we have visualized RDF graph which shows the semantic relations as shown in Figure 15.9. This graph represents the bot personality, in further this approach leads towards the explainable graph which can be return back to the user that shows the evidence of bot reply (The details of the graph are shown in Table 15.1).

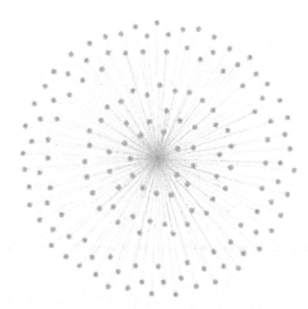

FIGURE 15.11　Network analysis and network graphs of Maria Wasti.

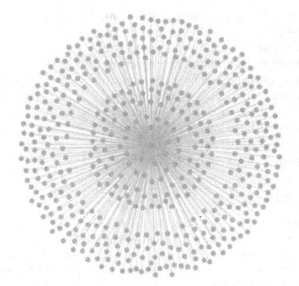

FIGURE 15.12   Network analysis and network graphs of Kamran Akmal.

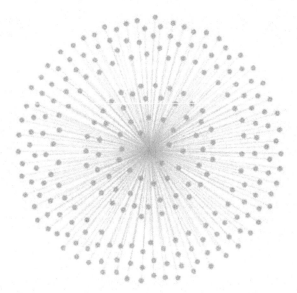

FIGURE 15.13   Analysing the topology and network graphs of Ayesha Omer.

FIGURE 15.14   Analysing the topology and network graphs of Umer Akmal.

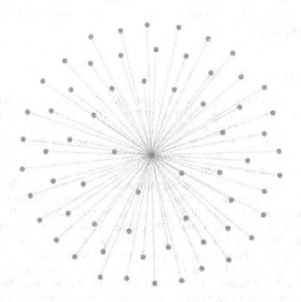

FIGURE 15.15   Network analysis and network graphs of Ayesha Khan.

TABLE 15.1
Details of the Graphs

Entity	Nodes	Edges	Pendants	Network Density
Maria Wasti	154	154	153	0.013071895
Kamran Akmal	481	481	480	0.004166667
Ayesha Khan	64	64	63	0.031746032
Ayesha Omer	249	249	248	0.004032258
Umer Akmal	8	8	7	0.285714286

15.6 CONCLUSION

This paper proposes a novel technique for visualizing of chat-bot knowledge graph. The paper also links the knowledge graph with open-linked data. A machine learning-based approach has been proposed for generating SPARQL queries. A novel social networking dataset has also been proposed in this research work. The proposed approach has been implemented using the Stanford NLP library in python. This research work can be extended by incorporating chatbot knowledge in other domains besides social networks. The knowledge graph can be modelled in other languages such as DARPA Agent Mark-up Language (DAML) and Web Ontology Language (OWL).

REFERENCES

1. Ait-Mlouk and Jiang, L., 'KBot: A Knowledge Graph Based ChatBot for Natural Language Understanding Over Linked Data'. *IEEE Access*, vol. 8, pp. 149220–149230, 2020. doi:10.1109/ACCESS.2020.3016142.
2. Krommyda, M., and Kantere, V., 'Visualization Systems for Linked Datasets'. *IEEE 36th International Conference on Data Engineering (ICDE).* 2020. doi:10.1109/icde48307.2020.00171.
3. Lei, Q., and Liu, Y., 'Constructing Movie Domain Knowledge Graph Based on LOD'. *Twelfth International Conference on Ubi-Media Computing (Ubi-Media),* Bali, Indonesia, pp. 54–57, 2019. doi:10.1109/Ubi-Media.2019.00019.
4. Szostak, R., Scharnhorst, A., Beek, W., and Smiraglia, R. P., 'Connecting KOSs and the LOD Cloud'. *Challenges and Opportunities for Knowledge Organization in the Digital Age*, pp. 521–529, 2018. doi:10.5771/9783956504211-521.
5. Lopez, V., Eisman, E., and Castro, J., 'A Tool for Training Primary Health Care Medical Students -The Virtual Simulated Patient'. In *2008 20th IEEE International Conference on Tools with Artificial Intelligence*, vol. 2, pp. 194–201. IEEE, 2008.
6. 'The Complete Beginner's Guide to Chatbots'. Chatbots Magazine, 2019. https://chatbotsmagazine.com/the-complete-beginner-s-guide-to-chatbots-8280b7b906ca. Accessed on Sep 09.
7. 'Natural Language Processing'. Wikipedia. Accessed Aug 10, 2019. https://en.wikipedia.org/wiki/Natural_language_processing.

8. Bickmore, T., Pfeifer, L., and Jack, B, 'Taking the Time to Care: Empowering Low Health Literacy Hospital Patients with Virtual Nurse Agents', 2009.
9. "Information Sciences Institutes". USC University of California. https://www.isi.edu/home.
10. Noman, I., Abbasi, A. Z., and Shaikh, Z. A., 'Semantic Web: Choosing the Right Methodologies, Tools and Standards'. *International Conference on Information and Emerging Technologies.* IEEE, 2010.
11. Islam, N., Siddiqui, M. S., and Shaikh, Z. A., 'TODE: A Dot Net Based Tool for Ontology Development and Editing'. *2nd International Conference on Computer Engineering and Technology.* vol. 6. IEEE, 2010.
12. Islam, N., and Laeeq, K., 'Salaat Ontology: A Domain Ontology for Modeling Information Related to Prayers in Islam'. *Indian Journal of Science and Technology*, vol. 12, p. 31, 2019.
13. Islam, N., and Shaikh, Z. A., 'Towards Ontology Editing, Querying and Visualization in. Net Environment'. *2019 8th International Conference on Information and Communication Technologies (ICICT).* IEEE, 2019.
14. Islam, N., and Sheikh, G. S., 'A Review of Techniques for Ontology Editor Evaluation'. *Journal of Information & Communication Technology (JICT)*, vol. 9, no. 2, 2015, p. 11.
15. Kondylakis, H., Tsirigotakis, D., Fragkiadakis, G., Panteri, E., Papadakis, A., Fragkakis, A., Tzagkarakis, E., Rallis, I., Saridakis, Z., Trampas, A., Pirounakis, G., and Papadakis, N., 'R2D2: A Dbpedia Chatbot Using Triple-Pattern Like Queries'. *Algorithms*, vol. 13, no. 9, 2020, p. 217. doi:10.3390/a13090217.
16. Wang, Y., Rong, W., Ouyang, Y., and Xiong, Z., 'Augmenting Dialogue Response Generation with Unstructured Textual Knowledge'. *IEEE Access*, vol. 7, 2019, pp. 34954–34963. doi:10.1109/ACCESS.2019.2904603.
17. Kocielnik, R., Avrahami, D., Marlow, J., Lu, D., and Hsieh, G., 'ACM Press the 2018- Hong Kong, China (2018.06.09–2018.06.13)'. *Proceedings of the 2018 on Designing Interactive Systems Conference 2018- DIS'18- Designing for Workplace Reflection*, pp. 881–894. doi:10.1145/3196709.3196784.
18. Ait-Mlouk, A., and Jiang, L., 'KBot: A Knowledge Graph Based ChatBot for Natural Language Understanding Over Linked Data'. *IEEE Access*, vol. 8, 2020, pp. 149220–149230. doi:10.1109/ACCESS.2020.3016142.
19. Yoo, S., and Jeong, O., 'Auto-Growing Knowledge Graph-Based Intelligent Chatbot Using Bert'. *ICIC Express Letters, ICIC International*, ISSN 1881-803X, vol. 14, no. 1, January 2020. doi:10.24507/icicel.14.01.67.
20. Marakakis, E., Kondylakis, H., and Papakonstantinou, A., 'APANTISIS: A Greek Question-Answering System for Knowledge-Base Exploration'. In *Strategic Innovative Marketing.* Edited by Androniki Kavoura, Damianos P. Sakas, and Petros Tomaras, Springer: Cham, Switzerland, pp. 501–510, 2017.
21. Diefenbach, D., Lopez, V., Singh, K., and Maret, P., 'Core Techniques of Question Answering Systems Over Knowledge Bases: A Survey'. *Knowledge and Information Systems*, vol. 55, 2018, 529–569.
22. Mutiwokuziva, M. T., Chanda, M. W., Kadebu, P., Mukwazvure, A., and Gotora, T. T., 'A Neural-network based Chat Bot'. *Proceedings of the International Conference on Communication and Electronics Systems (ICCES)*, Coimbatore, India, pp. 19–20, October 2017.
23. Unger, C., Bühmann, L., Lehmann, J., Ngonga Ngomo, A. C., Gerber, D., and Cimiano, P., 'Template-Based Question Answering Over RDF Data'. *Proceedings of the 21st International Conference on World Wide Web*, Lyon, France, pp. 639–648, 16–20 April 2012.

24. Soru, T., Marx, E., Valdestilhas, A., Esteves, D., Moussallem, D., and Publio, G., 'Neural Machine Translation for Query Construction and Composition'. *arXiv:1806.10478*, 2018.

25. Marx, E., Höffner, K., Shekarpour, S., Ngomo, A. C. N., Lehmann, J., and Auer, S., 'Exploring Term Networks for Semantic Search Over RDF Knowledge Graphs'. *Proceedings of the Metadata and Semantics Research, MTSR, Communications in Computer and Information Science*, Göttingen, Germany, vol. 672, p. 15, 22–25 November 2016.

26. Dubey, M., Banerjee, D., Chaudhuri, D., and Lehmann, J., 'EARL: Joint Entity and Relation Linking for Question Answering over Knowledge Graphs'. *Proceedings of the International Semantic Web Conference*, Monterey, CA, 8–12 October 2018.

27. Augello, A., Pilato, G., Machi, A., and Gaglio, S, 'An Approach to Enhance Chatbot Semantic Power and Maintainability: Experiences within the FRASI Project'. *Proceedings of the IEEE Sixth International Conference on Semantic Computing*, Palermo, Italy, pp. 186–193, 19–21 September 2012.

28. Zhu, Y., Wan, J., Zhou, Z., Chen, L., Qiu, L., Zhang, W., Jiang, X., and Yu, Y., 'Triple-to-Text: Converting RDF Triples into High-Quality Natural Languages via Optimizing an Inverse KL Divergence'. *Proceedings of the International ACM SIGIR Conference on Research and Development in Information Retrieval*, Paris, France, pp. 455–464, 21–25 July 2019.

29. Li, Z., Lin, Z., Ding, N., Zheng, H. T., and Shen, Y., 'Triple-to-Text Generation with an Anchor-to-Prototype Framework'. *Proceedings of the Twenty-Ninth International Joint Conference on Artificial Intelligence*, Yokohama, Japan, pp. 3780–3786, 11–17 July 2020.

30. Vougiouklis, P., Elsahar, H., Kaffee, L., Gravier, C., Laforest, F., Hare, J. S., and Simperl, E., 'Neural Wikipedian: Generating Textual Summaries from Knowledge Base Triples'. *Journal of Web Semantics*, vol. 52, 2018, pp. 1–15. CrossRef.

31. Moussallem, D., Speck, R., and Ngonga Ngomo, A. C., 'Generating Explanations in Natural Language from Knowledge Graphs'. In *Knowledge Graphs for eXplainable Artificial Intelligence*. Edited by Ilaria Tiddi, Freddy Lécué, and Pascal Hitzler, IOS Press: Amsterdam, NH, pp. 213–241, 2020.

32. Clarizia, F., Colace, F., Lombardi, M., Pascale, F., and Santaniello, D., 'Chatbot: An Education Support System for Student'. In *International Symposium on Cyberspace Safety and Security*, pp. 291–302. Springer: Cham, 23 September 2018.

33. Ronzhin, F., Maria, B., Beek, L., and van't Veer, R., 'Kadaster Knowledge Graph: Beyond the Fifth Star of Open Data'. *Information*, vol. 10, no. 10, 2019, p. 310. doi:10.3390/info10100310.

34. Ehrlinger, L., and Wöß, W., 'Towards a Definition of Knowledge Graphs'. *Proceedings of the SEMANTiCS Posters and Demos Track*, Leipzig, Germany, vol. 48, pp. 13–14, September 2016.

35. Battle, R., and Kolas, D., 'Geosparql: Enabling a Geospatial Semantic Web'. *Semantic Web Journal*, vol. 3, 2011, pp. 355–370.

36. Jensen, J., 'A Systematic Literature Review of the Use of Semantic Web Technologies in Formal Education'. *British Journal of Educational Technology*, vol. 502, 2019, pp. 505–517.

37. Smutny, P., and Schreiberova, P., 'Chatbots for Learning: A Review of Educational Chatbots for the Facebook Messenger'. *Computers & Education*, vol. 151, 2020, p. 103862.

38. Budiu, R., 'The User Experience of Chatbots'. Retrieved from Nielsen Norman Group: https://www.nngroup.com/articles/chatbots/.

39. Humm, B. G., et al. 'Machine Intelligence Today: Applications, Methodology, and Technology'. *Informatik Spektrum*, vol. 44, no. 2, 2021, pp. 104–114.
40. Stiftung zur Förderung der Hochschulrektorenkonferenz, 'Studieren und promovieren in Deutschland. Informationen über deutsche Hochschulen, Studiengänge, Promotionen', 2019. https://www.hochschulkompass.de/home.html. Accessed 30 Oct 2021.
41. Hitzler, P., 'A Review of the Semantic Web Field'. *Communications of the ACM*, vol. 64, no. 2, 2021, pp. 76–83.
42. Hepenstal, S., Kodagoda, N., Zhang, L., Paudyal, P., and Wong, B. L., 'Algorithmic Transparency of Conversational Agents'. *CEUR Workshop Proceedings*, 2019.
43. Anand, G., Shah, R. R., and Mutharaju, R., 'Knowledge Graphs in Medical Domain', 2020.
44. Gapanyuk, Y. et al., 'The Hybrid Chatbot System Combining Q&A and Knowledge-Base Approaches'. *7th International Conference on Analysis of Images, Social Networks and Texts*, 2018.

16 Toward Data Integration in the Era of Big Data
Role of Ontologies

*Houda EL BOUHISSI, Archana Patel,
and Narayan C. Debnath*

CONTENTS

16.1   INTRODUCTION AND MOTIVATION

Digital companies such as Amazon generate huge volumes of information: Petabytes and even exabytes of data commonly referred to as "Big Data" [1]. This term refers to all digital data produced, processed, and analyzed using new technologies for different purposes. Nevertheless, different applications in a distributed environment handle a large amount of data and may exceed the capacity of the available resources. So, it is difficult to predict this condition when dealing with data that circulates online. Yaqoob et al. [2] presented a robust study about big data and a comprehensive understanding of the big data scenario.

Let us analyze the following scenario: The Ministry of High education in Algeria manages more than 34 universities in the country and deploys big data architectures and some structures that operate with traditional databases. We suppose that the Ministry wants to integrate the activities of these universities toward operating centrally to offer better services to the students (Master application, etc.). Data integration involves the collection, storage, structuring, and combining of data to operate as a unified view. Data integration plays an important role in making it easier to exchange information and communicate across the enterprise, whether it is integrating core systems or integrating processes, administrative tasks, and databases. Data integration requires appropriate software tools that automatically gather and analyze real-time information from various online data sources. This is the place where researchers use ontologies because they represent knowledge in such a way that is understandable by both humans and machines. Ontologies describe the semantics of data and are widely used for semantic interoperability and integration.

Semantic interoperability includes structural and syntactic heterogeneity. Structural heterogeneity refers to the different data models of database systems and syntactic heterogeneity concerns the contextual data meaning and interpretation [3]. Structural heterogeneity occurs when the modeling schema is not in accordance with the Database management systems. A possible solution to these conflicts is to identify the structural problem. On the other hand, syntactic heterogeneity occurs when different terminology is used across companies in the same context. For example, if we want to query the address of a customer, we have to look for the general schema that contains the alias value as "address" (we find the same word with a different meaning). Syntactic heterogeneity can be addressed by using adapters that can translate user queries into different database query languages. From the literature review, we conclude that the methods and tools developed for dealing with the semantics of data can significantly contribute to overcome the challenges of big data.

In this chapter, we present an approach for data integration in the era of big data. We focus on ontology design and describe how ontologies contribute to data semantics. Our proposal is based on the NoSQL Column-Oriented databases, modular and global ontologies, and the SPARQL engine. The purpose of this chapter is to describe how Data Semantics use ontologies as a formal domain description to support big data management. The rest of the chapter is structured as follows. Section 16.2 presents some basic concepts. Section 16.3 overviews the related work regarding Data Integration in big data. Further, Section 16.4 describes the key elements of the proposed approach and provides more details on the functionalities. Finally, Section 16.5 gives some conclusions and describes future work.

16.2 BACKGROUND

This section provides an overview of some basic concepts relevant to this research area. It includes big data and its integration. In addition, we present the ontologies and the terminologies that describe the proposed approach.

16.2.1 Big Data Characteristics

We generate about 2.5 trillion bytes of data every day. Traditional data management tools cannot handle this massive amount of data. Data comes from different digital sources, like- online shopping, booking requests, blogs, tweets, and many more. The internet giants such as Yahoo, Facebook, and Google were the first to deploy the big data technology to manage their databases and make it easier to data access [4].

According to Ranjan and Foropon [5] "big data differs from traditional data in many dimensions called big data characteristics: Volume, Velocity, Variety, and Veracity".

- **Volume:** big data is associated with a huge volume of data, ranging from a few tens of terabytes (1 TB=212 bytes) to several petabytes (1 Po=215 bytes) in a single data set. The volume corresponds to the mass of information produced every second. Companies managing massive data are required to find the necessary techniques and means capable of managing the volumes of data collected each day.
- **Velocity:** It describes the frequency by which information is generated, captured, stored, and shared. It is also the processing of continuous data streams. Companies need to understand data creation, processing, and visualization speed. On the other side, they need to analyze and deliver the data to the user in a way that meets the requirements of user and real-time applications.
- **Variety:** Increasingly, the rate of structured data manipulated in relational database tables is declining relative to the expansion of unstructured data types. Big data technologies allow the creation, integration, analysis, recognition, and classification of different types of data.
- **Veracity:** In general, it concerns with the truthfulness and accuracy of the data. However, in big data environment, this characteristic is more important since it concerns not only the quality of the data but also the reliability and the credibility of the data source itself.
- **Value:** In the context of infobesity, it is a matter of being able to focus on data that has real value and can be used to justify their analysis.

16.2.2 BIG DATA INTEGRATION

In recent years, the amount of data generated by physical machines connected to the internet has grown exponentially with time, this complex and large data is called Big Data. It is not only voluminous but also diverse. This heterogenous data creates a problem in storage, analysis, processing, and mainly data integration, which emerges as a challenge for organizations that deploy large data architectures. Therefore, a global approach is primordial to negotiate the challenges of integration. Data integration means "to aggregate data from various sources to provide a meaningful and valuable data sharing" [6]. Traditional data integration tools follow the ETL process that involves Extraction (E), Transformation (T), and Loading (L) of data, depicted in Figure 16.1. ETL aims to load the data from the different sources into a data warehouse. Figure 16.1 shows the traditional data integration process.

In the era of big data, data integration is very challenging since the traditional techniques cannot handle it. The aim of big data integration is to deliver a

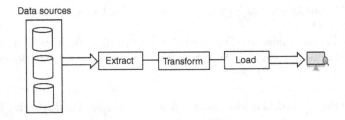

FIGURE 16.1   Traditional data integration process.

single trusted, accurate, and semantically enhanced view of data from a variety of sources. Data integration is a very important tool for boosting competitive and collaborative intelligence. Ziegler and Dittrich [7] highlight the reasons behind integrating multiple data sources, which are twofold :

- First, big data integration delivers simple and easy access to various information by providing a single view of a set of the existing information.
- Second, it is able to create a better data repository by combining data from different information systems which work in conjunction with each other.

In a big data environment, the applications use various types of data. Therefore, interoperability is a need for the applications to be able to exchange this data. Interoperability is a major strength as it facilitates the exchange and sharing of data between applications without any effort. However, implementing interoperability requires efficient solutions and techniques, mainly due to the heterogeneity and location of the information. Kadadi et al. [8] highlight the data integration challenges in a big data environment. These challenges are to determine the scope of data, query optimization, data inconsistency, inadequate resources, scalability, and ETL Process in big data. They stated that the currently available data integration architecture is not able to handle all the challenges.

16.2.3   NoSQL Data Bases

NoSQL is a type of database, which is non-relational, i.e., without schema or schema-less. NoSQL databases are categorized into four types: key/value, column-oriented, document-oriented, and graph databases [9]. Each type has advantages and disadvantages depending on the context of the application. NoSQL is used in big data environment and real-time web applications. These systems databases allow big data storage, analysis, as they handle a huge amount of data and support a wide variety of technologies. NoSQL databases are very flexible and are able to store large volumes of structured, semi-structured, and unstructured data. However, NoSQL databases can partially address volume, variety, and speed issues and do not address semantic heterogeneity.

16.2.4 ONTOLOGIES

Gruber defines ontology as "a specification of a shared conceptualization of a domain" [10]. Ontologies are used in many fields, such as information retrieval and knowledge extraction. The purpose of ontologies is to capture domain knowledge and provide a sharable representation of it. In other words, ontologies are a way to describe the structure of a certain domain, which enables the development of a common understanding and the reality of its concepts. The currently used ontology language is Web Ontology Language (OWL) [11]. The main components of an ontology are concepts, properties, relations, instances, and axioms:

- Concepts are also called classes. A concept is a set of entities within a domain that can be primitive or composite, abstract or concrete. We consider anything as a concept, like table, cat, task, action, etc. Attributes information (name and value) is used to describe an object.

  OWL classes are built on top of and add additional semantics to RDFS classes [12]. Whereas classes in most other languages only have heuristic definitions OWL classes have a rigorous formal definition. An OWL class is a set. A superclass is a superset, i.e., a set that is more general and contains more elements than its subset. Individuals in OWL are elements of sets. When an individual is an instance of a class then that individual is an element of the set represented by that class. The properties that describe sets in set theory also apply to OWL classes and can be asserted by various axioms. The intersection of two classes are all the individuals that are instances of both classes. The union of two classes is the set formed by combining all the instances of each class.

  In set theory, a special set is an empty set: the set with no elements in it. The empty set is represented in OWL by the class owl:Nothing. The empty set is important in set theory because it is a subset of every other set. Thus, owl:Nothing is a subclass of every OWL class. No individual can be an instance of owl:Nothing. If the reasoner determines that a class is unsatisfiable it will classify the unsatisfiable class as a subclass of owl:Nothing which means that class can have no instances. Similarly, there is a class that has *every* individual in it. This class is a superclass of every other class and is represented by the class owl:Thing. All classes have owl:Thing as a superclass. Every individual is an instance of owl: Thing [13].

  Properties: Relations between individuals are described by OWL properties. There are three types of OWL properties: object properties, data properties, and annotation properties [14].
- **Object properties**: Object properties are properties that have other classes as their range. I.e., the value of the property will be an instance of an OWL class.
- **Data properties**: Data properties are properties that have Literals such as strings and integers as their range.

Annotation properties: Annotation properties are used for meta-data and meta-data such as comments need to be used for everything in the ontology. Because annotation properties are for meta-data they can't be utilized by the reasoner.

OWL provides additional ways to describe a property that can then be utilized by the reasoner for automatic inferencing. The most commonly used definitions are:

Functional properties: Functional properties must have at most one value for each individual.

Symmetric properties: Symmetric properties point in both directions.

Inverse properties: Inverse properties go in opposite directions.

Transitive properties: Transitive properties propagate relations from one object to another.

Reflexive properties: Reflexive properties are properties that always apply from an individual back to that same individual.

Super and sub-properties: A super property is more general than its sub-property.

All properties can have super and sub-properties. Only object and data properties can be functional. The rest of the characteristics can only apply to object properties.

- Relationships describe the interactions between concepts or a concept's individuals in the domain. Functions a special type of relationship where the value of the last argument is unique for a list of values of the previous n-1 arguments [15].
- Instances are also referred to as individuals and represent the things of a given concept [16]. For example, "red" is an instance of the concept "color".
- Axiom assertions that are always true. They are included in an ontology for several reasons, such as constraining its information, verifying its accuracy, or inferring new information. One of the most powerful features of OWL is the ability to provide formal definitions of classes using the Description Logic (DL) language. These axioms are typically asserted on property values for the class. There are three kinds of axioms that can be asserted about OWL classes:
  1. Quantifier restrictions. These describe that a property must have some or all values that are of a particular class or datatype.
  2. Cardinality restrictions. These describe the number of individuals that must be related to a class by a specific property.
  3. Value restrictions. These describe specific values that a property must have.

First-Order Logic (FOL) has two types of quantifiers: existential and universal quantification [17]. Existential quantification means that there exists at least one value that matches a variable in a formula and makes it true. Universal quantification means that all possible values that match a variable in a formula will make it true. OWL implements existential quantification via the Description Logic (DL) operator *some* and universal quantification via the DL operator only [18].

Ontology is more than connected graphs: An ontology is a model of concepts. It defines the overall structure of the system: the class hierarchy, the properties, etc [19]. It represents knowledge in triple format. Let's take an entity, namely, Mona Lisa, and start to build its knowledge. First, we collect more entities about it from various sources. Suppose, this entity contains the following mentioned statements:

- Mona Lisa is in Louvre Museum
- Louvre Museum is in Paris
- Louvre Museum is the same as Great Louvre
- French government take care of Great Louvre
- Great Louvre is near the Tuileries Gardens

These facts are denoted in Figure 16.2a. The drawn graph denotes the statement in the form of A→B, where A and B represent the entities, the relationship that exists between the entities are explained by arrow (→). For example, the statement Mona Lisa $\xrightarrow{is\_in}$ Louvre Museum shows that the entities Mona Lisa and Louvre Museum are connected with is_in relationship. The connected graph is a powerful way for the representation of the domain entities as well as simulation of human knowledge. However, the meaning (semantic) of the relations that connect two entities is still missing. Consequently, the machine cannot do reasoning, track data provenance, and infer new knowledge as a human can do by applying logic. By such representation, a machine cannot answer simple questions like Is Louvre Museum in France? Is Mona Lisa in France?

By converting this graph into an ontology, the machine performs reasoning similar to human beings. The first step to transforming the connected graph into an ontology is adding is_a relationship to the concepts to mark them as a class and add the semantics later on. Figure 16.2b depicts six new classes namely Painting, Public Museum, City, Country, Public Garden, and Constitutional body. All entities are mapped to the class and are called an instance/individual of that class. For example, Mona Lisa is an instance of a class Painting. To make a class provides more meaning to the entities by associating relations that link a class to another class. All instances of connected classes must hold the same relationship that exists between the classes and vice versa. For example, suppose a class Painting and class Country is related as Painting $\xrightarrow{is\_in}$ Country, then from Figure 16.2b, the machine can deduce that Mona Lisa is in France because Mona Lisa and France are instances of class Painting and Country respectively. By applying the logic to the existing relations presented in Figure 16.2b, the machine can infer more relations, for example: Painting $\xrightarrow{is\_in}$ *City*, Painting $\xrightarrow{is\_in}$ Country, Museum $\xrightarrow{is\_in}$ *City* Museum $\xrightarrow{is\_in}$,

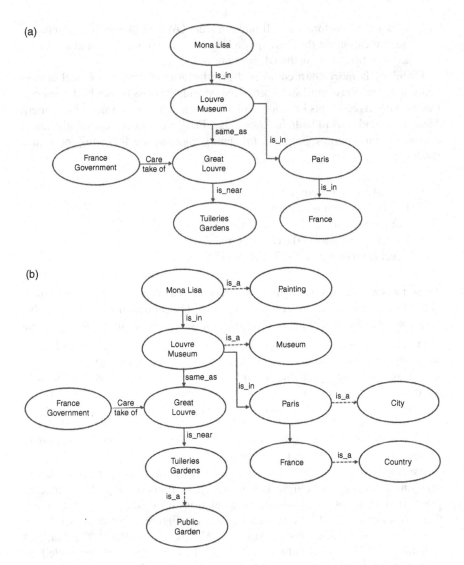

FIGURE 16.2   (a) Concepts with connections (b) an ontology with hierarchical relations.

Country, Constitutional body $\overset{\text{take care of}}{\rightarrow}$ Public Garden. Ontology also permits to add semantic to the relations, e.g. a relation can be defined as reflexive, transitive, disjoint, complementary, etc. By default, all the relationships that apply to the classes must be imposed on the instances/individuals of those classes. With the help of these semantics, ontology fetches more knowledge about the entities. The Web Ontology Language (OWL) is the latest ontology language. This language is defined on top of RDFS which adds powerful semantics based on set theory and logic. OWL is also supported by reasoners that are similar to the classifiers of previous languages such as KL-One and Loom.

16.3   RELATED WORKS

The continued rise in the volume and velocity of data, as well as the variety of data types, requires the use of data integration tools for combining data from various sources. Several approaches have been proposed for the integration of big data that comes from multiple sources and address the interoperability issues. In this section, we present the important approaches that meet our work.

Curé et al. [20] present a data integration solution based on ontologies whose aim is to merge data from document-oriented and column-oriented NoSQL systems. In this proposal, the authors produced a local schema for each system in the form of ontology in order to define a global schema from the matching between ontologies. In addition, the authors proposed a Bridge Query Language that translates SPARQL queries to the different query languages of local systems. In the context of big data, the author S. Ajani [21] presented semantic metadata and ontology-based semantic search technique for huge census domains. The purpose of this proposal is to create a census domain-specific ontology and use semantic search to extract desired results from the big datasets. According to the author, the proposed concept, may be useful for government and third-party agencies in improving semantic search on census data, as well as performing various actions and making decisions. An ontology contributes to the organization of scattered information pertaining to a particular topic of interest. Jirkovský and Obitko [22] presented a new integration approach regarding the automation of the industrial domain. The proposal addressed the structural and semantic heterogeneity for big data integration. The authors deal with different data, and to resolve the semantic heterogeneity, they created a shared ontology from the data sources. The approach involves several steps. In the first step, the authors focus on structural heterogeneity and create semi-automatically an ontology to support knowledge sharing. The process of processing different data sources differs depending on their category. The further step includes building a shared ontology from the previous steps. Knoblock and Szekely [23] proposed a novel approach supported by a software tool called KARMA for exploiting semantics for big data integration in the cultural heritage. The proposal deals with various data sources, including relational and hierarchical sources, to build new rich semantic models, useful for integration. The proposal involves four main steps. The first step imports data from multiple sources. In the second step, the authors performed data cleansing and normalization to prepare the data for the further step. The third step is to create a semantic description of each source, with the system attempting to automate the source modeling process. Finally, they integrated the data by converting it into a uniform format using the semantic descriptions and then integrating the data within this unified framework. The authors encountered a significant problem in rapidly importing, normalizing, modeling, and integrating data from a variety of museum data sources.

Bansal and Kagemann [24] proposed an ETL framework for the integration of big data. This framework generates a semantic model for those datasets that are being integrated and then based on the generated semantic model, semantic data is produced. They have used semantic technologies during the transformation phase to create a semantic data model in the form of RDF triples. The RDF triples are then stored in a data warehouse. The transformation phase is manual, and it involves dataset analysis, schema analysis, normalization, and data cleansing. Based on the results of the transformation step, an ontology is created from data based on the predefined schema. If the data sources belong to different domains, several ontologies will be created, and alignment rules are needed. The extraction and loading phases of the ETL process remain the same. Keller et al. [25] suggested a new methodology to merge different sources of air traffic data management by employing semantic integration techniques. The proposed system transforms data from multiple sources into a normalized semantic model based on ontologies. For this purpose, the authors used SPARQL as a query tool to retrieve information from the integrated triple store. The main architecture of the proposed system involves four steps. The first step selects and retrieves three data types for integration as "flight path data", "airport weather data", and "air traffic advisory information". The second step consists of using an ontology to add a semantic layer. However, users of the system, such as aeronautical actors, are expected to understand the ontology and learn the syntax of the SPARQL language while it is not obvious in practice. The third step enables to convert the initial data sources into Resource Description Framework[1] (RDF) triples. Finally, the last step includes querying and downloading the data with the help of the SPARQL language. The whole process is semi-automatic. Fang et al. [26] proposed a novel approach for integrating huge amounts of semantic data. The proposed methodology consists of four layers. A metadata layer uses different metadata rules to describe tourism domain data such as "geographic location". An ontology layer allows different forms of metadata to communicate semantically. The authors employed two techniques to achieve this. The first technique uses the OWL knowledge representation language to combine the attributes and concepts of the various metadata rules into the ontology. Moreover, the second is to convert the metadata format to RDF. Finally, the data application layer allows interactive search using a keyword search and a more user-friendly interface.

Vathy-Fogarassy and Hugyák [27] have proposed a meta-model based on data merging approach that provides uniform data access. This work described a user-friendly interface that enables query data from heterogeneous database systems without any programming skills. In this work, the authors tried to resolve structural, semantic, and syntactic heterogeneities by using a metamodeling approach. Abbes and Gargouri [28] proposed an approach whose purpose is to build ontologies from MongoDB databases for big data integration. They have provided a shared global model that works with heterogeneous data sources. The authors followed three- steps approaches. The first step performs the wrapping of data sources to MongoDB databases. The second step consists in generating local ontologies. Moreover, in the third step, the local ontologies are combined to get a global one. The approach used exploits a NOSQL database for wrapping data sources, namely MongoDB, and describes two tools that are used in the second. The third step generates local ontologies and merges them to create a global ontology. The challenge of the proposal is to extract coherent knowledge from the heterogeneous data and to store as well as manage this variety of data. Guerra et al. [29] presented a new method whose purpose is to describe how the data integration pipeline has been implemented to build a dataset. The proposal involves three main steps. The first step is "entity extraction", which aims to identify the data of interest from each input data source. The second step is "entity matching" which consists of detecting duplicate raw entities in the database. Finally, the last step is "data fusion" which solves possible conflicts due to the data. The main drawback of this proposal is that the schemas heterogeneity challenge needs to be handled by the user. Moreover, this approach does not deal with when a query needs data belonging to different source systems. Fusco and Aversano [30] have proposed an approach that integrates and mediates heterogeneous data and provides unified access to data. For the unified access to heterogeneous sources, they performed the following activities, namely source overlay, schema mapping, schema merging, and query reformulation. Overall, the approach is semi-automated and supported by only a specific software system. The system provides an abstraction of the activities at the first level, components involved in the execution, and component specialization at the second level. However, the acquisition of unstructured data sources was not yet considered by the proposal.

Table 16.1 outlines the main features of the related work, including the major commonalities and differences between these approaches regarding the semantic approach proposed in this chapter. We used seven comparison criteria that are mentioned:

- The column "Study" defines the underlying approach.
- The column "Automation Degree" specifies whether the approach is automatic (A), semi-automatic (SA), or Manual (A).
- The column "Data source" designates the data source used as input.
- The column "Output" indicates the integration purpose.
- The column "Used techniques" specifies the methods or techniques used for the data integration.
- The column "Advantages" underlines the advantages of the approach.
- The column "Semantic model" indicates the output model for the integration (OWL, XML, etc.).

These features specify whether the present approaches: emphasize data mining or optimization, offer proofs-of-concept, are concerned with big data, use OWL/ RDF in the semantic model, align with other ontologies, and describe workflow composition tasks. Table 16.1 identifies the actual contributions of the proposed semantic model with respect to the state of art.

The existing works attempt to solve data integration problems from various perspectives. Most of the proposals underline ontologies to enhance big data

TABLE 16.1
Related Works According to Big Data Integration

Study	Automation Degree	Data Set	Output	Used Technique	Advantages	Semantic Model
[11]	Manual	NOSQL databases	Data integration framework	Analyze a set of schema-less NOSQL databases to generate local ontologies. Generate a global ontology based on the discovery of correspondences between the local ontologies. Propose a query translation solution from SPARQL to query languages of the sources	Deals with the absence of a schema at the sources and of global ontology.	Ontology
[12]	Semi-automatic	Large census domain	Semantic metadata and an ontology.	Building the census domain ontology Performing semantic search	Semantic data model based on semantic web technology will be promising to provide a better solution in terms of costs and benefits.	XML file, OWL Ontology and RDF
[13]	Semi-automatic	Shared ontology Shared storage	Providing an end-to-end solution, where end-users will formulate queries based on ontology.	Data pre-processing Shared ontology creation Data transformation	Data heterogeneity can be reduced by means of shared ontology. The capability of handling various data sources stored in databases or files	OWL Ontology

(Continued)

TABLE 16.1 *Continued*
Related Works According to Big Data Integration

Study	Automation Degree	Data Set	Output	Used Technique	Advantages	Semantic Model
[14]	Automatic	Relational Databases, Web services, Excel files, XML,JSON, CSV	Semantic models	- Importing - Cleaning - Modeling -Integrating Data	Analyze known source descriptions to propose mappings that capture more closely the semantics	Schema mapping
[15]	Manual	Data available across the Web	A data mart or a data warehouse that serves BD	Semantic model generation of the data sets under integration Semantic linked data generation in compliance with the data model	Semantic framework, which produces linked data that supports innovative data-driven apps for smart living	OWL Ontology
[16]	Automatic	Csv files, html files, web services	RDF triples	Establish a common description of ATM data using an ontology model - apply semantic integration techniques to populate a triple store with data from Sherlock.	Serve a broad community of aviation researchers	Ontology
[17]	Semi-automatic	Metadata layer (MARC, DC, GILS)	Interface	Data collection. Data Backup Data Processing Maintenance and updating of information	To help lay the right foundations for intensive, intelligent, and unified tourism management	RDF Ontology

(Continued)

TABLE 16.1 Continued
Related Works According to Big Data Integration

Study	Automation Degree	Data Set	Output	Used Technique	Advantages	Semantic Model
[18]	Semi-automatic	Structured data. Semi-structured data	Integration tool	General diagram. Adapters for databases Migration of data from different source systems Processing strategy	New data integration methodology that does not require programming skills and can query relational and non-relational database	NoSQL
[19]	Automatic	Structured data Semi structured data Unstructured data	Ontology	Wrapping Build local ontology Build global ontology	Ontologies for reuse in knowledge representation	OWL Ontology
[20]	Automatic	CSV relational databases, and crawled websites	Dataset with removing duplicated entities belonging to different entity classes	Matching Fusion	-Effectiveness -Robustness to missing values - Deal with missing values - Preserve the hierarchical relationships we found in the data	Entity mapping
[21]	Semi-automatic	Structured data Semi-structured data Unstructured data.	Global ontology	Development of data sources Pattern matching Merge schemas. Reformulation of queries.	Transparency is guaranteed by translating the queries placed on the virtual view into queries that can be directly executed from local sources.	OWL Ontology OWL.

analytics and are mostly oriented to the specific domain of health system applications and e-learning in education. However, the analysis of these works indicates a few critical problems, which are unsolved in existing technologies:

- Lack of a solution for real-time integration: Most of the solutions are based on Hadoop and its ecosystem, which largely address the data volume issue but do not proceed in real-time.
- Lack of an integration solution that takes data quality into account: Most techniques do not focus on data quality, which is a critical barrier to data integration success. Integration of diverse data can result in messed-up or meaningless data.

In conclusion, big data still needs an efficient data management strategy. Future works in the big data environment must focus on the management of this massively distributed data. A challenge that concerns the entire big data process, i.e. security, documentation, data quality assurance, data verification, and semantic data integration, which are crucial tasks to ensure the efficiency of data technologies. Semantic data integration can provide a solution to the heterogeneity problem and provide interoperability between applications that use big data [31]. In fact, regarding the integration domain, the use of ontologies as a formal description knowledge model is crucial for smart processing and retrieval of big data since it introduces new insights into the information systems and applications.

16.4 PROPOSED APPROACH

The presented literature survey revealed that the research work on knowledge extraction from databases generally used structured data, i.e., relational data. However, few works deal with semi-structured or unstructured data. In general, ontologies can contribute as a semantic layer and formal knowledge description for:

- Data discovery and interpretation.
- Data management.
- Data exploration and integration.

Semantic big data integration focuses on the data variety by enabling the solution of several interoperability conflicts, such as representation, structuring, matching, and naming conflicts. These conflicts arise because distinct data sources may utilize different data models and representation strategies. Furthermore, the data may have a similar syntax with a different meaning (mouse may be an animal and a computer device), so it is difficult to distinguish that data. As a result, data integration approaches are required that can handle all interoperability issues while we take into account the data complexity.

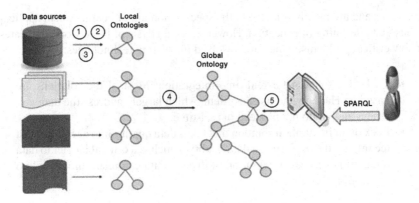

FIGURE 16.3   System architecture.

This paper proposes a process that deals with multiple heterogeneous data sources and involves several steps. The proposal's main features are:

- **First**: unstructured data sources are wrapped in relational databases, while structured data sources are wrapped in NoSQL databases.
- **Second**: a mapping of relational and NoSQL databases to ontology modules is performed via transformation rules.
- **Third**: creating a global ontology module by combining ontology modules.
- **Fourth**: to respond to user queries, the SPARQL language is used to query the global ontology.

Our approach extracts knowledge from different sources and semantically improves these massive data with ontologies for better understanding and decision making. The proposed approach employs both relational and NoSQL databases. Figure 16.3 gives an overview of the proposed system.

The first step, "Data collection" imports heterogeneous data from disparate sources. The second step, "Data cleansing" filters the imported heterogeneous data by removing duplicate, missing, or outlier data. The third step, "Data analysis" loads each cleaned data source into NoSQL databases, and for each data source, creating a local modular ontology. The fourth step creates a global ontology from the built ontologies and finally, the fifth step, "Query construction" queries the global ontology with SPARQL. Below, we describe each step in detail.

   Step 1. Data collection: This step aims to collect data that is both meaningful and trustworthy. In software, this step is called the loading of the data sets. It enables the extraction of data from an existing data source to construct the local modular ontologies (through concepts, relations, axioms, and individuals). These input sources include several data types: structured, semi-structured, or unstructured. Each data source is wrapped, and we collect valuable data for content analysis.

Step 2. Data cleansing: Data cleansing is the process of eliminating or changing data that is erroneous, incomplete, irrelevant, duplicated, or poorly formatted in order to prepare it for analysis. This data is often not necessary or beneficial for evaluating data, and it can slow down the management process or produce erroneous results.

Step 3. Data analysis: In a massively distributed data environment, data comes from multiple sources (both SQL and NoSQL databases). Hence, there is a lack of semantics in these databases. We propose an ontology-based methodology to recover the hidden semantics of big data as a potential solution to the problem. This methodology aims to build a global ontology for big data that comes from different data sources. Initially, local ontologies are built for each data source. The proposed system delights Cassandra databases as input data and involves two phases for the construction of the local ontology; these phases are data homogenization and the definition of mapping rules. We describe each phase as follows:

a. Data Homogenization

The proposed system uses a NoSQL database, namely Cassandra. Cassandra is *"an open source NoSQL, non-relational, column-oriented database and largely distributed"* [9]. A collection of rows is encapsulated within a column family. Each row has a set of columns that are arranged in a certain order. It even features unique columns known as super columns that store a map of sub-columns. We bind each retrieved data source after the cleansing step to a Cassandra database as a middle representation formalism. To load the data into the Cassandra databases, we use the data integration tool Talend. Talend includes big data analytics, data integration, data security, and more [32]. This software collects data from multiple sources and integrates it into NoSQL databases.

b. Data Mapping

To generate semantics from the Cassandra Database, we build the ontology according to a specific domain. Next, we use the following association rules to produce the OWL ontology. This phase is inspired by our previous work [31]. The association rules for extracting ontology components from the Cassandra database are given below. These rules are divided into four parts: (i) rules for extracting classes, (ii) rules for extracting datatype properties, (iii) rules for individuals, and (iv) rules for extracting axioms.

– **Ontology classes**: We extract the ontology classes from the collection as well as the subsumed relations in order to define the ontology tree structure. For this purpose, each column family in the Cassandra Database is mapped into an OWL class.

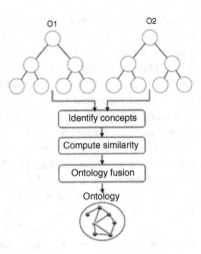

FIGURE 16.4   Ontology fusion.

- **Datatype property**: Each column in the Casandra Database can contain a variety of data types, including primitive data types (int, text, etc.), referenced data types (map, set, list), and user-defined data types. That column is mapped into a datatype attribute for basic data types. Each other type is linked to an object attribute.
- **Individuals**: The content of each column in the row (real value) is mapped to individuals in the OWL ontology.
- **Axioms**: If two separate rows in the column family have an identical set of data values for all the columns, the classes corresponding to these two rows are assumed similar. However, we assume that these classes are unrelated.

Step 4. Global ontology building: In this step, we build our global ontology from the modular ontologies that are built locally (Figure 16.4) from the same or related domains. This ontology encloses the knowledge of the initial ontologies.

An ontology is a tuple $<C, R, H^C, H^R, I>$ where C is the set of classes, R is the relationships, H^C is a hierarchical structure of classes, H^R is a hierarchical structure of relations, and I is the set of instances [33]. Ontology matching/fusion is the process of accomplishing mutual understandings by detecting semantic relationships called alignment between entities of source ontology (O_S) and target ontology (O_T). The relationship can be equivalence (=), less general ($\sqsubseteq$), more general ($\sqsupseteq$) and disjoint ($\perp$). Alignment is the actual result of an ontology matching process. Many ontologies refer to the same domain and the same objects; hence it is necessary to merge and organize them. Indeed, the ultimate goal of fusion is to represent a better perspective of the knowledge of a domain. In general, ontology fusion is used in the field of data integration; however, it can also be considered as a technique used in the domain of ontology enrichment. It facilitates the transformation of the related knowledge into the ontology in less time and cost. However, it is still unclear how the merging process is achieved. Indeed, there is no consensus on the methodology of ontology fusion. The only common phase is the initial phase, which takes a set of ontologies (two or more) as input. Some start directly with all ontologies to be merged (non-incremental method); others start with a selected initial group of ontologies (usually one ontology) which is then progressively extended by the other ontologies (incremental method).

The ontology proposed fusion technique is described in Algorithm 1, which employs a semantic similarity measure between two concepts taken from two different ontologies [34,35]. Using the similarity metric, we do a peer-to-peer fusion of the ontologies. For the similarity metric, we additionally defined a threshold, which is a real value in the range [0, 1]. The use of a threshold in computations represents the permitted tolerance. Furthermore, the threshold represents the required accuracy. The accuracy improves as the threshold value approaches "1".

- If the threshold value is "1", it means that the two concepts must have total semantic equivalence (both concepts are identical). In this case, we use only one concept in the global ontology.
- If the threshold value is "0", it means that the two concepts are not equivalent. In this case, we assign a subset or superset relationship between both concepts in the global ontology [36].

Algo- Ontology fusion
Input: O1, O2: domain ontologies
Output: O: domain ontology
Begin
Load(O1)
Load(O2)
N1← O1concepts number
N2← O2 concepts number
        For each c_i (i=1 to N1) do begin
        For each c_j (j=1 to N1) do begin
        Sim=Calculate similarity between (c_i, c_j)
If sim >= threshold value then c_i and c_i are similar, replace c_i by c_j
Add c_i to O with individuals
Save(O)
End
    End

Once the global ontology is created, we ensure that it is complete and accurately represents all available data sources. We examine the accuracy, consistency, and completeness of the generated ontology. The verification includes the structure and content of the ontology. To accomplish this, we use the tool "protégé²" to validate the generated ontology. This tool checks the logical consistency and builds a suitable OWL file needed for the next step.

Step 5. Query construction: In this last step, according to the user needs, we build a query by using the Simple Protocol And RDF Query Language (SPARQL) [37]. Query response is important in the era of big data integration since it provides a mechanism by which users and applications can interact with the data sources using ontologies. SPARQL is a query language developed and recommended by the World Wide Web Consortium (W3C). SPARQL is an RDF query language and supports querying of multiple RDF graphs [38]. The SPARQL engine evaluates the query and returns the required responses once it is created. The query is built directly using the language constructors at this level, which necessitates prior knowledge. We intend to support natural language queries that will be converted into SPARQL in the future.

16.5 CONCLUSION

With the proliferation of information systems and the distributed, heterogeneous nature of data, ontologies become a necessity as they enable the understanding of different types of data by both humans and machines. Indeed, ontologies help to address semantic interoperability, make large-scale systems more comprehensible and allow data integration. In addition, ontologies improve data adaptability and re-usability. Currently, ontologies are increasingly used in decision support systems and contribute to knowledge management and discovery. Overall, ontologies play an important role in different big data scenarios as they provide transversal meanings to terminologies. In this chapter, we have tried to give a global overview of the different challenges related to big data integration and techniques. Then, we presented the important approaches related to big data integration. Furthermore, we discussed our approach that aims to build a framework for big data integration based on modular ontologies and the SPARQL language. In future work, we plan to consider other NoSQL databases and automate the composition step of ontology modules. Another interesting perspective is the dynamic update of the resulting modular ontologies according to the updates made on the data sources.

REFERENCES

1. Gandomi, A., & Haider, M. (2015). Beyond the hype: Big data concepts, methods, and analytics. *International Journal of Information Management*, 35(2), 137–144.
2. Yaqoob, I., Hashem, I.A., Gani, A.B., Mokhtar, S., Ahmed, E., Anuar, N.B., & Vasilakos, A.V. (2016). Big data: From beginning to future. *International Journal of Information Management*, 36, 1231–1247.
3. George, D. (2005). Understanding structural and semantic heterogeneity in the context of database schema integration. *Journal of the Department of Computing, UCLAN*, 4(1), 29–44.

4. Dahdouh, K., Dakkak, A., Oughdir, L., & Messaoudi, F. (2018). Big data for online learning systems. *Education and Information Technologies*, 23(6), 2783–2800.

5. Ranjan, J., & Foropon, C. (2021). Big data analytics in building the competitive intelligence of organizations. *International Journal of Information Management*, 56, 102231.

6. Yu, L., Duan, Y., & Li, K. C. (2021). A real-world service mashup platform based on data integration, information synthesis, and knowledge fusion. *Connection Science*, 33(3), 463–481.

7. Ziegler, P., & Dittrich, K. R. (2007). Data integration—problems, approaches, and perspectives. In *Conceptual Modelling in Information Systems Engineering* (pp. 39–58). Springer, Berlin, Heidelberg.

8. Kadadi, A., Agrawal, R., Nyamful, C., & Atiq, R. (2014, October). Challenges of data integration and interoperability in big data. In *2014 IEEE International Conference on Big Data (Big Data)* (pp. 38–40). IEEE.

9. Abramova, V., & Bernardino, J. (2013, July). NoSQL databases: MongoDB vs Cassandra. In *Proceedings of the International C* Conference on Computer Science and Software Engineering* (pp. 14–22).

10. Gruber, T. R. (1995). Toward principles for the design of ontologies used for knowledge sharing? *International Journal of Human-Computer Studies*, 43(5–6), 907–928.

11. Patel, A., & Debnath, N. C. (2021). Development of the InBan_CIDO ontology by reusing the concepts along with detecting overlapping information. *Inventive Computation and Information Technologies: Proceedings of ICICIT 2021*, p. 349.

12. DeBellis, M., Aasman, Jans., & Patel, A. (2022). Semantic technology pillars: The story so far. In *Data Science with Semantic Technologies: Theory, Practice, and Application*, Wiley, Scrivener Publishing.

13. Mishra, A. K., Patel, A., & Jain, S. (2021). Impact of Covid-19 outbreak on performance of Indian Banking Sector. In *CEUR Workshop Proc.*

14. Patel, A., & Jain, S. (2021). *Ontology Versioning Framework for Representing Ontological Concept as Knowledge Unit*. ceur-ws.org.

15. Guarino, N., Oberle, D., & Staab, S. (2009). What is an ontology? In *Handbook on Ontologies* (pp. 1–17). Springer, Berlin, Heidelberg.

16. Noy, N., & McGuinness, D. L. (2001). Ontology development 101. Knowledge Systems Laboratory, Stanford University, 2001.

17. Amati, G., & Ounis, I. (2000). Conceptual graphs and first order logic. *The Computer Journal*, 43(1), 1–12.

18. Kalyanpur, A., Parsia, B., Horridge, M., & Sirin, E. (2007). Finding all justifications of OWL DL entailments. In *The Semantic Web* (pp. 267–280). Springer, Berlin, Heidelberg.

19. Patel, A., Debnath, N. C., Mishra, A. K., & Jain, S. (2021). Covid19-IBO: A Covid-19 impact on Indian banking ontology along with an efficient schema matching approach. *New Generation Computing*, 39(3), 647–676.

20. Curé, O., Lamolle, M., & Duc, C. L. (2013). Ontology based data integration over document and column family oriented NOSQL. *arXiv preprint arXiv:1307.2603*.

21. Ajani, S. (2014). An ontology and semantic metadata based semantic search technique for census domain in a big data context. *International Journal of Engineering Research & Technology*, 3(2), 1–5.

22. Jirkovský, V., & Obitko, M. (2014). Semantic heterogeneity reduction for big data in industrial automation. ITAT, p. 1214.

23. Knoblock, C. A., & Szekely, P. (2015). Exploiting semantics for big data integration. *Ai Magazine*, 36(1), 25–38.

24. Bansal, S. K., & Kagemann, S. (2015). Integrating big data: A semantic extract-transform-load framework. *Computer*, 48(3), 42–50.
25. Keller, R. M., Ranjan, S., Wei, M. Y., & Eshow, M. M. (2016). Semantic representation and scale-up of integrated air traffic management data. In *Proceedings of the International Workshop on Semantic Big Data* (pp. 1–6).
26. Fang, Y., Jiaming, Z., Yaohui, L., & Mei, G. (2016). Semantic description and link construction of smart tourism linked data based on big data. In *2016 IEEE International Conference on Cloud Computing and Big Data Analysis (ICCCBDA)* (pp. 32–36). IEEE.
27. Vathy-Fogarassy, Á., & Hugyák, T. (2017). Uniform data access platform for SQL and NoSQL database systems. *Information Systems*, 69, 93–105.
28. Abbes, H., & Gargouri, F. (2018). MongoDB-based modular ontology building for big data integration. *Journal on Data Semantics*, 7(1), 1–27.
29. Guerra, F., Sottovia, P., Paganelli, M., & Vincini, M. (2019). Big data integration of heterogeneous data sources: the re-search alps case study. In *2019 IEEE International Congress on Big Data (BigDataCongress)* (pp. 106–110). IEEE.
30. Fusco, G., & Aversano, L. (2020). An approach for semantic integration of heterogeneous data sources. *PeerJ Computer Science*, 6, e254.
31. Mhammedi, S., & Gherabi, N. (2021). Heterogeneous integration of big data using semantic web technologies. In *Intelligent Systems in Big Data, Semantic Web and Machine Learning* (pp. 167–177). Springer, Cham.
32. Sreemathy, J., Brindha, R., Nagalakshmi, M. S., Suvekha, N., Ragul, N. K., & Praveennandha, M. (2021, March). Overview of ETL tools and talend-data integration. In *2021 7th International Conference on Advanced Computing and Communication Systems (ICACCS)* (Vol. 1, pp. 1650–1654). IEEE.
33. Patel, A., & Jain, S. (2021). A novel approach to discover ontology alignment. *Recent Advances in Computer Science and Communications (Formerly: Recent Patents on Computer Science)*, 14(1), 273–281.
34. Bouhissi, H. E., Salem, A. B. M., & Tari, A. (2019). Semantic enrichment of web services using linked open data. *International Journal of Web Engineering and Technology*, 14(4), 383–416.
35. El Bouhissi, H., Malki, M., & Cherif, M. A. S. A. (2014). From user's goal to semantic Web services discovery: Approach based on traceability. *International Journal of Information Technology and Web Engineering (IJITWE)*, 9(3), 15–39.
36. Otero-Cerdeira, L., Rodríguez-Martínez, F. J., & Gómez-Rodríguez, A. (2015). Ontology matching: A literature review. *Expert Systems with Applications*, 42(2), 949–971.
37. Zhang, C., & Beetz, J. (2017). Querying linked building data using SPARQL with functional extensions. In *eWork and eBusiness in Architecture, Engineering and Construction* (pp. 27–34). CRC Press.
38. Quilitz, B., & Leser, U. (2008, June). Querying distributed RDF data sources with SPARQL. In *European Semantic Web Conference* (pp. 524–538). Springer, Berlin, Heidelberg.

NOTES

1 https://www.w3.org/RDF/
2 https://protege.stanford.edu/

Index